D0296736

Technology Trendlines

Technology Trendlines

Jessica Keyes

VAN NOSTRAND REINHOLD
I(T)P ™ A Division of International Thomson Publishing Inc.

New York • Albany • Bonn • Boston • Detroit • London • Madrid • Melbourne
Mexico City • Paris • San Francisco • Singapore • Tokyo • Toronto

I(T)P™ A division of International Thomson Publishing Inc.
The ITP logo is a trademark under license.

Printed in the United States of America.

For more information, contact:

Van Nostrand Reinhold
115 Fifth Avenue
New York, NY 10003

Chapman & Hall GmbH
Pappelallee 3
69469 Weinheim
Germany

Chapman & Hall
2-6 Boundary Row
London
SE1 8HN
United Kingdom

International Thomson Publishing Asia
221 Henderson Road #05-10
Henderson Building
Singapore 0315

Thomas Nelson Australia
102 Dodds Street
South Melbourne, 3205
Victoria, Australia

International Thomson Publishing Japan
Hirakawacho Kyowa Building, 3F
2-2-1 Hirakawacho
Chiyoda-ku, 102 Tokyo
Japan

Nelson Canada
1120 Birchmount Road
Scarborough, Ontario
Canada M1K 5G4

International Thomson Editores
Campos Eliseos 385, Piso 7
Col. Polanco
11560 Mexico D.F. Mexico

1 2 3 4 5 6 7 8 9 10 QEBFF 01 00 99 98 97 96 95

Library of Congress Cataloging-in-Publication Data

Keyes, Jessica 1950-
Technology trendlines / Jessica Keyes.
p. cm.
Includes index.
ISBN 0-442-02022-8
1. Information technology. 2. Technological innovations.
I. Title.
T58.5.K49 1995
004—dc20 95-12048
CIP

Project Management: Raymond Campbell • Art Direction: Jo-Ann Campbell • Proofreading: Joan P. Radin •
Production: mle design • 562 Milford Point Rd. Milford, CT 06460 • 203-878-3793

This book is most appreciatively dedicated to
my clients and friends, old and new,
to my family, and to my editors.

CONTENTS

FIGURES

FOREWORD

I started out in the field of information technology when it was considered a good day if you wrote a program that ran on the mainframe in under 128K. Today's computing environment is much different. Only a decade and a half ago running accounts payable, billing, and payroll while simultaneously keeping track of your mailing list would have required a machine that took up yards of square footage and cost more than you probably make in a dozen years. But that's exactly what we have on the desktop today.

Things have changed. And that's what this book is about.

The motivation for this book is quite simple, really. It is to make the reader aware of the information technology trendlines that will affect his or her life in the years to come.

Technology Trendlines is really a book for two sets of people. On the one hand, users of the technology will be interested in what's down the road for them in terms, of intelligent agent software, video conferencing, video servers, and the Information Superhighway. On the other hand, IS management and developers will be interested in finding out the ramifications of building such systems.

This book is a guide to the future of technology. It's a guide to your future as well.

This book would not have been possible without the help and encouragement of many people. First of all, I'd like to thank my husband and parents without whose unwavering support this book would never have been finished.

I would also like to thank my editor at Van Nostrand Reinhold. Jeanne Glasser has been an inspiration as well as a great source of organization, style, and content advice from the very first outline of this book. Her help shaped this book in more ways than one.

If there's one other person who was as involved in this project as I, it was my editorial assistance and research guru. Debra Nencel is not only a whiz with the electronic quill pen, having written and edited a number of publications in her time, she also ably assisted me in tracking down the trendlines that were—and were not—of interest. Thanks Deb.

Many other people were involved in this book as well—people like Joe Segel of QVC, Margaret Hamilton of HTI, and Information Builder's John Senor. These people gave willingly of their time and information to help me better understand the trends that are shaping our brave new world, and I offer them my grateful thanks.

JK, New York City

PREFACE

It wasn't that many years ago when keeping up the pace of technological change was an easy task. It changed at a rather slow—even languorous—pace. In fact, things didn't change much in the days of the IBM 360 and *Saturday Night Fever*.

Steve Jobs, et al. changed all that. The PC put into place more than just a new technology. It put into place a new way of thinking about the possibilities of technology. To be sure, IBM and other mainframe manufacturers were no slouches when it came to innovation. But the mainframe, or Big Iron as it has come to be known, is symbolic of a particular type of mindset: Big, centralized, and expensive.

The world was a different place before the 1980s: Before the fall of the Berlin Wall and NAFTA. The pre-1980s world was still the era of the hierarchical organization characterized by all-knowing bosses and truly subordinate—and often kept-in-the-dark—employees. These employees received information in only two ways; either distilled and filtered from their bosses or non-real time from the corporate mainframe (usually on a 24-hour time delay).

The PC changed all that. Within the space of a half decade, the PC began to appear nearly everywhere. On these PCs was the software that hastened the revolution.

If one piece of software can be credited with ushering in the age of the PC-empowered employee, it is the spreadsheet. When Dan Bricklin, now president of Software Garden, wrote VisiCalc he did not know that he would be the mentor of a way of thinking as well as the creator of an automated way of working with spreadsheets. But that's exactly what Bricklin did. He single-handedly made people realize that a PC is more than just a souped-up typewriter, it is an information processor.

No longer would the "average Joe" have to wait for the boss to trickle down those nuggets of important information. No longer would the "average Sue" have to wait for the mainframe to chug out its precious cargo. Information was on the desktop now. And that desktop had the power to chug away with every bit of intelligence of the mainframe that preceded it. But this time, Sue and Joe controlled it.

VisiCalc's success opened the door—some say the Pandora's box—to new ways of thinking about solving problems with technology. That door has never been closed.

When I wrote *InfoTrends* in 1992, I concentrated on several hot technology trends that were shaping our industry back then. My contention then was that more and more firms were beginning to use technology for strategic purposes rather than just for tactical purposes. In this vein I tackled the heady topics of downsizing, partnering, integration, outsourcing, and knowledge-based systems, among others, to demonstrate their competitive value.

In *Trendlines* I expand this theme to include the broad impact that technology is having on not only business but everyday life. Make no mistake about it—technology is changing every aspect of life. For example, by the time this book is published AT&T will have brought out Sage which has the smarts to tie together voice mail, faxes, and e-mail so that it deliverable to you via one smart phone in one smart place at one smart time.

In the first few chapters of this book I discuss some of the macrotrends that seem to be shaping the information industry. Among them are: intelligent agents, smart development, the information superhighway, and the personal politician.

Some of the industry's most provocative thinkers gather together to get into the nitty gritty of the massive technological change that *Trendlines* is about. These are the microtrends. Rick Faletti and Mike Frame of Northern Telecom write about delivering multimedia services on what they refer to as the Information Network. 3Com's Lionel de Maine explains how to develop and maintain that information network. And 3Com's David Helfrich describes how to use that network for an intriguing way of working—teleworking. But Joseph M. Segel, QVC's Chairman Emeritus, may burst the bubble with his chapter on separating hype from reality on the information superhighway.

There's more. Information Builder's John Senor uses his chapter to provide some insight into middleware which may be the best way to laying the foundation for open, distributed computing. DBStar's Dale Way predicts nothing short of a radical transformation in the way we do information management because of client/server.

Mark Youngblood, who recently departed from Micrographx to become a principal at the Renova Corporation, provides a holistic way of performing total process management; Margaret Hamilton, who managed the software development for the Apollo and Skylab missions, describes an error-free way of systems engineering and software development.

Trendlines has more than 25 chapters contributed by the "thinkers and doers" in the industry. From topics such as why you need a repository to enterprise multimedia systems' to PC card technology, this book should hold your interest, and change the way you think about technology.

CHAPTER 1

Technology and Change

by Jessica Keyes

Technology = Change. The faster the pace of technological advancement, the faster the pace of business change.

Indeed technology is changing. From mainframes in the seventies, to standalone PCs in the 1980s, to networks of servers in the 1990s. Today we are posed to enter a new age, an age where businesses, like the PCs they connect together, form cooperative, open networks of skillsets. Essentially, we are steadily moving away from an era of hierarchical organizations where information must force its way down a labyrinth of special interests and power clicks to a more networked structure where information is the only master.

A structural change such as this requires the organization to be well-aware of the technological trends empowering these massive changes. Organizations must be reactive to these trends as well. This is not to say that they must jump on every high-tech bandwagon in spite of advice to the contrary from the popular press. Instead organizations must be aware of trends to the point at which they are able to select the one, two, or three trends they will hang their corporate hats on.

Knowledge is power. In this case, knowledge equates to an understanding of what new technologies are becoming available, how they can be useful to you, what their costs are, and, ultimately, how to turn it into a competitive advantage.

...ORLD AND TECHNOLOGY—A SYMBIOTIC ...TIONSHIP

...hat came first, the chicken or the egg? This is actually quite a meaningful question if you replace chicken with technology and the egg with need. So what comes first? The technology or the need for the technology?

Both actually. There have been many cases when technology was introduced into the marketplace where there was no real need for it. Until it was introduced, that is. And suddenly everyone began using it. These are the "market shapers." Visicalc, the first automated spreadsheet, was a market shaper. Nobody asked for it. But when it was introduced it jumped into the marketplace—and ultimately changed the way we think about computers. Essentially, a market shaper is a piece of software, or hardware, that is pivotal in changing the face of business.

On the other hand, the introduction of the word processor was not a pivotal event in the history of business computing. For years before the introduction of the first word processor, the typewriter industry had been successfully making and selling millions of souped-up typewriters. These typewriters were electronic, contained some memory, and had built-in functionality such as spell checks. The PC-oriented word processor was nothing more than the next logical step in the technological advancement of typing. It wasn't dramatic. It didn't shape the industry. It was simply a nice easy progression from mechanical typewriter to electronic word processor to PC word processor. So, as you can see, technology creeps into the door in two ways: One as an egg, and the other as a chicken.

While nobody can guess at what the next Bill Gates is cooking up in his or her garage, it would be worthwhile to look at what some forecast our future to hold. The World Future Society, which is located in Bethesda, Maryland, has long gained notoriety for its succinct, and often accurate look into the crystal ball.

In their 1994 white paper entitled, "SOCIAL AND TECHNOLOGICAL FORECASTS FOR THE NEXT 25 YEARS," the World Future Society sometimes chilling forecasts may be providing the seeds of market shaper ideas. While the following bullet points are not all technology-related, a savvy entrepreneur can take each of these forecasts and find a way to push information technology into the mix.

- On the average, by the year 2000, cars will be constructed mainly of plastic and last approximately 22 years.

- Japan, now a major player in the economic sphere, will suffer turmoil in the 1990s and never recover.

- Outsourcing technological projects overseas to third world countries will become a hot international trade issue by the end of this decade. Workers who telecommute across borders will be competing against their compatriots in affluent countries.

- 52% of the world's population will be residing in urban centers by the year 2000; a number that may leap to 90% by the end of the 21st century.
- A complete collapse of the global economy is quite possible in the very near future.
- By the end of this decade, NASA hopes to launch the first flight of the National Aero-Space Plane. According to NASA scientists, air travel in the future will include planes that will travel at Mach 12, transporting commuters halfway around the world in under two hours.
- Health wars loom on the horizon: Activists of the major diseases will fight each other for scant research dollars.
- Personal-preference cards (smart cards that will contain such data as your suit size and the color of the last tie you purchased) should make shopping in the future less of a trial.
- Hydroponic greenhouses, where shoppers will be able to pick their produce right off the vine, may soon be installed in the produce section of your favorite market.
- Coral reefs, home to a variety of species—potential sources of new medicines and food supplies—may not survive. As much as 60% could be destroyed in the next 20 to 40 years.
- As species continue to vanish from our world, zoos will not only collect the remaining few but also catalog their genetic materials to preserve them—and their botanical counterparts—for possible re-establishment or medicinal purposes.
- As the availability of transplantable organs continue to fall well short of the need, organ robbers may definitely be in our future. Criminals who will steal kidneys, hearts and other organs to sell on the black market.

From population control software to virtual offices. From virtual zoos to kidney tracking software. From direct marketing software that can track down to the actual consumer to software that checks and controls the world's economy, future world trends give rise to future technology trends. Of course, this is some years down the road.

FORECAST FOR THE NEAR FUTURE

So, what's in store in the near future? Fortune Magazine looked at the desktop of 1996 and came away with an interesting variation on what we are using today.[1]

[1]Deutschman, Alan, "Your Desktop in the Year 1996," *Fortune*, July 11, 1994. pp. 86-98.

Although many innovative products are on the market today, such as PlusTek's (Sunnyvale, CA) multi-lingual scanner, which scans and then translates documents from one language to another, and 3D Systems' (Valencia, CA) 3D printer, which prints three-dimensional models of physical objects, the PC of tomorrow is sure to be unrecognizable to today's users. For one, the appearance will be quite different. Larger screens will be required to take full advantage of the flashier graphics and the simultaneous running of several applications. Peripherals will include micro video cameras for teleconferencing and optical scanners for digitizing images. One thing will remain the same—despite the maturation of technological recognition of human speech and handwriting, the typewriter-style keyboard will remain.

The most dramatic difference between today's PC and tomorrow's in the ways the hardware will be utilized. Networks will enable instantaneous access to global information. Innovative software packages, designed not to speed up the output of isolated individuals but to ease collaboration and the pooling of knowledge, will be brought on board. The PC will become a mini-post office, organizing your voice mail, faxes and E-mail, including messages sweetened with video clips. Even the telephone will be absorbed by the PC. In addition, for those who are Internet phobic, traveling through cyberspace will become friendlier.

Over time, as hardware costs continue to fall, innovative computing practices will move into the PC realm from the more rarefied world of workstations. Today, even moderately-priced workstations from Silicon Graphic and HP include their own digital cameras for desktop video conferencing.

The optical scanner, currently a much relied-upon tool for graphic designers hooked into a workstation, will soon move from workstation to desktop PC. For business executives who work specifically with text and numbers, scanners provide a way of capturing large amounts of data for future manipulation. Although today's computers have difficulty handling the power and memory needs of this tool, tomorrow's versions should not find these requirements at all troublesome.

Collaborative computing will reign supreme. The interconnecting of computers has caused business executives to rethink the way they do business—from facilitating virtual meetings between on-site and off-site participants to quickening the complicated process of consensus building and group decision-making.

E-mail will continue to grow. David Ferris, a San Francisco consultant on messaging systems, estimates that the average number of E-mail messages sent and received per person will increase from 20 a day in 1995 to about 40 by 1996. BIS Strategic Decisions of Norwell, MA predicts that by 1998 the number of LAN-based E-mail users will reach 38.9 million, more than twice the total of 16.9 million users in 1993. E-mail will not be used only for the electronic transmission of text. As the memory capabilities of computers continue to increase, users will be able to transmit video clips, graphics, and large spreadsheets as well.

By 1996 you will be able to continually leave a window open running your communications program. There will be no need to close down the files and program you're working on; you'll have on-going conversations with your fellow workers and exchange messages with people in other companies.

Usage is currently still very limited, mostly due to the pricing policy of online services. Charging customers by the minute for extended usage, these services do not encourage the fiscally responsible to utilize the service to its maximum potential. But change is in the air. For example, Metricom says it can offer unlimited access to online communication—wireless, amazingly—for as little as $3 a month, albeit at a very slow rate of data transmission.

Groupware, software that enables people to work collaboratively and share ideas, will be the big prize in the networking crackerjack box. Presently, when one thinks groupware, one thinks Notes, a sizzling product from Lotus Development Corp. Notes serves as the company's message box, a central place for employees to post all kinds of information. Among the 750,000 who use the software, accounting giant Peat Marwick, has 15,000 employees hooked up.

E-mail and Lotus Notes will provide the foundation for a new generation of network software. The programs that were designed for solo usage—word processing, spreadsheets, databases—will soon be replaced by software that will be able to tie everyone together—no matter how remote the location to allow easy collaboration on projects, linking workers together, no matter how many miles separate them.

An example of this new wave of networked software comes from Pillar, a privately owned company in Foster City, CA. Pillar's FYPlan program is designed to manage specific tasks that require input from many people—creating corporate budgets. It spreads the input from the company's accountants to managers throughout the organization. Pillar enables the executives to record and share with their colleagues the information and thought processes behind the calculations. According to Marc Benioff, vice president for client/server systems at Oracle Corp, groupware is about amassing the total mental, physical, and spiritual energy of a corporation. The more thoughts are shared the higher the corporate IQ.

Groupware is also enabling people in different locations to work together in real time. A technological trend that will enable two or more people connected on a network to simultaneously edit or annotate a shared document, view presentations, or participate in meetings has been showcased at a significant number of trade shows. This development, rich in potential for manufacturers of high-end PCs, has already propelled Intel to develop its own program, ProShare Personal Conferencing Software ($99), which lets two people in remote locations jointly edit a document. Competing against ProShare is IBM's Person to Person software ($99) are offerings from such entrepreneurial firms as Future Labs of Los Altos, CA, which publishes Talk Show ($199).

Ultimately, companies are seeking to merge software and hardware for video conferencing. By 1996, it will probably be an everyday happening that pairs of knowledge workers will routinely review printed materials on screen and simultaneously see each other via video. Intel offers a program similar to ProShare that includes a small video phone window on screen, although the PCs must be hooked up to a digital ISDN phone line. AT&T and InVision Systems of Vienna, VA are considering selling similar packages.

Right now the pictures are fuzzy and tend to flicker, but as more and more consumers gravitate to PCs powered by microprocessors such as Intel's Pentium and Motorola PowerPC, users will see a much improved video image.

In the very near future, intelligent filters will be assisting companies who rely on shared information. With BeyondMail, published by Banyan Systems in Westboro, MA a worker is enabled to program his or her computer to check the company file server regularly for information relevant to his or her position. The software searches through the files for keywords and snags the appropriate documents.

In the near future users will have easier access to worldwide information. Presently the only way to travel through cyberspace is to book passage on an online service or on the Internet. Link Resources, a New York market research firm, estimates the number of U.S. households that subscribe to online services will increase substantially by 1996. These online services are more user-friendly but they lack the depth of knowledge available on the less user-friendly Internet.

By 1996 this should change. America Online is proceeding rapidly with its plans to connect subscribers to popular Internet bulletin boards. According to CEO Steve Case, by 1996 America Online hopes to be the Internet access provider. Also, rumor has it that IBM Chairman Lou Gerstner is ready to pull into the fast lane with a new service line offering online computing, communications, and transaction services, all created and delivered by Big Blue. With network connections in 90 countries and 700 cities, IBM is well-positioned to achieve its goal.

James Gleick, a former *New York Times* reporter and author of a best selling book on chaos theory, now operates the Pipeline, an Internet concern in New York city that offers a simple Windows interface that allows subscribers to "mouse" their way to computers far and wide.

One sour note, though, today's virtually free Internet superhighway will soon give in to tolls to support that network. The backbone, now run by the government, will soon be turned over to private industry. Price hikes invariably follow privatization.

On top of this, network traffic is increasing beyond the point of tolerance. Many in the academic and research communities are recommending tolls to ensure adequate response time. Scott Shenker and his colleagues at the Xerox Palo Alto Research Center noticed that there are at least two distinct types of Internet services: real-time services, such as video conferencing, which require a steady connection with minimal delays, and best-effort services, such as E-mail, which can tolerate slower delivery. For both to work well, real-time data must be given priority over ordinary best-effort traffic. But what is to prevent users from cutting in online by mislabeling their E-mail as a video? Shenker believes that a priority pricing scheme would solve this problem.

In spite of predicted escalating costs, many more companies are riding on the highway than waiting for the fallout. J.P. Morgan has provided Mosaic, a graphical navigation program that is available at no-charge on the Internet, to its employees. Mosaic allows users to access audio and video clips and text without having to be educated as to which remote database is housing the information.

Mosaic is just the beginning. Jim Clark, the founder and former chairman of Silicon Graphics, started a company this spring to create and market a sort of super-Mosaic. Through a venture called CommerceNet, a group of 50 companies including Apple, Intel, and Sun Microsystems, will soon be using Mosaic to conduct business with one another.

As the business community's interest continues to grow, customer support will continue to grow and improve. Several entrepreneurial firms are already developing systems to centralize customer service. The businesses that are successful in the future will be the ones whose workers know how to share information.

THE INFORMATION APPLIANCE

The common denominator for most of these innovative trends is the advent of client/server computing. In the early 1980s, PCs became ubiquitous. This sort of computing was largely disconnected from the mainframe (or any central server) and thus may have actually contributed to what many refer to as a productivity paradox by creating "islands of (mis)information" around the organization.

When the knowledge worker discovered the power of the PC, he or she began (as one would expect) to grow his or her own applications. While this ultimately empowered him or her to do things not thinkable a few years before, it also de-powered the organization from controlling its information assets. Essentially, users began to hoard data.

With the advent of the server and introduction of the LAN, organizations won back some if their hard-earned data. While the PC-based worker still controlled his or her own destiny by virtue of the desktop, the organization was better able to control the flow of information to the worker.

It took approximately half a decade, for the PC to circumvent the organization and another half a decade for client/server to set the compass for the organization to properly set sail. But in spite of the momentous change set into motion by these events, this was merely the prelude to the massive change yet to occur, the move from the PC through the client/server organization to the information appliance.

A convergence of technologies has made the PC an able assistant to the knowledge worker. Using the server as an information "spigot," the network as the information flow, and the PC as the information appliance, a worker can now choose from among a plethora of options to massage and view his or her data.

Saul Wurman wrote *Information Anxiety* (Doubleday) in 1989. His premise is that information was increasing too rapidly for a human worker to process and absorb. While the rate of information flow was accelerated, today's workers have tools to help them sort the information available to them.

Intelligent agents are smart software programs that act as intermediaries between the user and the information flow. Take a simple example: CompuServe, an online service that boosts somewhere over a million subscribers, sells a service called the Executive News Service. One need only enter a few filtering keywords, select one or more newspapers, and/or wire services (as shown in Figure 1.1) and you're off and running.

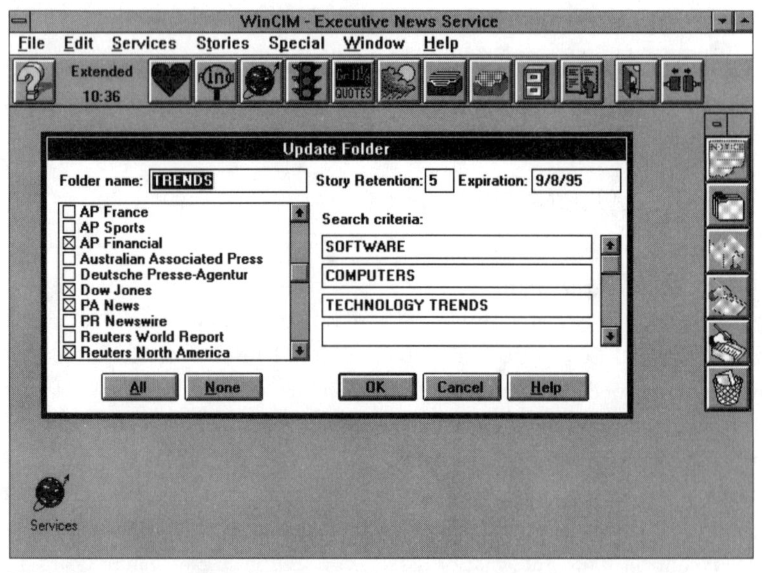

FIGURE 1.1. CompuServe's Executive News Service—smart agent software

In this example, I am collecting information about technology trends. There is no way I can read each of the trades and business magazines, so I have ENS do it for me. Every day I log on to CompuServe and see what the network has brought me in the way of news overnight. It helps me stay current in a timely manner.

Hertz Corporation, the three-billion dollar car rental company based in New Jersey, uses intelligent agent software to help it in its quest to win the pricing wars with companies such as Avis and National.[2] Car rental rates are not constant. Today a car might

[2]Wilder, Clinton. "Important Information—Fast," *InformationWeek*. November 13, 1994. pp. 77-82.

rent for $60 a day, and tomorrow the same car might rent for $50. Because Hertz needs to make 50,000 price changes a week, it depends on technology to lead the way.

Working with consulting giant Arthur Andersen and decision-support vendor Comshare Inc, of Ann Arbor, Michigan, the Hertz team developed an intelligent agent system called Commander Exception Monitor. Running on 486-based IBM PS/2s linked to an ES/9021 mainframe located in Oklahoma City, Hertz's system sifts through huge mainframe databases as well as client/server databases to alert users to variations that impact on the bottom-line.

Intelligent agents come in small boxes, too. Sony's new Magic Link handheld personal digital assistant has a base price of $995 and offers E-mail, fax, and paging capabilities. Magic Link is the first machine to work with AT&T's PersonaLink communications service and the first machine based on the General Magic's software that uses "agents" to search the network and follow an individual's instructions to perform such tasks as shopping and making airline reservations.

Virtual and Not So Virtual

The PC as an information appliance is an apt description and the use of the PC in this way is not limited to large businesses like Hertz. In fact, the PC has been instrumental in enabling small-to mid-sized businesses to compete with the big guys.

Some call it the virtual office. While that name is becoming somewhat of a cliche, it's a cliche with real meaning for business. No longer is it necessary to set up shop in the downtown business district, hire a secretary, and go to a printer for letterhead and business cards. Armed with a PC a company can produce these things in-house. Many do.

For example, regarding this book, I've never met many of its contributors in person. I invited them to work with me on this project through the Internet and it was through the Internet that they sent me their chapters—as well as their figures. This was the same tactic I took for my last several books, and it's the way I run my business. I solicit assignments and communicate with customers all online through the Internet (newart@panix.com) and through CompuServe (72470,2111).

The Virtual Hospital

Cuba Memorial Hospital, a typically small-town health facility, in a typically small western New York town, has seen the technological future—and seized it![3] Cuba Memorial has made a choice that will enable it to save even more lives in its bucolic

[3]Talan, Jamie. "The Virtual Hospital: Telemedicine Brings Cutting Edge Health Care to Small Town Patients," *New York Newsday*. October 3, 1994. p. B29.

community. Already being utilized as a tool to connect its patients with some of the top specialists in the country, Cuba is now expanding telemedicine to its emergency room. There its health care professionals will be able to turn on a video camera and computer, dial another hospital miles away, and seconds later a trauma specialist will appear on the screen—an off-site physician who will then, with continued use of the camera, and other high-tech tools, be able to observe and diagnose treatment for the emergency admittance. Later, via video, conferencing be available for consultations.

The virtual hospital has arrived. Without leaving their own community, patients can now consult some of the country's top specialists. Relying on a multitude of high-tech communication devices, such as video cameras that can zoom in on microscopic as minute skin melanomas, electronic stethoscopes that magnify a heartbeat and transmit it hundreds of miles away through phone or cable lines, virtually instantaneous computer access to a patient's medical charts, and large-screen TVs that permit doctors to see and chat with patients and other medical professionals, the small-town hospital has increased its value to its patients.

Dr. Francis Tedesco, president of the Medical College of Georgia, a pioneer in this growing communications network, passionately boosted telemedicine, stating that doctors can look into the eyes, hear the heartbeat, and listen to the lungs of patients, and they can adapt this technology to any instrument.

According to Dr. Jane Preston, president of the American Telemedicine Association in Austin, TX, the emergency room is definitely going to be where this technology shines. James Toler, co-director of the bioengineering center at Georgia Institute of Technology, adds that the ability to administer treatment to patients during the "golden hour" saves lives and results in less serious medical complications. Heart attack or stroke, natural disaster or vehicular accident victims, telemedicine will enable patients to receive the quickest help possible.

Though the technology has already been instituted on a limited basis in 20 states, it is not generally accepted by Medicare, Medicaid, or any existing insurance plan—even though it cuts costs. It is decidedly more cost-effective for a patient to stay at a small hospital than to be admitted to a higher-priced health facility.

This technology has brought with it countless considerations: Can specialists provide out-of-state medial care via video links without having a license to practice in that state? Will practitioners feel comfortable using a system that creates a video diary of their every move and decision? Who will be responsible for the patients' care—the video doctor or primary care physicians? Will patients feel comfortable with this new technology? And lastly, who will invest the money to start the system in the first place?

The costs are substantial. $130,000 is required for the compilation of computers, video cameras, special scopes for internal exams, and a device that looks like an electronic notebook that permits doctors to measure minute changes on a cardiogram or to zoom in on patient's haphazard stockpile of prescription drugs, and this equipment must be installed at each site. Additionally, a hospital has to employ and train specialists to operate the equipment.

Although already funding some telemedicine programs on a limited basis, the federal government is cautious. Speaking for the government, Helen L. Smits, deputy administrator of the federal Health Care Financing Administration, stated that since this technology is virtually untested, a certain reserve on their part is necessary. A total understanding of both its benefits and limitations are desired before bestowing government money and commitment.

New uses are already occurring. The Medical College of Georgia is testing home hook-ups with 25 of its patients. Success along this line could ultimately lead to installation at nursing homes, thus decreasing emergency room transfers. A major benefactor of this technology would certainly be rural hospitals. Able to bring the specialists to them by video links, they will be better able to retain those patients who in the past would have had to be transferred for specialized care.

In some areas the transition has not been smooth. The human element has produced kinks. Doctors are having some difficulty adjusting to being televised. It has not been easy to train themselves to look through the camera's eye on the monitor, not the patients eyes in person, as well as to remember to guide the technicians operating the system, advising when and where to zoom.

Aside from a treatment tool, telemedicine has also produced benefits in terms of education. Many states have already approved programs allowing doctors to obtain continuing education credits working with a specialist on telemedicine systems. If Dr. Barney Kenet, a dermatologist at New York Hospital-Cornell Medial Center and his brother Dr. Robert Kenet, a cardiologist and engineer, are successful with their experimentations, earlier detections of skin cancer may be another positive result of this technology. Programming a computer to peer through a scope that shines light just beneath the surface of the epidermis, the doctors can study the pigmentation pattern and determine the risk factor of a mole. The pattern is stored on a computer disk and can be relayed anywhere electronically.

The tactile relationship between patient and doctor may be suffering at present by this technology, but Toler and his colleagues at the Georgia Institute of Technology are hard at work on ways to break the barrier of long-distance touch. They are experimenting with a sensor-laden glove that would enable a physician to "feel" a patient without actually being on site. The rural doctor dons the glove and examines the patient; the information is then transferred to the Medial College of Georgia, where a doctor inserts his hand in a sensory generator and senses the texture and temperature of the body area. Though still in the prototype phase, the tool should be available in 1995, according to Toler.

Military uses are also conceivable in the near future. A backpack system for the military that would use two lipstick-size cameras inserted into a helmet has been designed to enable a medic to transmit information in actual-time from the battlefield. Doctors at military bases could then examine the wounded soldier and advise treatment and prepare for his or her arrival at the hospital.

Dr. Jay Sanders of Massachusetts General Hospital, an early booster and user of this technology (in the early 1970s he set up one of the first telemedicine systems between his hospital and Boston's Logan Airport), would like to see it employed in schools, as a teaching tool, and in people's homes to enable doctors to make housecalls—whatever the time of day. A true believer, he sees telemedicine as one of the factors that will radically change health care.

The Virtual Bank

New York's Citibank is typical of those banks looking to cash in on the virtual business. Citibank recently made a giant leap toward virtual banking by making stock quotes and securities transactions available at its 1,800 proprietary automated teller machines (ATMs) nationwide. This service was inaugurated, in the fall of 1994 in New York, Chicago, Miami, San Francisco, and Washington.

Citibank's ATM users are able to see up to 10 stock quotations, buy and sell stocks, or invest in money market funds simply by following the instructions on the screen. This is the first of many virtual banking programs under development at Citibank. The bank is also improving its phone option with sophisticated screen phones and expanding its PC banking through software giveaways.

In addition to the convenience offered their customers, there are cost-saving benefits of this new technology. According to Mark Hardie, an analyst with the Tower Group, a bank technology consultancy in Wellesley, MA, electronic transactions are more economical than those processed by personnel, and the high operational costs of branch offices can definitely be decreased.

IBM video conferencing kiosks is offering the $150 billion Royal Bank of Canada in Toronto the chance to centralize its intelligence and services. According to Dennis Graham, manager of advanced concept networks for Royal Bank, these kiosks will make more sense than having licensed, experienced mutual salespeople stranded in seldom-visited branch offices,

Allowing the bank to situate their experts in one central location, lessening these employees' downtimes and increasing their availability to a wider range of customers via the video conferencing kiosks, this new technology has definitely improved their customer service and management of personnel. A pilot kiosk, installed in June of 1994 at the bank's main Toronto office, had already serviced 500 people and processed over 100 transactions by March of 1995.

Another financial institution that has made the foray into the world of virtual business is Huntington Bancshares, Inc. of Columbus, OH. This $16.5 billion institution is currently converting its branch offices to accommodate these new customer-intensive technologies. The move allows the bank to expand without building or leasing new office space.

To make automated check cashing services possible, Huntington has hooked up with AT&T Global Information Solutions to add imaging technology to its teller machines. More detailed transactions, such as opening deposits accounts, are handled separately through interactive video conferencing terminals or by tele-banking. The reason for this, according to Jan Tyler, research and development manager at Huntington, is the desire to keep ATM lines and transactions short.

VIDEO CONFERENCING—THE KEY TO THE VIRTUAL BUSINESS

The key driver behind the virtual business is video conferencing which allows groups not only to share a common PC blackboard, but to see or hear each other over long geographic distances. PictureTel a Danvers, MA-based company, is one enterprising organization that is marketing such a technology. LIVE Share (Figure 1.2) operates using affordable dial-up digital lines. A small video camera is mounted to the top of the PC that transmits the individual's image while audio is transmitted through a PictureTel telephone. Software enables users on the both end of the connection to share a PC session. So, one person can be in New York, and another can be in San Diego and both can pour over a spreadsheet in real-time, talking and seeing each other the entire time.

FIGURE 1.2. Videoconferencing with PictureTel

Not to be outdone, AT&T and Intel Corporation have produced a very user-friendly desktop video conferencing product has been made available through a joint venture.

The subsidiary, created by the joint venture in the summer of 1994, will link Intel's ProShare video conferencing equipment and AT&T's Worldworx networking services. The joint product, produced to operate over digital phone lines, includes software and a reasonably-priced video camera and headset, and will be priced between $999 and $1,500. AT&T's Worldworx network service will manage ProShare calls for between 50¢ and $1 per minute.

The AT&T an Intel subsidiary will also be working with the Personal Communications Workgroup to develop a desktop video standard that will allow systems from various vendors to communicate fully. But suppliers are at least a year away from adopting design configurations, which were originally expected by the end of 1994. ProShare employs Intel's proprietary Indeo technology, while other vendors utilize H.320, a standard videocompression algorithm.

Amy Pearl runs the Enterprise Communication Initiative (ECI) at Mountain View-based Sun Microsystems. ECI is a cross-functional collaboration between Sun's MIS department, Sun's training organization, Sun Laboratories, and SunSoft. Pearl believes that the advent of E-mail has not replaced the need for interactive, synchronous communication, even in computers. For example, for as long as computers have supported multiple users, people have found simple text-based interactive "talk" programs useful.

Video conferencing is a technology that has been emerging over the past thirty years, motivated by how much of our face-to-face communication is visual, or nonverbal. For example, efforts have been made to augment telephones to include transmission of visual images. Recently, the Computer Supported Cooperative Work (CSCW) community has been exploring the use of computer-controlled analog audio and video transmission to support group work. Recent advances in media compression technology, along with shrinking component size and cost, are making digital video feasible. Unlike analog video, digital video can take advantage of the growing number of digital communication networks, including phone (e.g., ISDN) and institutional networks (e.g., LANs and WANs). It is also possible for programs to manipulate digital video, the effects of which are just now being explored. An example of this is the morphing of video images.

Sun funded a project called Videoconf to explore these issues. Like the PictureTel system, Videoconf enables users on multiple ends of a geographic spectrum to share information as well as see and hear each other. Although Sun is very much a vendor, Videconf is an internal project. Given the company's size and diversity, video conferencing was seen to be a way to simultaneously decrease costs and improve productivity.

According to Sarah Dickinson, an analyst with Personal Technology Research, a Waltham, MA consultancy desktop video products (including hardware and software for voice, data, and video services, but not including the PC) will be about $2.1 billion in revenue by the year 1998.

Of course, just beyond video conferencing is the idea of the use of video in business.

THE VIDEO CONNECTION

One need only peruse the trades—by using your intelligent agent software, of course—to find numerous articles about the coming boom in video.

A video camera that connects to mobile computers via a PCMCIA card was released in November 1994 by the Scottish company, VLSI Vision Ltd.[4] The company sees as its target market such on-the-go professionals as real estate agents, site inspectors, and graphic designers. Also considering a move into the mobile digital video market are such firms as MRT International of Hollis, NH, and Logitech Inc. of Fremont, CA.

Named PC Card Camera by VLSI Vision, the device consists of a handheld camera and mounting apparatus and a PCMCIA Type II interface card with cable and documentation. Camera exposure can be set either automatically or manually, and optional lenses are available. To enable the camera to capture still-image and motion-video, VLSI Vision has created VVL Snap, a still-image and motion-video application that runs under Microsoft Windows.

According to VLSI Vision, the collected images are displayed on the computer in real-time, enabling the user to control what is being captured precisely. Then, for ease of integration into other applications, these images are stored as TIFF files.

The selling price of the camera is under $500. The company also plans to make available a series of accessories such as a PCMCIA/AT bus adapter, for using the system with a non-PCMCIA-equipped desktops. Alister Minty, product group manager for VLSI Vision, sees the PC Card Camera being utilized in presentations, document imaging, real-estate applications and conferencing. Video is such a hot trend that companies like Microsoft and Oracle are putting their money where their mouths are.

Though business applications for video are still in the very beginning stages, some companies are already experimenting with the best usage for this new technology.[5] For example, NBC News is using the Oracle Media Serve to log video clips.

This is the side road, but the main path chosen by most companies for video servers is training. As vast libraries of interactive courseware continue to be accrued by corporations, they are beginning to stress the existing standard LANs. Bruce Yon, multimedia analyst at Dataquest at San Jose, CA, concurred, stating that the ever-increasing software is surpassing the ability of businesses to absorb servers or CD-ROM jukeboxes. Video servers enable employees to train themselves at their PCs as their time allows instead of completing a whole course or jamming the network.

[4]Clancy, Heather. "VLSI to Connect Video Cameras, Mobile Computers Via PCMCIAS Link," *Computer Reseller News*. September 19, 1994. p. 88.

[5]Knibbe, Willem. "Video Servers on Trial in Businesses," *InfoWorld*. October 10, 1994. p. 22.

Video Servers have definitely found a home in manufacturing. Jeff Capeci, advanced manufacturing engineer at defense contractor Nordon Systems Inc. says his company is using the technology to help educate its workers to perform certain manufacturing processes, feeding the video instruction live to their terminal for convenient review.

Analysts foresee businesses such as advertising and video production firms as desirous of such products as Microsoft's Network (originally code-named Tiger), and Oracle's Media Server. Storing sales data and archiving footage are the problems that will be solved by video servers for such media-intensive organizations.

Kitty Cullen, Oracle vice president of new media product marketing, strongly believes that the major empowering force behind this technology will be the business community. This technology is Oracle's strength, and playing to this strength, Oracle has joined forces with companies, including Lotus Development Corp. and 3Com Corp., to form a consulting group that will help companies plan for their interactive future and work with content providers.

The business community may be the major focus of these companies, but Microsoft is not ignoring the consumer market. The company recently signed an agreement to provide the software components, including its Tiger continuous video server, for a Southwestern Bell Corp. interactive test in Richardson, TX. This test is scheduled for 1996. A comparable system will be tested by Microsoft and Tele-Communications, Inc. starting early next year.

On a side-note, Microsoft's aggressive push into every market that we've discussed so far will see the breakup of the company (ala AT&T) by the end of the decade. Aside from the fact that Microsoft is the software company people most love to hate, even pro-Microsofters are beginning to think that it is just not right to have one company dominate an industry.

THE ELECTRONIC CONSUMER

A few short years ago Lotus embarked on a mission to sell information on consumers. While this project failed due to public outcry over the loss of privacy, this trend is here to stay (unless Congress steps in to stop it). This may indeed happen. In October of 1994, Representative Edward Markey, chairman of the House Telecommunications and Finance subcommittee, blasted America Online's decision to sell its subscriber list to direct marketing groups. In his letter, Markey wrote, "comprehensive privacy protections must become part of the electronic ethics of companies doing business on the information superhighway and a fundamental right of all its travelers."

The era of the electronic consumer is here to stay. A cash transaction is blissfully anonymous, but as soon as one uses a credit card more, and more data is compiled on his or her spending patterns. In the beginning there was just too much of a mass of data to do anything with, but with the advent of "smart" systems, this mass of data began to be shaped into something worthwhile.

With this information, retailers can track who its customers were, and what they are buying. With credit information and medical information being digitized, these databases are making their way into the commercial arena. Today there is nothing except public condemnation to stop a savvy organization from pinpointing diabetic customers who vacationed in Aspen and who make over $120,000 a year. Even the casinos are getting into the act. Caesar's Palace is using information compiled from its hotel, restaurants, and gaming tables to personalize its marketing efforts.

This automated market segmentation has given rise to individualized direct mail. Where one used to receive the generic catalog or sales letter, now the marketer can tailor a letter or catalog just for you. Big Brother is indeed watching. But there are two sides to the electronic consumer. Companies are working hard on using technology to deliver product catalogs directly to the consumer—bypassing paper altogether.

Of course, we're all familiar with the now-ubiquitous home shopping channel. Joe Segel, chairman emeritus of QVC, is one of the contributors to this book. QVC and the Home Shopping Channel have been selling goods over the TV for years. Now they've begun to go one step further.

Those of you who navigate the Internet with Mosaic know exactly what I'm talking about. The Internet Shopping Channel (ISN), as shown in Figure 1.3 is a wholly owned subsidiary of the Home Shopping Network. CommerceNet, shown in Figure 1.4 and mentioned above, is a consortium comprised mostly of software companies who, like ISN, are out to make a buck on the Information Superhighway. And ISN and CommerceNet are not alone. The telecom companies are fast getting into the act as well. In November MCI announced a new product entitled internetMCI that lets Windows clients access the Internet's World Wide Web.

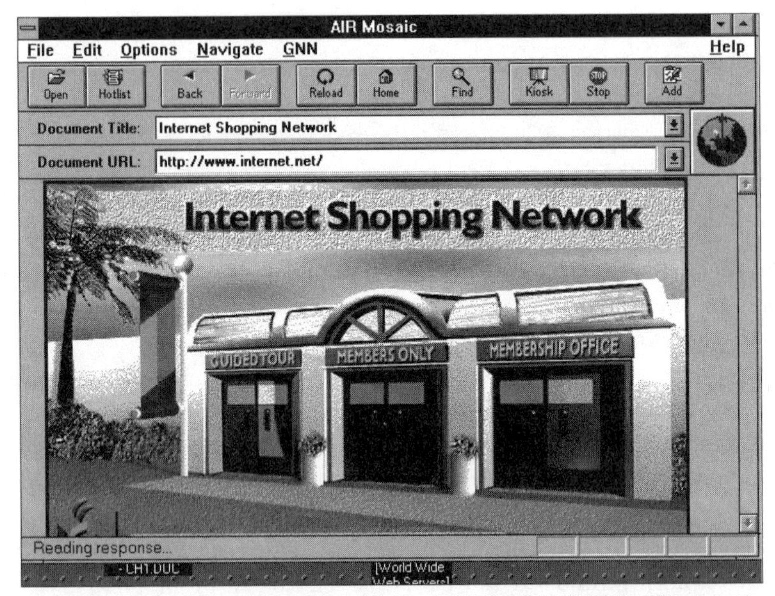

FIGURE 1.3. The Internet Shopping Network (ISN)—World Wide Web and Mosaic

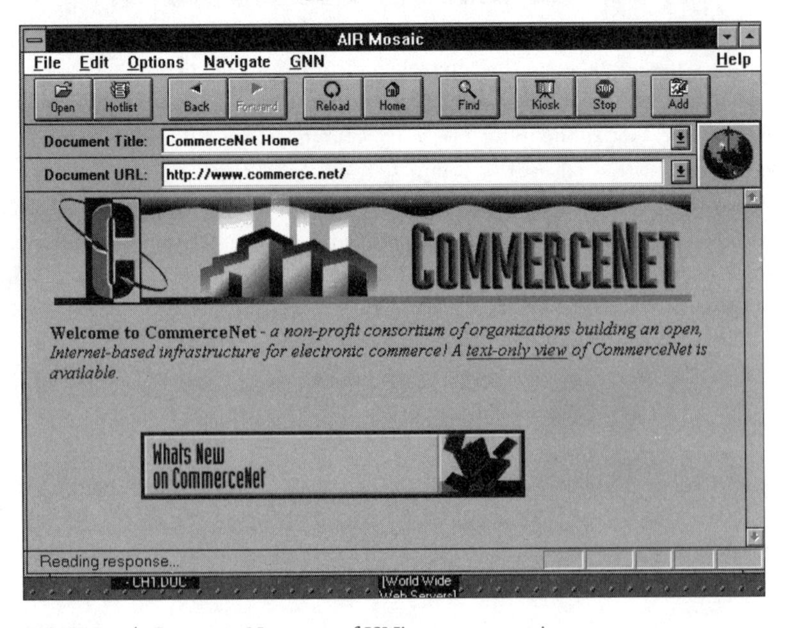

FIGURE 1.4. CommerceNet—one of ISN's many competitors

As of August 1994, over 18,000 commercial entities were registered as Internet service providers, whether that service was World Wide Web, FTP (File Transfer), or Gopher. This includes companies as diverse as Ketchum Public Relations, Con Edison, The Body Shop, Fidelity Investments, Master Card, and the London Stock Exchange.

Add the Internet information to the long-time ability to run a classified on such services as Compuserve and America Online and an American need never get on an airplane to shop in London again.

The Internet, Compuserve, America Online, Prodigy, and the hundreds of Bulletin Board Services (BBS) springing up around the world are examples of interactive consumerism. Of course, the market would not be complete without the presence of Microsoft. Aside from outfitting Windows 95 with an Internet connection, Gates and friends unveiled their own online service at the November 1994 Comdex trade show. Dubbed Microsoft Network, instant access to millions of Windows-enabled PCs will usher in what Gates expects to be an era in which all businesses are online and interconnected.

Lately, a number of companies have been sending out marketing diskettes and CDs for a more passive form of consumerism. Instead of interacting with the manufacturer or retailer real-time, a catalog on a CD-ROM provides the ability to reduce costs by replacing paper and sending out a decidedly more sexy sort of catalog.

San Francisco-based Magellan Systems was one of the first to see the possibilities of this type of marketing. Creators of the Merchant CD-ROM, they have managed to convince the likes of JCPenney, Speigel, L.L. Bean, and twenty others to produce their catalog via CD-ROM.

Late in 1994, 10,000 PC users tested another of these novel CD-ROMs.[6] ProductView contains pitches for cars, stereo equipment, vacations, financial services, and sporting goods. By clicking icons and hypertext, users are able to comparison shop as well as to receive detailed information about each advertised item. ProductView Interactive carries no charge to both consumers and advertisers, and consumers will obtain discounts for advertised products. In exchange, consumers will freely supply demographic information that ProductView Interactive will then resell to advertisers. It is the means by which advertisers can be aware of precisely who is viewing their ads. Robert Young, CEO of ProductView Interactive, sees this as advertising's future.

In agreement is Edwin Artzt, chairman and CEO of Procter & Gamble Co., who shined a very bright light on this new technology during his keynote address at the American Association of Advertising Agencies' annual convention in May, 1994. Though grateful for his very strong endorsement of this medium, many of Interactive's trail blazers wished it had come sooner.

Martin Nisenholtz, former VP of Interactive marketing at Ogilvy and Mather Direct, the direct-marketing division of Ogilvy and Mather believes that in the not too distant future the entire marketing process will be digital. With Ameritech in Chicago as director of content strategy, Nisenholtz added that marketers, consumer research firms and

[6]Wilder, Clinton. "Interactive Ads," *InformationWeek*. October 3, 1994. pp. 25-29.

hardware and software vendors are developing the back-end of marketing analysis systems through customized databases. But they must also keep an eye on the front-end. They must never fail to grasp the latest in technological approaches to consumers.

Such major retailers as FTD, Hallmark Cards, JCPenney, Land's End, Sony, Tower, and Victoria's Secret are already dipping their toes in the interactive waters, agreeing to participate in the initial testing of such interactive services as En Passant, Interactive Catalog, and Shopping 2000.

Shopping-2000 a CD-based catalog, developed by Contentware of New York, utilizes text, photos, audio, and full motion video to sell their client's goods. Ken Koppel, president of Contentware, zealously promotes the interactive environs, stating that it's really the only place an advertiser can be assured of a customer's full attention. People go online to get information—the more the better.

Concurring is Paula George Tompkins, president of the SoftAd Group, a young and growing California agency that produces interactive content for such clients as Ford Motor Co and Abbott Laboratories, Inc. According to Tompkins, this technology exerts a tighter grip on a potential customer's attention, something that traditional advertising methods—with the channel surfacing and magazine page flipping tendencies of consumers—can no longer guarantee.

A company needn't go to the expense of full-motion video. If they can export data and graphics at 9.6 kbps or produce a CD-ROM title, then they are more than ready for interactive advertising. But creating the right interactive content is the test. The mantra will probably be "Nix the sales pitch, bring on the bells, whistles, giggles, and facts."

An example of a successful interactive diskette is The Ford Motor Company's, Ford Simulator. Targeting the highly-desirable 25- to 45-year-old, college-educated, upwardly mobile computer users, this program that lets them compare colors, engine sizes, options and payment plans for Ford's entire consumer line of 25 cars and light trucks. According to Larry Dale, a Ford marketing specialist in Dearborn, MI, information, not hype, is the goal. Their Ford IQs will be high when they go to purchase.

Advertising on diskettes, CD-ROM, or online services is separate and apart from interactive advertising—a sales tool that has its own unique language and culture. With this new data comes a new challenge to the marketing information industry. Nielsen Marketing Research in Northbrook, IL, and Information Resources Inc. (IRI) in Chicago, foresee a future where their customers will definitely want this market analyzed. According to an IRI spokesman, they will want to know if it's a real market or just techies at play. Viewing agencies as critical to the process, ProductView Interactive's Board of Directors consists of the former CEO's of Young & Rubicam, Leo Burnett, and Foote, Cone & Belding.

A September 1994 Harris poll shows the interactive services market may not be all its cracked up to be, however. Only 40% of those questioned were interested in ordering movies-on-demand or sports events, and only one third wanted interactive shopping. On the other hand, 63% said they would want health-care information, lists of government services, a phone directory, and product reviews. Almost 75% wanted a customized news report, and about 50% wanted E-mail.

The question is: will this market boom or dry up?

THE COMING BOOM IN I-TV?

Of course, the television industry is not about to be overtaken by the computer industry. Interactive TV (I-TV) is very much a contender.

Last fall, a few thousand cable subscribers in Seattle, Omaha, and Orlando participated in an interactive TV experiment.[7] They were able to call up movies on demand, order products and services using their remote control devices, and play video games in actual time against other connected viewers. It was an experiment that was viewed with a great deal of interest.

But a widespread number of industry players and observers are casting a very weary eye on this technology. They advised caution: it's still in its infancy: it's too soon to rush into mass distribution. Some of the world's top technology and television companies might regret their haste; they might wind up suffering a loss of respect and dollars.

Fortune 500 companies, well aware of the financial risks, are taking a wait-and-see attitude with regard to this technology as an advertising vehicle. According to Larry Dale, a marketing specialist at Ford Motor Co., Ford doesn't want to create software for a lot of different platforms. It would mean a different set-top box for every system; they would rather wait to see what develops before they commit.

The roadblocks to achieving success in interactive TV will not be dismantled easily. Success in this new medium will come in understanding each of these barriers:

- **Complexity**. The required infrastructure of set-top boxes and servers is described by John Malone, CEO of cable TV giant Tele-Communications Inc., as more technically-contrived than anything that has ever been designed. As high-speed, asynchronous transfer mode (ATM) switching technology becomes more financially accessible and provides the network transport piece of the puzzle, many of those involved agree that developing the software to control delivery of digital content in a daunting challenge.

- **Cost**. It's guesswork as to what the consumer price tag will be. Malone sees a $500 digital set-top box, which cable providers could amortize over five years by an estimated charge of $10 per month. But this all will depend on market desire driving the profit motive. According to James Barton, president and general man-

[7]Wilder, Clinton. "I-TV: Hit or Myth," *InformationWeek*. August 29, 1994. pp 12-14.

age of Silicon Graphics Inc. (Interactive Digital Solutions joint venture with AT&T), there will be some out-back areas that may never see this technology; the delivery costs will be too high.

- **Lack of Standards**. It is a brutal war. In media services it is Oracle vs. Microsoft; in set-top boxes it is Scientific-Atlanta vs. General Instrument and others; in the quest for the hardware platform standard it is virtually all major computer vendors vying for this valuable prize. This war makes potential content suppliers such as Ford not rush to supply. Even the most seasoned technology providers see the risks of providing such services to millions of viewers. Jerry Held, Oracle's senior VP of interactive multimedia products, states that Oracle's Media Server databases will be some of the largest systems they have ever compiled.

The delivery of video on demand and interactive shopping services is just one part of the package. The back-end systems that fill orders, run credit checks, and mail monthly bills are, according to Barton of Silicon Graphics, the largest slice of the "pie." A new operational framework will be necessary before home shopping applications will be seen on interactive TV.

Silicon Graphics' Challenge server is the power behind Time Warner's interactive TV trial in Orlando, the first major test to be delayed due to unexpected technology. Originally set for start Spring 1994, Orlando is now set for a fourth-quarter launch. Such major corporate players as Ford, Hallmark Cards, FTD, JCPenney, and Nordstrom are participating in US Avenue, an interactive shopping service from Bell Company US West Inc.'s interactive video enterprises unit.

Despite the aforementioned drawbacks, interest from the entertainment industry continues to be strong. August 1994 saw Walt Disney align itself with such communications as Ameritech, BellSouth, and Southwestern Bell to bring Disney's content—including home shopping—to interactive TV networks. Also Pacific Telesis stated on August 16, 1994, that it will immediately start building a state-wide interactive TV network in California at an estimated cost of $16 billion over seven years.

But the fact remains: even if the technology succeeds will consumers really want to interact with their TV sets. And will they pay for the privilege? These are the questions that remain to be answered. An interesting side note here is that sales of non-interactive TV sets (and the bigger the better) are booming. "We're in the midst of a record year for the industry," says Zenith executive VP. One person calculated that the number of TV screens purchased by Americans would fill 750 acres, the size of Montana's Little Big Horn Battlefield.

These are interesting questions, but many think that this is a more likely scenario than consumers interacting with their PCs for shopping. Although today's businesses are very much PC-centric, the question is—if a TV can be outfitted to provide the information and the resolution of today's PCs and offer the extra advantages of video and high resolution, will business migrate to the TV as its information appliance of choice, or will it stick to the PC?

The fact that we will have a choice at all is largely due to this thing the industry calls convergence. While Kevin Brauer of Sprint addresses this in depth in Chapter 26, it is worth gives it a special mention here. Very little of what we have discussed thus far, and what we will discuss in later sections of this book, will be possible without what has been termed convergence.

In late 1994, Bell Atlantic, NYNEX, Pacific Telesis, joined forces to pursue the multimedia-network marketplace. The goal of this jointly funded venture is to support construction and maintenance for their information superhighway efforts. Earlier in the year, Silicon Graphics and AT&T funded their own joint venture.

Ultimately, it is predicted, all things informational will merge with all things electronic. The future, should be quite interesting.

THE ELECTRONIC VOTER

One of the benefits of "going electronic" is that the consumer/user is in more of a position to obtain objective news. It comes as no secret that newspapers and magazines are very much biased.

Not too long ago, the TV news program *Dateline* came under fierce attack for jury-rigging a car explosion. Their motives were innocent enough—to expose a safety hazard. But their methodology was anything but. As a result, the *Dateline* disaster has become synonymous with what's wrong with journalism today.

Howard Rosenberg of the *Los Angeles Times* called the fiasco an "electronic Titanic—an unprecedented disaster in the annals of network news and perhaps the biggest TV scam since the Quiz Scandals." What *Dateline* had the audacity to do was to stage a crash test that was rigged to get a particular outcome. They purposely concealed from the public's view the hidden rockets, the over-filled tank, and the loose gas tank that precipitated the crash. If many thought the *Dateline* scandal was a freak or bizarre departure from accepted network standards, they were wrong. Both CBS and ABC had previously run the same sorts of grossly misleading crash videos and simulations while also withholding the same sorts of material facts about the tests.[8]

But perhaps the most serious of misjudgments on NBC's part was in trusting "experts" who were deeply involved in litigating against the target of the expose.

A 1993 issue of the *National Review* devotes the majority of its publication to one salient topic, "The Decline of American Journalism."[9] Although a publication with a

[8]Olson, Walter. "It Didn't Start With Dateline NBC," *National Review*. June 21, 1993.

[9]*National Review*. June 21, 1993.

bias admittedly right of center, the article it devotes to this topic is rare in its self-exposure. The essence of what this series of articles said is that journalists themselves are biased, and they write through the haze of this bias. In other words, since most journalists are quite liberal in their political stance, articles about pro-liberal causes are written about supportively while articles about conservative topics are written about less so—if at all.

Those astute readers who have noticed the seeming embrace of the PC (politically correct) movement by the press will understand immediately what I'm referring to. How else can one explain the dozens of magazine lead stories and TV specials on AIDS? Granted it's a most dreadful scourge, but if the truth be known, on a monthly basis many more women are diagnosed with breast cancer than are people diagnosed with AIDS. How else can one explain the press' seeming fascination with people like Donald Trump or JFK, Jr.? The press rarely writes stories that reflect our interests and needs.

Criticism of the press is not limited the right-of-center publications. A literature search in the public library on the subject turns up these gems.

Junk stories are slowly eroding public's faith in the media. The press is being used as a dupe for lawyers and other charlatans who want to implicate public figures. Seeking to satisfy what they perceive to be the tastes of the American public, the press is printing first and asking questions later. Mark DiIonno, *The Sporting News*. April 20, 1992.

Obviously, the love affair with the media is just about over. Although not everyone seems to have gotten the message, the press has as many foibles as the rest of us. They make errors. They take shortcuts.

Unfortunately, somewhere around third grade or so, it was hammered into us that the printed word was gospel, "if it's in writing it must be so." And this is a habit that's hard to break. But break it we must if we are to make informed decisions.

Our trade publications are not much better. Writers have natural biases, so what gets covered is sometimes a matter of what the writer likes (and who he or she is friendly with) than the actual merit of the product. But there's a far more serious charge leveled against our trade journalists: in a survey my company performed in January of 1993, a majority felt that those wielding the big advertising bucks garnered more and more favorable press than those with smaller budgets.

The advent of electronic news allows users to bypass the subjective for the objective. No longer at the mercy of two or three sources of information, there are virtually hundreds to choose from. Using the intelligent agent software that we spoke of in the beginning of the chapter, it is possible to pick out only the topics one is interested in, from a multitude of sources. There's nothing like reading several reports of the same topic to find the real truth. Intelligent agents beget intelligent users. It's not too much a stretch of the imagination to predict that tomorrow's consumers will do a lot more with their information appliance (whether PC or I-TV) than reading, research, and shopping. Tomorrow's workers just might start voting online as well.

We are using the same voting technology that we've been using for decades. We wake up, get dressed, and go out to vote in a mechanical booth. But what if we could use our PCs or ITVs to vote? How would America change?

This is question worth consideration. Of course, electronic voting by itself will not alter American life, but what if we stretch this to include interactive government?

Picture a senator who no longer needs to go to Washington. He or she "meets" with the Senate through video conferencing and confers with constituents through E-mail. Office-bound, this senator is no longer operating in a vacuum. No longer is he or she just a suit on the Hill unaware of what is really happening in the neighborhood he or she represents.

Around election time, our Senator can E-mail constituents for free. No longer the slave to special interests, the Senate could become more reactive to the American people.

Conclusion

We started this chapter talking about change. Thanks to the industry's innovators, who include many more than just Bill Gates of Microsoft, technology is changing at a rapid clip.

The macro Trendlines addressed in this chapter form the basis for this massive change. Whether that be I-TV, or commerce on the Internet, or intelligent software, there will be no segment of our society that will be unaffected. The real question is, will you be ready?

Author Biography

Jessica Keyes is president of Techinsider/New Art Inc., a technology consultancy/ research firm specializing in productivity and high-technology applications. She is the publisher of TECHINSIDER REPORTS and COMPUTER MARKET LETTER. An organizer and leader of TECHINSIDER seminars, Keyes has given seminars for such prestigious universities as Carnegie Mellon, Boston University, University of Illinois, James Madison University, and San Francisco State University. Prior to founding Techinsider, Keyes was Managing Director of Technology for the New York Stock Exchange and has been an officer with Swiss Bank Co. and Banker's Trust, both in New York City. She has over 15 years of technical experience in such diverse areas as AI, multimedia, CASE, and reengineering. She holds a M.B.A from New York University.

A noted columnist and correspondent with over 150 articles published in such journals as *Software Magazine, Computerworld, AI Expert* and *Datamation*, Keyes is also the author of four books including *Infotrends: The Competitive Use of Information* (McGraw-Hill), which was chosen as one of the best business books of 1992 by the Library Journal, *The McGraw-Hill Multimedia Handbook* and the *Productivity Paradox*.

CHAPTER 2

The Smart Business

by Jessica Keyes

The term competitive intelligence is very much on the tip of the tongue in today's economic maelstrom. The majority, if not all, of American companies collect some sort of information about the direction that their competitors are taking. What few realize, though, is that competitive intelligence is really only part of a larger view of the business world we live in. This view is called business intelligence.

Competitive Intelligence is a subset of Business Intelligence. And as a subset it is not the complete picture. Herbert E. Meyer, a noted author and consultant, as well as past vice chairman of the U.S. National Intelligence Council, calls business intelligence the corporate equivalent to radar. Similar to radar, the business environment must be continually scanned to avoid danger and to seize opportunities. Meyer describes business intelligence as the other half of strategic planning. Once a plan is completed, business intelligence monitors its implementation and assists in making strategic course corrections along the way.

Business intelligence is not just looking at the competition but seeing all the changes that occur—including politics, consumer affairs, and even environmental issues. All of these influence the long-term future of any company—in any industry. The use of competitive intelligence transcends industry boundaries and, if used correctly, provides the organization with an immediate advantage.

No two companies will implement a competitive intelligence system in the same way. The rest of this chapter will be split between case studies of the best of these systems and explaining how to build your own competitive intelligence system.

QUAKER OATS EATS DATA

Perhaps the department with the biggest need to digest huge amounts of information is corporate marketing. Aside from the product or service being marketed, the efforts of the marketing department make the biggest impact on the profitability of a company. It stands to reason that providing marketers with the appropriate tools and techniques will most certainly enhance their efforts, and, in doing so, enhance the bottom line. These tools and techniques make up the foundation for a trend that I like to call *info-marketing.*

The Quaker Oats Company™ is one of a growing legion of companies that sees the relationship between smart marketing and increased profitability. The Quaker Oats marketing decision system was built specifically for use by the marketers as an assistant in making marketing decisions and is indicative of the company's understanding of the importance of the role of technology in their marketing efforts.

The Quaker Oats Company, located in Chicago, IL, is a leading multi-billion dollar international food manufacturer of cereals, pancake mixes, snacks, frozen pizzas, and pet foods. Over the past several years, Quaker has embarked on an aggressive acquisition campaign with the net result of establishing several new sales forces and introducing new product lines.

From a marketing perspective, this diversity of products makes performing the necessary marketing analysis difficult. All of these forces led Quaker to build an automated decision-support tool. With its plethora of functions and ease of use "Mikey" makes a good example of the first level of info-marketing tools—the executive information system.

Mikey

Quaker Oats built a PC-based system that they dubbed "Mikey." The system gets its named from the famous Quaker Oats commercial where a finicky little boy finally tries a Quaker cereal and likes it. At corporate headquarters, and at sales offices around the country, Mikey is currently being used for online access and ad hoc query capability of large corporate marketing bases. Due to its distributed nature, Mikey has become Quaker's central coordinating facility for the creation and distribution of marketing plans, production requirements, and financial estimates.

Using pcExpress and Express MDB, from the Chicago-based marketing tool vendor Information Resources Inc, Quaker built a robust marketing system whose components include business review reporting, marketing planning, ad hoc reporting, general information, and utilities. The business review reporting component produces standard reports based on the company's historical sales and comparisons with competitors. The ten standard reports can be generated for an extensive range of market, brand, time, and measure selections or time aggregations.

The ad hoc reporting component allows marketing users to essentially write their own marketing analysis programs. This permits them to look at data and create their own brand or market aggregates.

The market planning module permits the marketing department to review the marketing performance of any particular product. Data stored and able to be analyzed here includes such things as package weight, cases, the cost of the product to the company, price, and advertising budget. Mikey understands the relationships between all of these items. In a planning mode, if marketing staff decide to alter one of these variables, for example package weight, the system will automatically change the other relevant components in the mix.

USING EXECUTIVE INFORMATION SYSTEMS

Executive Information Systems are a big part of the trend towards intelligent agents that we spoke about in Chapter 1. They are hands-on tools that focus, filter, and organize information to make more effective use of that information. The principle behind these systems is that by using information more effectively and more strategically, a company can ultimately increase profitability.

The goals of an EIS should be:

- To reduce the amount of data bombarding the executive
- To increase the relevance, timeliness, and usability of the information that does reach the executive
- To focus a management team on critical success factors
- To enhance executive follow-through and communication with others
- To track the earliest of warning indicators: competitive moves, customer demands, and more

Comshare is an EIS-vendor located in Ann Arbor, MI. Its executive information system tool, appropriately named EIS, is used by hundreds of companies involved in info-marketing. Since Comshare's EIS is representative of this class of info-marketing tool, it serves as a good example and I discuss its attributes below.

Comshare's EIS is actually a series of tools, first and foremost of which is the Briefing Book™. Basic to many paper-based executive reporting systems is what is known as the monthly briefing book. Some companies call this the monthly project control report; others call it an executive summary. Whatever the semantics, its purpose is the same—to advise the senior executives of critical issues and the status of key projects within the organization. The problem with the paper reports is that it usually arrives too late for preventative or corrective action. In addition, the format of the report is too rigid for an in-depth investigation. The worst problem is that the executive cannot ask it questions to get more detailed information. Thus, the briefing book is supplemented in most companies with monthly status meetings to respond to these deficiencies.

This lack of information results in an organization's executives spending approximately 80% of their time in attending these status meetings. If one adds in the trickle-down requirement of rolling information down to the staff level, these additional "one-on-ones" and roll-down *meetings* add overhead to a firm that forces it to spend more time meeting about being competitive than actually *being* competitive.

In many companies, at least part of this problem is being addressed by automating the briefing book. Comshare's Briefing Book can be implemented in a few different ways. One way involves a menu of items corresponding to tabs in a paper briefing book or goals in a budget write-up. Another way starts with a top-level executive report using the system's color-coded exception reporting facilities. With this approach, "bad" variances can be colored red. Touching the red variances results in another screen being displayed—either of component variances, textual explanations, trend graphs, or a combination of any one or more of these.

There are many advantages to an automated briefing book. Each executive can receive a personalized selection of reports and charts, reducing the amount of irrelevant information he or she sees. In addition, each executive can set up variances responding to acceptable tolerances. For example, an automobile executive wishes to flag any line of car in which the company's market share falls below 25%. Since executives often like to see facts in context, automated briefing books should have the capability of comparing information on a competitive or historical basis. So, looking at current sales data as the latest event in a continuum indicates whether sales volume is heading up or down. Similarly, executives usually want to compare information with goals, budgets, and forecasts as well as with information stored on the competition.

Perhaps the biggest obstacle to upwards reporting is that the senior executive is only privy to information on a monthly basis. In an age of stiff competition, but easy access to distribution channels, a fast reaction time to an event may make a major difference to the bottom line. This "information float," as it is sometimes called, the time it takes for information to wind its way up through the channels to senior management, can be nearly eliminated by use of an automated system.

Comshare's Execu-View™ runs in conjunction with the Briefing Book. Once a significant variance is identified within a Briefing Book report, the executive must be able to investigate the variance with much more detail and from multiple perspectives. It's not enough to know that at the consolidated level, profits have deviated from goal by 7%. Some business units might be over goal and others well under goal. The executive must have some facility to answer questions such as: What makes up the deviation? Is it a faltering distribution channel? Are there competitive problems in an established product? Is it a failure in a particular geographic area? In a particular product line? Or in one customer grouping?

To do this, a multi-dimensional information base designed to support the managerial perspective of financial performance must be made available somehow. Execu-View permits the encoding of definitions of economic behaviors tailored specifically for this viewpoint. For example, in taking sales and expense data from the general ledger into a product profitability model, allocation rules are used to prorate cross-product costs like sales and overhead. The result is financial data with a layer of strategic perspective.

In addition to this filtered information flow, the Comshare model of an executive information system comes equipped with a calculator which permits the addition of new columns, calculation of ratios, trends, and even variances. Another feature is a connection to the Dow Jones Newswire through Comshare's Newswire™.

Sun Goes for the Golden EIS

Few realize how large a company Sun Company is. As a major domestic refiner of oil and natural gas, Sun has been a supplier of fuel needs to Americans for over 100 years. But this $11.1 billion company also produces and markets petroleum products worldwide.

With five major refiners and six divisions, senior management's goal is to get this wide range of products to market while being as flexible as possible to maximize their service and profits. To do this, Sun's steering committee, consisting of the top executives from its six divisions as well as executives from Finance, Planning and Systems, and Administration understood that they needed to develop a common model of the entire enterprise so that division executives could be aware of the impact of their decisions on other divisions.

Sun built its "Enterprise Information System" using Comshare's EIS. Sun's EIS, which allows cross-functional sharing of information across organization boundaries, is a veritable fountain of filtered information. It includes sections showing operational information for each of the six divisions; daily operational information supplied by the general manager of each of the five refineries; analysts' interpretations combining competitive, market, and internal information. There is also an information feed from external information including commercial listings of oil prices and news summary stories about the industry containing insights not likely to be found in a general trade journal.

According to Ken Fulmer, Manager of Product Management in the Planning and Systems Group, "The ability to compare ourselves to the competition in the external world is absolutely a key part of our system. In order for our executives to stay abreast of rapidly changing market conditions, they want to see not only the company's internal data, but also industry data that's external."

The biggest benefit of the EIS and the whole enterprise concept at Sun has been that the executives have a better understanding of what is going on in their business. Because each executive sees the same information, including the reports used by the top executives in each of the other divisions, they have better trust, better communication and a better understanding of the situations that exist in the other parts of the company. As a group, management can better identify which challenges to react to and how best to deploy resources for the greatest efficiency and profitability for the organization as a whole.

du Pont Uses EIS to Go Global

With over $35.5 billion in revenues generated outside the U.S. alone, E.I. du Pont de Nemours recognized early on the need to operate effectively in global markets. Thus, globalization has become a major objective for the entire company. Part of that push has been the development of global information systems. "Key to du Pont's management philosophy is the belief that the effective utilization of global information technology is fundamental to maintaining a competitive position today and even more so in the future," says Pete Trainor, Manager of du Pont's Global Financial Database Project.

Most companies, including du Pont, store information in departmental databases. In a company like du Pont this compartmentalizes, or fragments, information in such a way as to render it unusable as viewed from a global perspective. This is where Trainor's Global Database comes in.

Trainor devised a three-prong approach to develop du Pont's global system. Trainor first developed the infrastructure by establishing the networks necessary to support the global system. Perhaps the hardest task was the second component, getting the information in order. This meant setting standards to achieve data consistency and information resource management. The third component was to develop an interface that was insightful and easy to use.

The very first system put into production was du Pont's Executive Information System (DEIS). Using Comshare's EIS, this global system is now in use by over 150 managers, directors, vice presidents, as well as the CEO. DEIS provides global access to over 18 critical areas of information including key new strategic plans.

As was the case at Sun, executives at du Pont began to think on a organizational level rather than at a departmental level.

FRITO-LAY HANDS ITSELF USEFUL DATA

Quaker Oats' use of a powerful decision support tool is indicative of a trend that is catching on in the competitive arena of packaged goods. Although Quaker is probably one of the first companies to have introduced the concept of info-marketing, others are right behind them. Kraft USA, Proctor and Gamble, and RJR Nabisco are reportedly installing massive systems to track sales. But among all of these Goliaths, it is a David that has produced one of the most innovative info-marketing systems.

Frito-Lay, a subsidiary of Pepsico Inc., makes corn chips, potato chips, and tortilla chips. This is an extremely competitive market, with much of that competition coming at the local and regional level. With 100 product lines in 400,000 stores, a mountain of data is produced on a daily basis. In order to gain an advantage over the competition, Frito-Lay needed to be able to collect, digest, and then act on that mountain of information quickly.

Frito-Lay's solution was a combination of advanced technologies. This includes scanner data from Information Resources Inc. (IRI) combined with sales information from field staff all combined on a sophisticated network, encompassing hand-held computers, the newest of IBM PC computers, and even a private satellite communications network connecting the 300 distribution sites. This system is ultimately accessible through a decision-support system developed by the Ann Arbor-based, Comshare Inc.

The idea of electronic data entry from the sales force dates back to the late 1970s; but it wasn't until 1989 that Frito-Lay saw their vision become a reality. By then the company had equipped more than 10,000 sales representatives with hand-held computers developed with Fujitsu at a cost of more than $40 million.

These hand-held computers, called "bricks" by Frito-Lay, are used to track inventories at retail stores as well as to enter orders. The bricks are connected to mini-printers in all delivery trucks. The sales reps use these printers to prepare invoices which indicate which items are to be restocked at each stop on their route, as well as to indicate current prices and any promotional discounts.

At day's end, the bricks are connected to an IBM minicomputer at a Frito-Lay distribution center, and all sales data collected at the stores that day are sent to the central data center at the company's Plano, TX headquarters. Here the central computers digest the information, run it through a series of edits and finally put it into an understandable format. Next, the data is shipped out to the regional and divisional levels for further analysis by marketing and other staff. At the same time, the central computer sends pricing and promotional updates back to the hand-held computers. Finally, each Monday it gives the sales reps a review of the previous week's results on their routes.

A Micro-marketing Approach

Frito-Lay built a total information system that could be customized at every level of the company. As the network developed, the company realized it could shift from a national marketing strategy to one that targeted local consumers. This is known as micro-marketing.

Employees using the system run the gamut from marketing support to the chief executive officer, Robert Beeby. On one occasion Beeby noticed spotted red numbers on his Comshare EIS screen. This indicated that sales were down in the central region. Quickly calling up another screen, Beeby located the source of the problem in Texas. Continuing on his hunt, Beeby tracked the red numbers to a specific sales division and finally to a chain of stores. Apparently a local company had successfully introduced a new white corn chip which was eating into Frito-Lay's market share in that location. Frito-Lay immediately put a white-corn version of Tostitos into production which appeared on shelves a few months later.

Mary Ellen Johnson typifies the Frito-Lay sales manager. In order to do her job, Johnson was often required to pull together sales information from a variety of sources that sometimes took many weeks to assemble. In some cases, the information had to be obtained from telephone calls: in other cases the data was simply not available.

With Frito-Lay's info-marketing system, Johnson now is able to get reports on her accounts by brand, by type of store, or by package size. She is also able to obtain results from the best and the worst of the 388 sales representatives in her territory as well as pricing moves by competitors. She can compare the results of her sales reps performance with the previous week, the previous year, and with current targets. She can even compare a product's sales in different markets, such as supermarket versus convenience stores.

The original motivation behind Frito-Lay's unique marketing tool was to cut down on the amount of time it took sales representatives to keep their records and free them to make new sales calls. In reducing this overhead, Frito-Lay also realized a benefit of minimizing the accounting discrepancies between the sales force and headquarters which had risen to $4 million annually.

The significance of these systems is that they achieve positive returns on a multitude of levels. Frito-Lay's info-marketing system has affected every level of staff, and every layer of the corporate hierarchy. What originally started out as a mechanism to reduce the overhead of the sales rep in the field has turned into a bonanza of information for marketing staff back home, has cut down on administrative overhead at the home office, has fine-tuned the production cycle by providing timely information, and has even provided an executive information system for the 40 or so senior executives in the company. This system is pervasive.

The executive information system allows key executives to retrieve status reports by region or product. Using Comshare's EIS, these executives merely touch the computer's screen to obtain desired information. For example, by pressing an image of a book labeled "Briefing Book," executives can select maps and charts which will provide them with more detailed information. An image of a green monster calls up information about competitors.

Using this system Frito-Lay has been able to maintain a prestigious annual sales growth, about 6%, which is twice the industry average. Their earnings are impressive as well, currently at the double-digit level.

GATHERING BUSINESS INTELLIGENCE

Frito-Lay and Quaker Oats use a first-class set of business intelligence tools and procedures to collect, collate and re-assemble internal and external data to provide enough information to perform competitive decision making.

The creation of the Frito-Lay and Quaker info-marketing systems, which have been refined to the point of providing these companies tailored, filtered, and usable information, required an intensive two-level effort. On the first level was the development of the underlying technological infrastructure permitting the information to be distributed to and analyzed by the appropriate parties. On a higher level was the effort required to determine the depth and the breadth of the information that would be required.

Sitting a hundred people in a room with access to every newspaper, journal, magazine, and book that has ever been produced will produce only disconnected tidbits of information. Virtually every piece of information in those journals, magazines, books, and newspapers is available online through any of a myriad of "information vendors." Using a mere PC and a modem, it is possible for a farmer in Idaho to dial-up long-term weather conditions and for a businessman to dial-up and download information on products or competitors or trends.

The list of what information is available online covers everything from Who's Who to the Moscow news. But even with easy access to this wealth of information, you still do not have intelligence. Take the case of a major pharmaceutical company. Over a one year period they spent $12 million in online downloads. According to the CEO, virtually all this expensive information was worthless. That's because they did not have the ability, or know-how, to turn this plethora of information into intelligence. It is a discipline that requires the use of tools and techniques to be able to coordinate and correlate discrete bits of information into intelligence.

Perhaps the biggest stumbling block to the process of creating business intelligence is where it is done. Since the creation of this intelligence relies on the downloading of information, the task is often assigned to the information technology department. This was the case with the pharmaceutical company mentioned. It turned out to be an expensive mistake for them.

A profile of a department most likely to succeed in this endeavor would include the abilities of being able to work with technology, to gather the raw information, to have the writing talent to present it in an understandable fashion to management, and to have the socio-political skills to get the conclusions drawn from those analysis accepted by the diverse, and often conflicting, groups that make up a modern corporation.

This is the tact that Herbert Meyer, and his partner Mike Pincus, take when advising their corporate clients on how to build an intelligence gathering department. They look for people with library skills, technical skills, and a familiarity with the company's business. But they also look for someone who has an "in" with the CEO, so the results don't get politicized.

According to Pincus, "Part of the problem with intelligence is that when it's done well, it tends to offset bad judgment which often comes from executive support people. These executive support management people feel threatened by intelligence, since it tends to offset their own bad opinions. Where you have bad management advice, you have to bring someone into the unit who has the social/political/corporate capability to be able to move around in that environment without upsetting people or causing them to feel threatened." In other words, the role of this person is internal public relations.

The first step involved in building an intelligence unit is to build a profile of a company. This is really a needs analysis which documents the products and services that a company manufactures or performs, its goals and priorities, as well as requirements for competitive information. Basically this will be a comprehensive list of categories of information that the company must monitor if it is to be competitive. Examples of categories are suppliers, markets, customers, and so on. The profile also uncovers irrelevant information that the company is tracking. In addition to all of these, an assessment must be made of the cultural climate of a company. How is information passed up and down the corporate hierarchy? What political machinations are in place that could possibly affect, or even impede, the information flow?

With profile in hand, Pincus and Meyer perform what they call an "intelligence audit." This is the process that will determine if the right people are getting the right information. This is really a two-step process. First, as one would expect, the information needs uncovered during the process of developing the profile are satisfied by locating the proper online source that contains that piece of information. As demonstrated above, virtually anything ever written can be located in an online database. The trick is in being able to first locate it, which is what Pincus and Meyer do, and then be able to download it, and finally to be able to process it. This is where the second step of the intelligence audit comes in. Pincus and Meyer examine the company's "technological mentality." What kind of technical expertise does the company have? What are the company's personnel comfortable with? From the information collected in this process Pincus and Meyer are able to develop a technological solution that would best satisfy the needs, and capabilities, of their client.

Probably the most crucial step in this entire process is the training of selected personnel in how to convert the information obtained to business intelligence. This is actually done on two levels. On the technical level, one or more people must be trained to develop skills in correlating information in support of staff who will ultimately turn this raw information into intelligence. This top tier of staff are the ones who will need to develop and hone skills to coordinate, correlate, analyze and ultimately convert raw streams of information into useful business intelligence.

A Checklist for an Information Audit

Information audits are tailored to specific companies and their individual needs. The goal of this process is to pinpoint the information requirements of a company and then proceed to recommend solutions to satisfy these requirements.

Basically the process, if one follows the Meyer approach, is composed of four steps:

- Selecting what needs to be known
- Collecting the information
- Transforming this collected information into finished products
- Distributing the finished product to the appropriate staff

Selecting what needs to be known. According to Meyer figuring out the right things to know is one of the trickiest, least understood, and underrated jobs. To perform this feat requires not so much of an expertise in one or more fields, but the ability to recognize what factors will influence that particular issue or area of concern.

The process is begun by reviewing the objectives that have been outlined by the CEO or management committee. Meyer provides an example of this process in a business environment:[1]

> The CEO of an aluminum manufacturer wants to improve sales of the company's pop-top beverage cans. To do this requires an assessment of the prospects for growth in the beverage industry. This is the obvious information that would be required. A person with experience in performing these audits would most certainly look beyond the obvious to, say, assessing the prospects of Third-World aluminum producers into the canning business. Even this might be obvious to some, so we need to go deeper into the assessment and evaluate producers of other materials that could perhaps replace aluminum cans. In essence, this example demonstrates the need to think about issues in a multidimensional way.

Collecting the information. Once what needs to be known has been decided, we can turn to collecting the appropriate information. There are several categories of information.

[1]Meyer, Herbert E. *Real-World Intelligence*. Grove and Weidenfeld, New York. 1987.

First, there is *information that is already available in-house* either residing on some corporate database, on a distributed database (perhaps on some PC) and paper files. The next category of information can be referred to as *public information*. This is information that is on the public record, available in the form of magazines, newspapers, and information from public agencies.

The next category is *private information*. This is information that is not publicly known but is available for a fee. Much of this information is available online through one of the many sources listed in the appendix of this book. But Meyer goes one step further in this category. He suggests that a lot of this information can be obtained through old-fashioned legwork. If a company wants to know whether Singapore or Taiwan would be a good place to locate a manufacturing business, someone should be sent to scout around Washington, New York, London, Zurich, and Tokyo meeting with consultants and political figures who could share their understanding of Singapore's and Taiwan's economic and political prospects. Meyer also suggests that this person should also seek out universities to try to locate an expert on the topic. Ultimately this person will visit the countries in question and talk with as many people, from as many walks of life, as they can to find out what is likely to happen in these countries years from now.

The final category of information is what is known as *secret information*. This is information privately held by competitor companies. Unfortunately, most of this information is impossible to obtain legally.

Transforming the collected information into finished products. Deciding upon and then collecting the information is only half the battle. For the information to be truly useful, it must be presented in analytic reports which provide, according to Meyer, "the best judgments, conclusions and projections based on this information."

Transforming this data into useful information is a multi-step process. These steps require a team to study the material and then to debate what the material actually means, whether it is accurate, and whether it harbors any inconsistencies. At this point, all facts will be verified, experts consulted, and thesis developed and tested.

At this step we depart from the Meyer approach. Although Meyer actively uses technology in this process, his multi-step approach relies on manual efforts of intelligence officers. These people argue over the facts, and then make a decision as to the correct interpretation of the data to be delivered to the CEO or other staff member.

In the info-marketing approach, this step is replaced by loading all collected information into a technological toolset by automatic analysis and distribution. The toolsets described in this, and other, chapters in this book have the power to make these types of judgment calls.

Distribution of the finished product to appropriate staff. Information should be presented to the staff members appropriate to that staff members level within the organization. Certainly executive information systems, as described in a previous section, provide this capacity.

THE METAMORPHOSIS OF LINCOLN NATIONAL

Lincoln National Corporation is a $23 billion insurance and financial services company located in Fort Wayne, IN. As is the case in any company of Lincoln's size, the process of collecting, interpreting and disseminating information was time consuming at best and hit-and-miss at its worst.

As is true for nearly any financial services company, one of the most demanding business problems Lincoln National faced was its need to digest large amounts of information. In the past this had been done by issuing a daily news digest, as is done in the majority of companies. This paper report basically formed the baseline of information around which Lincoln executives and managers made their strategic decisions.

Understanding that the method of creating this daily report left large gaps of business intelligence unaccounted for, the Information Services group began the development of a corporate-wide information retrieval and telecommunications system using information auditing techniques, called the Office Productivity Network. Today this network is used to collect and disseminate business intelligence company-wide.

Consisting of word processing, spreadsheet, database management systems, desktop publishing, financial modeling, and more, the Office Productivity Network's most important component is its electronic mail feature. This software actually disseminates business intelligence to the appropriate staff.

Given the proliferation of productivity software (i.e., word processors, spreadsheets) on the network, one would suppose that a large part of the business intelligence disseminated to staff was internally generated. This turned out not to be the case. Lincoln discovered that most executive needs were not for information in the corporate database. In general, the executives at Lincoln got their business intelligence from the various news sources. This, then, was what they wanted in their Morning Report.

The Information Services group has been able to create an automated morning report that retrieves, searches, and correlates textual external data to assist staff. The Morning Report's intelligence comes from a variety of external sources including *The New York Times* and *BusinessWeek*. It has given Lincoln the ability to analyze information from a wide variety of other sources as well. At times, data in an automated format is not available. This is especially true for information from internal documents, paraphrased documents, and anecdotal information.These are all obtained using high-speed scanners which can convert printed material into the automated format that is required.

The Morning Report is viewed, by Lincoln, as a business intelligence gathering tool that feeds information into their executive support system. Even at this preliminary level, Lincoln has seen some significant productivity improvements. Prior to attending meetings, staff can review the pertinent information, negating the need to brief meeting attendees so that meetings can move forward more quickly. Perhaps the greatest benefit of all is improved communications within the company permitting key executives to make better decisions and allowing the company to avoid the inevitable filtering effect that so often happens as information makes its way through the corporate hierarchy.

Lincoln uses its executive support system for strategic planning and competitive analysis. Along with external information, Lincoln management can analyze internal sales data, competitor activities, field reports from sales staff, as well as competitor's financial data to determine the best way to compete.

Electronic mail forms the basis with which Lincoln can collect and disseminate data. It is used to communicate with the sales force in the field.The sales force, in turn, collects competitive data and enters it onto preformatted screens providing quick feedback to the strategic planners.

Competitive analysis is a major component of Lincoln's planning process. Using the information entered by the salespeople in the field, Lincoln builds a profile of each competitor's strengths and weaknesses. This is done by identifying the factors that are considered critical for each line of business and then ranking each competitor's capabilities in the same area. At the same time, the same criteria is used to rank Lincoln's own capabilities in those same areas. Using a side-by-side comparison of competitor versus itself, Lincoln can evaluate whether or not it is weak in the critical factors needed for success in any particular product line. If a perceived weakness is noted, Lincoln formulates a plan to strengthen itself in that particular area. At the same time Lincoln's marketing plan is modified to focus on its key strengths, while minimizing its weaknesses, as uncovered during this competitive analysis.

This marketing plan, as well as plans from the other eleven lines of businesses, is sent through the electronic mail system to the corporate office where it is consolidated and then sent to the CEO for final approval. One of Lincoln's greatest strengths is its ability to track and process competitor's data and then relate that data to its own data to further strengthen their own product and marketing plans. Being able to monitor what a competitor is doing requires a combination of technology and techniques.

WHAT'S THE COMPETITION DOING?

For those that wish to embark on a program of gathering competitive intelligence, luckily one innovation permits even the smallest of companies to drag important information out of hiding. This was the invention of the PC. Along with the PC came a plethora of information services that can potentially offer a firm all of the competitor intelligence it needs.

In order to perform the task of searching for competitive intelligence alone, it is worthwhile to first review several techniques that are performed in industry that will assist in putting the found information into perspective.

Combustion Engineering's Competitor Analysis

The philosophy behind Combustion Engineering's technique, is that information coupled with the experience of a seasoned industry manager is more than adequate to take the place of expensive experts in the field of competitive analysis.[2]

The goal behind Combustion Engineering's technique is to analyze one competitor at a time to identify strategies and predict future moves. The key difference between this technique and others is the level of involvement of senior managers of the firm. In most companies research is delegated to staff who prepare a report on all competitors at once. Combustion Engineering's method is to gather the information on one competitor and then use senior managers to logically deduce the strategy of the competitor in question.

Combustion Engineering uses a five-step approach to performing competitive analysis. Each will be discussed in turn.

Step 1—Preliminary Meeting. Once the competitor is chosen, a preliminary meeting is scheduled. This meeting should be attended by all senior managers who might have information or insight to contribute concerning this competitor. This includes the CEO as well as the general manager and managers from sales, marketing, finance and manufacturing. A broad array of staff attending is important to this technique since it serves to provide access to many diverse sources of information. This permits the merger of external information sources as well as internal sources collected by the organization such as documents, observations and personal experiences.

At this meeting attendees should agree that they will spend a specified amount of time collecting more recent information about the competitor. At this time, a second meeting should be scheduled which will review this more recent information.

Step 2—Information Meeting. At this meeting each attendee will receive an allotment of time to present his or her information to the group.

The group will then perform a relative strengths/weaknesses analysis. This will be done for all areas of interest uncovered by the information obtained by the group. The analysis will seek to draw conclusions about two criteria. First, is the competitor stronger or weaker than your company? Second, does the area have the potential to affect customer behavior?

Combustion Engineering rules dictate that unless the area meets both of these criteria, it should not be pursued further either in analysis or discussion. Since managers do not always agree on what areas to include or exclude, it is frequently necessary to appoint a moderator who is not part of the group.

Step 3—Cost Analysis. At this point, with areas of concern isolated, it is necessary to do a comparative cost analysis. The first step here is to prepare a breakdown of costs

[2]"Calculating Competitor Action: Combustion Engineering's Strategy," *Management Briefing: Marketing*. October-November 1988. The Conference Board.

for your product. This includes labor, manufacturing, cost of goods, distribution, sales, administration, as well as other relevant items of interest as necessary.

At this point, compare the competitor's cost for each of these factors according to the following scale:

- Significantly higher
- Slightly higher
- Slightly lower
- Significantly lower

Now translate these subjective ratings to something more tangible, such as slightly higher equivalent, say 15%. By weighting each of these factors by its relative contribution to the total product cost, it is possible now calculate the competitor's total costs.

Step 4—Competitor Motivation. This is perhaps the most intangible of the steps. The group must now attempt to analyze its competitor's motivation by determining how the competitor measures success as well as what its objectives and strategies are.

During the research phase the senior manager (and/or his or her staff) gathered considerable information on this topic. By using online databases, it is possible to collect information about promotions, annual reports, press releases, and the like. In addition, information from former employees, the sales force, investment analysts, supplier, and mutual clients is extremely useful and serves to broaden the picture.

Based on the senior managers understanding of the business it is feasible to be able to deduce the competitor's motivation. Motivation can often be deduced by observing the way the competitor measures itself. Annual reports are good sources for this information. For example, a competitor that wants to reap the benefits of investment in a particular industry will most likely measure success in terms of ROI.

Step 5—Total Picture. By reviewing information on the competitor's strengths and weaknesses, relative cost structure, goals, and strategies a total picture of the firm can be created. Using this information, the group should be able to use its own insight into the process of running a business in a like industry to determine the competitor's next likely moves.

For example, analysis shows that a competitor is stronger in direct sales, has a cost advantage in labor, and is focused on growing from a regional to a national firm. The group would draw the conclusion that the competitor will attempt to assemble a direct sales effort nationwide while positioning itself on the basis of low price.

Phantom Analysis. Combustion Engineering has also devised an approach to dealing with the situation where an outsider enters the market place. Here, the strategy above obviously wouldn't work.

Using the same group of people gathered to analyze competitor strategy, this exercise requests the group to look at the market as an objective third party would. The task is to design a fictitious company that would be able to penetrate the market successfully. The group would then compare this fictitious company with the competitor firms in the industry to see whether any of the traditional competitors could easily adopt this approach.

When Combustion Engineering's phantom analysis uncovers a strategy that traditional competitors might adopt easily, they adopt this strategy as a preemptive move. When this same analysis reveals that an outsider could penetrate the industry by following this strategy, Combustion Engineering attempts to create additional barriers to entry. This includes forming an alliance with an outside company to pursue the phantom strategy itself.

Missing Piece Analysis

A complementary strategy to Combustion Engineering's competitor analysis methodology is one developed by F. Michael Hruby, founder of the Society for Competitor Intelligence Professionals.

Hruby's *Missing Piece Analysis* also attempts to anticipate competitor moves, but it does this by identifying key weaknesses in the competitor.[3] By concentrating on the competitor's weakness, the great wealth of information on that competitor can be turned into usable, action-oriented intelligence.

The methodology for performing Hruby's missing-piece analysis is to analyze the strengths and weaknesses of a competitor in six areas. In each of these areas, the competitor is compared to the company doing the analysis.

Product. Compare the strength of the competitor's product from the consumer's point of view.

Manufacturing. Compare capabilities, cost, capacity.

Sales and marketing. How well does the competitor sell the product? Compare positioning, advertising, sales force, so on.

Finance. Compare financial resources and performance. How strong are these relative to a requirement for launching a strong competitive thrust?

Management. How effective, aggressive, and qualified are the competitor's managers?

Corporate culture. Examine values and history to determine whether a competitor is likely to enter or to attempt to dominate a market.

[3]Hruby, F. Michael. "Missing Piece Analysis Target's the Competitor's Weakness," *Marketing News*. January 2, 1989.

The goals of this exercise are to identify weaknesses in each of these areas as well as to see whether any one of these weaknesses stands out as a major vulnerability. According to Hruby, most companies have a key weakness—or "missing piece"—that can be exploited.

To perform this technique requires that the competitor be rated in each of the six areas listed. Ratings are done on a scale of 1 to 5 with a 1 being very weak, 2 is weak/noncompetitive, 3 is adequate/average, 4 is very strong/competitive, and 5 is excellent/superior.

Hruby summarizes these scores in a competitive-strengths matrix which lists the names of the competitors down the right hand side and the competitive areas of interest across the top. Scores are entered in the appropriate cells. The worst score for each competitor should be highlighted. This is the competitor's weakest point, and it should be monitored accordingly.

For example in a list of five companies, Company A and Company B may both be weak in the finance area. This means that they do not have enough strength to launch major advertising campaigns to bolster a new product. What this means is that if the company doing this analysis is ready, willing, and able to spend a lot of money, a new product launch would most probably be successful.

Company C scored low in the product category. This means that its product is not as good as the company doing the analysis. In this case, an advertising campaign emphasizing product differences would serve to grab some market share from Company C.

Company D, on the other hand, scores strong in all matrix areas. Given a strong product and an aggressive management team, this company is likely to make an aggressive move. Perhaps a new product launch or major advertising on an existing product. They might even reduce costs. Company D certainly bears watching.

Company E has a weak product but a good financial position. It might launch a new product. However, its weak management structure might defer any product launch.

In summary, upon analysis of the competitive strengths, matrix one would deduce that a combination of a strong financial position and competent management are a mix that indicates a strong likelihood of aggressive action on the part of the competitor. By using this analysis on information obtained from various sources, in particular information obtained through the use of information technology techniques, it is quite possible to keep tabs on what the competition is up to.

Monitoring the Competition's Moves through Information Technology

These and other techniques rely on information. Although much information can be obtained through salesmen, ex-employees of rivals, suppliers, and customers, this is merely a trickle, compared to the wealth of information available to a information-savvy company.

Meyer and Pincus' firm, Real-World Intelligence, described above, performs the intelligence audit to make sure that the right information is retrieved and delivered to the right person. Then it proceeds to modify the stream of information already retrieved by the company—adding new data streams and deleting inappropriate or useless streams.

But with the proliferation of PCs and access to some of the hundreds of databases, a savvy company which perhaps cannot afford the services of a consultant specializing in this business, or wants to do the analysis in-house can learn about its competition's sales, size, profit and loss, credit history, officers, organization, R&D efforts, intellectual property, and a host of other informational tidbits.

A word of advice is warranted here: Having a PC and subscriptions to half a dozen online services does not guarantee good intelligence. Before embarking on this journey solo, a company should make sure that the person doing this work is experienced in searching the online database(s) in general. Searching for information through any of the subscription services is often expensive. Spending time online "playing" around with the information trying to find some meaningful intelligence is even more expensive. Most professional online searches spend more time off-line than online performing their research. Off-line time (i.e., free time) is spent in determining which of the online databases should be searched. Next, time is spent formatting queries so that precious and expensive online time can be spent being productive immediately.

To cut costs, a company should not perform research on all the competitors in the field. Choosing the top three to five is usually sufficient. The next step is determining the most relevant areas of competition. For example, car rental companies compete on several factors, including pricing and location. Performing a search on the types of cars that they are buying yields intelligence that is of little use, but finding out that a rival just obtained counter space in hotels in major cities is very valuable.

The following steps are recommended by Kirk Tyson who is president of the Competitor Intelligence Group. Tyson recommends a tiered approach: daily intelligence briefings; profiles of your competitor's essential finances, market, and product lines; and focused data such as sales strategies.

Daily Briefings. Few managers do not know what a clipping service is. Before computers, these services would scan thousands of newspapers, magazines, and journals and literally clip out anything of interest to a particular customer. The customer in question usually indicated his or her parameters of choice, for example, all information about the aerospace industry, and the clipping services searched for this information daily.

With the advent of online subscription services, clipping has become automated. Not only can this information be viewed online, it can be printed on a PC printer and even downgraded to a variety of PC software.

Kirk Tyson thinks of this as the bedrock of a competitor intelligence program. Luckily there are a host of easy-to-use services out there that will assist the company that decides to "go it alone."

Probably the most popular of services is CompuServe's Executive News Service. CompuServe is one of the oldest and most popular of online services. Used by consumer and business person alike, it is both economical and easy to use. Once logged onto CompuServe the simple command, GO ENS, gets the user access into this service. CompuServe's eminently readable manual makes it easy to get into the service. In general, the Executive News Service puts the resources of The Associated Press, Reuters Financial Report, The Washington Post, OTC NewsAlert, and McGraw-Hill News to work.

There are several options for the user. One is simply to review any of the stories from the wire. There are many. For example, the Associated Press database carries over 7,000 stories daily. Since the purpose of this exercise is to target the exact information needed and to bypass irrelevant details which save time and money, the best way to approach the situation is to create a clipping folder.

Under CompuServe's Executive News Service, that is exactly how it is done. Here the thousands of stories are scanned on the company's behalf, and only the ones that meet the specific criteria pre-set by the company are stored in an automated clipping folder.

Specifying areas of interest is usually done by entering keywords. For example if a company were interested in saving all information about APPLE COMPUTER the keywords might be APPLE COMPUTER, INC. Since there are other online clipping services, such as GEnie's NewsGrid Executive News, time must be spent on each one honing the keywords for the most effective and economical use of computer time.

Although automated clipping services are alike, they differ in their strengths. For example, it has been said that GEnie's NewsGrid covers the international business market slightly better than does CompuServe. On the other hand, Dow Jones/Retrieval offers the widest range of business wires in a single package.

For the most part, most companies will soon discover that variety is indeed the spice of life and will sign up for more than one service simultaneously.

Rival Profiles. Clippings will provide the raw news that is required to keep abreast of competitor moves. But at some point, some analysis must be done to predict competitor moves (see the last few sections for techniques). To perform these analysis it will be necessary to collect information of another kind to create a rival profile.

The profile should contain information to track trends in market share, profits, management information, products as well as other criteria desired by the company.

CompuServe's IQuest is known as a gateway to hundreds of databases. Dialog, an example of its many categories discussed above, provides a similar array of information. BRS, Dow Jones, Nexis, and NewsNet are also good bets.

Knowing what information to search out is an art unto itself. It pays to study carefully the manuals that each service provides. In each, a list of databases and the type of data contained within is spelled out. For example, a good place to find out information about a public competitor is through its filings with the Securities and Exchange Commission

(SEC). Through the 10Ks and various other filings, it is possible to uncover such things as officers, ownership, income statements, and even a five-year business summary.

Private firm information can be obtained through Dun & Bradstreet. Through Dunsprint, available on CompuServe, it is possible to track information on sales, balance sheets, and even credit on some two million private companies. Additionally, through NewsNet it is possible to access TRW's Business Profiles for the same information. Information on foreign firms is available through many of these services as well.

Perhaps the key database in researching a true market picture is the Investext database available through CompuServe. This database stores over 200,000 securities analyst's reports on more than 12,000 public companies—including some foreign firms. Investext reports come from more than 90 of the most well-regarded securities analyst firms and includes such information as market share, gross margins on sales, research and development expenditures, cash flow, and other important financial statistics.

Industry newsletters are available online through NewsNet. Also available through NewsNet is the PR Newswire which tracks press releases. These two sources will assist in tracking trends and releases of newer technologies into the industry.

Sharp Focus. The information obtained now was general competitive data. How can these online services be used to answer more specific questions such as "is your competitor going to launch a new product?"

Several databases are available which provide more in-depth coverage of a particular firm's actions. One of these is Insider Watch, which is available both on Dialog and Dow Jones. This database gives up information concerning open market sales, as well as purchases and stock options exercised by over 100,000 corporate insiders in more than 8,500 firms.

Dialog also offers a wealth of biographical information about a competitor's officers. The Biography Master Index contains more than seven million references for over 40,000 companies. It also provides information such as number of employees and corporate structure.

Since corporate structure is such a tell-all concerning how a firm is organized, this information is also available on Business Dateline and Corporate Affiliations. It is available through Dow, BRS, Dialog, and Nexis. Other databases that tell-all are Financial Industry Information Service (FINIS) on Dialog, Nexis, and BRS. Those interested in merger and acquisition activity would best consult M&A filings on Dialog. Of course, no research would be complete without looking at Standard & Poor's Register—Corporate.

UNCOVERING THE GOLD IN THE CORPORATE DATABASE

While executive information system tools provide a baseline for analyzing business intelligence, there will come a time when higher order software tools will be required to sift through collected data and uncover relationships that can be mined to profitability. When developing an information technology infrastructure, these tools should not be overlooked.

While choosing an architecture for the company's corporate database standard is important and should be funded as well as supported by management, this process should not overshadow the more important issue of being able to navigate effectively through the data contained within, so that corporate data can be turned into competitive gold as in the case of Lincoln National.

Standard database query languages easily retrieve information, but only if the user knows specifically what he or she is looking for. If the request for information is vague one needs to use a more heavy duty tool.

A more difficult class of information to deal with is what can be termed unknown information. Upon analysis of a large database, it is possible to discover patterns, rules, and unexpected relationships between data items that were previously unrealized—the hidden gold in the corporate database.

This is an intriguing possibility, but not one that many companies are embarking on. This is unfortunate, since these same companies would most certainly benefit from discovering some interesting correlations between sales data and customer financial data. Before we launch into a discussion of how this technology works, a simple example of how useful this technique could be is warranted.

The New York Stock Exchange's regulatory department is charged with ensuring that the brokerage firms that are members of the Exchange are financially sound. This is accomplished by requiring these firms to file huge amounts of financial data which financial analysts who work for the Exchange review.

The preeminent tool for this purpose is a software program called the Exception Disposition Report (EDR). This report is produced by comparing the data filed by the brokerage firm to a set of statistical algorithms such as, "*If the firm's excess net capital is greater than 25% of its profits then flag this exception.*" These exceptions, in the form of rules, were developed by the financial analysts by comparing and reviewing one item of information against another item of information. The rules, called EDRs, were actually coded statements concerning the relationship of financial data item to financial data item. The financial analysts were continually meeting to improve on these rules. Since these rules were developed manually, there was always a chance that the relationship, as defined by the rule, was incorrect. A worse problem was omitting a rule—especially about a potentially volatile financial situation—altogether. What if software could be used to determine these rules automatically? Would not the Exchange's product then be a better one? A more competitive one?

An Automatic Discovery Program

The number and size of operational databases such as the one used by the Exchange's are increasing at a progressively quickened rate. Because of the number of these databases, their size, and their complexity, a tremendous amount of valuable knowledge locked up in these databases remains undiscovered. Since the tendency of most modern organizations is to cut back on staff, it follows that there will never be enough analysts to interpret the data in all the databases.

Fortunately information technology has spawned a new concept that has the ability to perform automatic analysis of large databases. The name coined for this technology is automatic discovery.

IntelligenceWare, located in Los Angeles, CA, is the market leader in automatic discovery software. IXL, an acronym for **I**nduction on e**X**tremely **L**arge database, is a unique system which analyzes large databases and discovers patterns, rules, and often unexpected relationships. IXL uses statistics and machine learning to generate easy-to-read rules which characterize data, providing insight and understanding.

The president of IntelligenceWare, Kamran Parsaye, coined the term "intelligent databases." The goal of intelligent databases is to be able to manage information in a natural way making the information, these databases, easy to store, access, and use.

The prototypical intelligent database, according to Parsaye, would have some robust requirements. An intelligent database would need to provide some high-level tools for data analysis, discovery and integrity control. These tools would be used to allow users not only to extract knowledge from databases but would also allow users to apply knowledge to data. So far, it is not possible to scan through the pages of a database as easily as it is to flip through the pages of a book. In order for the label intelligent database to be valid, this feature is necessary. Users should be able to retrieve information from a computerized database as easily as they can get it from a helpful human expert. Finally, an intelligent database must be able to retrieve knowledge—as opposed to data. To do this, requires the use of inferencing capabilities to determine what a user needs to know.

In developing the theory behind intelligent databases, Parsaye enumerated three basic levels in dealing with a database:[4]

- *Collect* data, e.g., we maintain records on clients, products, sales, etc.
- *Query* data, e.g., "which products had increasing sales last month?"
- *Understand* data, e.g., "what makes a product successful?"

[4]"IntelligenceWare." *What Can IXL Do That Statistics Cannot?* 1990. Internal paper.

In general, most current database systems passively permit these functions. A database is a static repository of information which will provide answers when a human initiates a session and asks a set of pertinent questions. IXL attempts to change this point of view by turning the database into an active repository of information, automatically posing queries to the database and uncovering useful, and sometimes unexpected, information.

This was the case for a well-known computer manufacturer who suffered sporadic defect problems in its disk drive manufacturing process that it just could not locate. Using the IXL program against a database which consisted of the audit logs of the manufacturing process, this company was able to pinpoint the particular operator who was causing the problem. The defect was then traced back to a lack of proper training.

An even more interesting case study deals with lead poisoning data from the University of Southern California's cancer registry. Analysis of this data, using the IXL program, uncovered a relationship between gender and the level of lead in the blood leading to kidney damage. Before IXL's analysis this relationship was unknown and potentially deadly.

Software such as IXL amplifies our ability to navigate and analyze information such that it can be rapidly turned from discrete and disconnected pieces of data into intelligence. But tools such as these address only one side of the competitive coin. As much as we would like to rely purely on analytics, the truth of the matter is that creativity provides a counterbalance that we cannot eliminate. Perhaps, however, there is a way to make people more creative.

COMPUTERIZED BRAINSTORMING

Human performance, enabling a person to perform at his or her full potential, perhaps that's what it all boils down to.

In the beginning, we developed technology appliances to make the drudgery of clerical work less burdensome, and even to replace humans. Later, technology began to be used to help humans sort through the massive information datastores. The age of the personal productivity appliance, the PC, began in the early 1980s and during that decade, and on into the 1990s, spurred an avalanche of productivity-enhancing tools that nearly boggle the mind. But still, the emphasis was on productivity-*enhancing*. What is readily needed by companies searching for that elusive silver bullet of competitive leadership is some sort of tool that is productivity-*producing*.

Marsh Fisher may have found that silver bullet. You may have heard of him. Fisher was the original founder of the Century 21 Real Estate empire. Any business person would take advice from Fisher. After all, the real estate business he founded is worth billions. But Fisher wanted to offer more than advice. He wanted to offer ideas. Actually, he wanted to offer competitive advantage through creativity. Fisher calls this type of software Human Performance Technology.

Fisher got the idea for creativity boosting back in the days when computers were large, monolithic, mainframes stuck away in the basements of office buildings and providing only a smattering of the functionality that has become standardly available in the 1990s. In 1964 Fisher was studying comedy writing. He noticed that most of the other students in his class were much better at being fast on their feet that he was. They seemed to ad lib a lot better than he did. So he started looking for some sort of crutch with which he could at least become competitive.

He began to study the art of ad lib-ing, and comedy in general, and he found that there is a unique association between the punch line and the set-up line. Related to both of these phrases is an assumed word or phrase. It is this word or phrase that associates the set-up line to the punch line.

When Fisher retired from Century 21, he began to study cognitive sciences which is a combination of linguistics and computer science. One of the goals of cognitive sciences is to determine whether the mind can be mimicked in the mysterious task of problem solving.

Fisher describes problem solving as the three R's: Recording, Recall, Re-associating. Recording of information is done spontaneously. Everything we see, hear, smell, or touch is stored in the grandest of all databanks—the human brain. Of course, once it's stored inside, it's sometimes quite difficult to get it back out. This is the task of Recall, or remembering. We have this massive warehouse of information stored in our subconscious and trying to find something buried away is usually quite difficult. Once an item of information is recalled, the third R is deployed. We Re-associate, or recombine, one or more items of information to produce an original creative idea.

Of course, if we had instant access to everything tucked away in our memories the road to creativity would be much less arduous. Unfortunately, as we are often reminded, as we search in vain for the name of the person that we just met in the hallway, this is usually not the case. Even if humans were all possessed of the gift of instant recall, there's still that third R to contend with: Re-association, the creative R. The R that gives us creative leverage.

In the 1960s Fisher wanted to give humans a creativity shot in the arm by publishing a book of associations. By the time he was ready to do it, the PC had become so ubiquitous that he decided to write it in software. This is when IdeaFisher was born.

IdeaFisher claims it can help us make something quite novel out of fragmented, and seemingly useless, bits of information. Here's how it works.

In this example, we join the Product Planning Group of a Sock Company. They are developing a plan to sell more socks in the summer months. In undertaking a challenge of this nature, it is important that the strategy team understand that this process actually consists of four processes: understanding the goal, defining the strategy, naming the product, and finally identifying the key attributes of the product for advertising and product positioning. We now take a quick tour through IdeaFisher to show how software of this type can assist the creative process.

Understanding the Goal

Our first step is to understand the specific challenge fully. In order to do this, we normally get a group together to brainstorm. To pick the goal, and the resultant ideas apart and piece them back together into a solution. Brainstorming relies on a series of questions and answers. But what if you can't come up with the right questions? Fortunately, IdeaFisher comes with a question bank, called the QBank, which is pre-loaded with some 3,000 questions to spur the process on. Questions are categorized along several lines including developing a story or script, developing a new product or service, developing a name, title, theme or slogan, and developing a marketing strategy or promotional campaign. Since our goal is to develop a new line of socks, we'll choose developing a new product or service. Here we look through a series of questions and pick the ones most appropriate to our goal. Questions such as: Does the audience or customer fit a particular category—a distinct type of thought and behavior (a stereotype)? What are the customer's relevant physical traits in addition to age and sex? List the person's relevant pyschographic traits. What product or service characteristics are most important to this customer?

After each question is selected the strategy team enters its responses directly online. The team brainstorms answers such as adult males and females of all ages, people at home and outside, people who like to be outdoors, gardening, bird-watching, as well as requirements such as: socks should be fashionable, socks should be useful in outdoor activities, socks should be in a fabric that does not hold moisture and is not hot.

What we have in our automated notepad, called the Idea Notepad in IdeaFisher, is a set of rather prosaic responses to some very-well directed questions. Once this prose is filtered into a series of key concepts through the Filter Question process, the team is ready to target the most relevant key concepts and move on to the next step.

After much debate our team finally targets the key concepts of bird watching, color coordinated, gardening, moisture, and useful in outdoor activities. This then is the breakdown of the key elements in their marketing strategy. Defining a specific strategy is the next activity.

Defining the Strategy

To develop a feasible strategy to sell more socks in the summer, our team will use IdeaFisher's IdeaBank. This is a fascinating repository of 28 major categories broken down into 387 topical or subcategories containing over 60,000 words. By associating these words and phrases it is possible to wind up with a staggering 705,000 direct associations. Our team wants to begin with the socks key concept. By highlighting this word in the notepad, the program will display all of the topical categories that contain the word socks. Apparently, IdeaFisher has 10 topics that deal with socks including black/gray, cleaning/dirty/clean, clothing/fashion/style, and push/pull/attract/repel. This last topic intrigues one of the members of the team so it is highlighted to see the section titles on the next level. It turns out that there are 945 idea words or phrases associated with push/pull/attract/repel neatly categorized into 11 groupings such as things/places, things that repel, things that attract, and abstractions/intangibles. The team decides to pursue things that repel. Highlighting this, they find 26 intriguing items such as anti-icer, body armor, car wax, and mosquito repellent. Certainly these are things that repel.

Marsh Fisher describes the act of creativity as one involving coming up with new ideas whose revelation excite the creator so much that he or she exclaims, "A-Ha!" He calls this the "A-Ha experience." The members of our fictitious team experience this feeling when they realize the interesting possibilities in mosquito repellent on socks.

Naming the Product

Now that the team has decided on their novel product, they need to come up with a good name for it, a good hook. Selecting a name for a product or service has many elements: It must be easily remembered. It must be descriptive. And it must tie in with the customer's perceived needs and values.

The team decides to compare two topical categories using IdeaFisher's Compare feature to create a unique name for the socks. They want to associate disparate ideas to merge them together into a single word or phrase that creates a novel hook for their new product. Picking socks as the first key concept to compare, the team is prompted to pick one of the many topical categories containing this word. The head of the team recommends that the team pursue limbs/appendages. Outdoors is the second word that the team wants to use in the comparison. Again, a list of topical categories containing the word outdoors is displayed. This time the team picks camping/hiking/mountaineering. IdeaFisher takes over at this point and produces a listing of words and phrases found in both of the topical categories selected. This list serves as a jog to creativity. The team looks through the list bypassing blister, footing, footpath, and 50-odd other words and phrases. One word on that list jumps out at them as the perfect name for the new line of socks. Surefooted.

So far, all in one sitting, the team has brainstormed the meaning of their challenge, defined their strategy and named their product. In a space of hours, rather than days. All that is remaining of their task is to identify key attributes for advertising and product positioning. In order to do this the team decides to explore the key concepts stored in the automated notepad in more detail.

Identifying Key Attributes

The team decides that the key attribute they want to emphasize is summer uses of socks. They select and highlight the word "summer" listed in the automated notepad. Ultimately the team winds up with a host of summertime activities and hobbies that people in the target market might enjoy more with Surefooted socks.

The final results of the IdeaFisher session, which began just a few hours before, is as follows:

1. Socks that keep bugs away.
2. Color-coordinated to current athletic clothes and incorporating reflective material in some models.
3. Lightweight material that doesn't hold heat or moisture inside.
4. An insect repellent fabric that could be used for clothing, sleeping bags and tents.

Computerized Brainstorming in Use

When Pabst Brewing Company customers began calling out PBR when ordering Pabst Blue Ribbon Beer, Pabst knew they were on to something hot. But they needed more than just PBR to create a hot jingle. They needed some inspiration.

That's when they turned to IdeaFisher to assist in writing a jingle that is based on abbreviations. This is what their fishing caught, "I'm gonna give my thirst some TLC, just PBR me ASAP." Pabst is not alone, IdeaFisher has been used to write proposed copy for everything from Bud Beer to the Discover card.

Even the world's foremost advertising agency uses it. Saatchi & Saatchi Advertising rarely runs out of ideas. But sometimes, even in the most creative of firms, the well runs dry. That's when this New York City-based company turns to IdeaFisher which Saatchi's Perry Davis calls a "creative springboard."

One way in which Saatchi uses IdeaFisher is as an idea-generator for focus groups. The traditional method of conducting a focus group is to get 8 or 10 people around a table and to go through a structured session of key ideas or seminal words that people will react to. The problem with this method is that it is very possible to travel down a path and ultimately lose sight of the forest because you keep examining the trees. In other words, it is quite possible to reach a very dead end. Using IdeaFisher changes all that and permits Saatchi to create the most rigorous of focus sessions, and at the lowest possible cost, for the client. According to Davis, "IdeaFisher gives us the broadest range of opportunities to wrap one's hands around. It's one of the all-time great tools."

Info-Marketing Redux

Without information technology, organizations would be reduced to searching for the proverbial needle in a haystack. In fact, without information technology many of the products or services companies offer for sale today simply would not be available.

Mauna La'i guava drink is a case in point. When Ocean Spray Cranberries performed test marketing for this tropical fruit drink, initial results were less than promising. What this translates to is that the product would not meet Ocean Spray's sales objectives. But Ocean Spray decided to recheck these results through Information Resources' electronic test market service. Surprisingly, the IRI analysis showed that the depth of repeat purchases would make up for lack of volume in the trial.

A plethora of package-goods companies are finding that the use of information technology to assist in uncovering golden nuggets amongst an avalanche of data is key to expanding market share. At Cadbury Beverages, much of this market information comes from scanner data sold by companies such as IRI. Using IRI's decision support software, Dataserver, Cadbury wafts through the data for a first pass. More information technology is used for a refined view of the data in the guise of IRI's Coverstory. Not only do these innovative software tools permit Cadbury to find new openings in the competitive marketplace, they enable Cadbury to perform intra-prospecting.

By far, the perfect "new" product is one that exceeds sales projections and whose time to market as well as developmental cost are low. Cadbury first looks for new opportunities in its own product categories (i.e., intra-prospecting). In these analysis, Cadbury scrutinizes subcategories of products that perform better than average. For the most part, though, new opportunities are found outside of current product boundaries.

Holland House cooking wines is a case in point. Upon researching the marinade category, Cadbury discovered that this was a market niche that was underpopulated by the competition. This was discovered by subjecting the data to intensive analysis in hopes of finding potential buyers. When the market researchers combined this information with information on product sales and competition, Cadbury management gave the green light.

Conclusion

Knowledge is power. Ultimately, *this* secretive gathering of information is what keeps a corporation abreast of its competition. A business' very breath depends on the bits of data it can gather on markets, technologies, customers, and competitors, as well as external political trends. It's information easily obtained through libraries, online services, new employees hired away from your competition, and industry gossip at trade shows.

Professor Benjamin Gilad of Rutgers University in New Brunswick, NJ compares competitive intelligence to Cerberus, the three-headed dog guarding the gates of the underworld in Greek mythology.[5] Gilad breaks down the distinct types of intelligence models into three categories: one, a non-intrusive monitoring system for early warning; two, tactical field support; and three, support strongly aligned with top-management strategy.

A case in point can be illustrated by 3M. By keeping an eye on foreign mission reports, the company discovered—six months before reportage by the electronic and print media—that one of its key customers was building a refrigerator compressor plant in Lithuania. At MassMutual, intelligence assists the field sales force with monitoring deals offered by competitors on new life-insurance products. At Southwestern Bell Telephone, mock-battle exercises enable strategic planners to be on their guard for multiple contingencies in the quick-paced telecommunications industry.

CIOs can play a major role in this intelligence gathering—but its a role that may take the CIO outside of traditional role play. David Drew, vice president of information technology for 3M in St. Paul, stated that in intelligence it is more important to have the right people in the organization analyzing the information that it is to have a system organizing information. He adds, that the more we try to institutionalize competitive intelligence, the less we seem to accomplish.

Acknowledged by intelligence experts to be one of the best of the big companies, 3M, according to Leonard M. Fuld, an intelligence consultant from Cambridge, MA, are further along than most other companies because they are teaching people throughout the organization the value of information.

Surprisingly, according to Herbert E. Meyer, former vice chairman of the National Intelligence Council, smaller and medium-size companies are embracing intelligence at a quicker pace than the Fortune 500. The larger companies are just too muscular, and there are too many layers of bureaucrats.

When an organization is small, explained Jan Herring, vice president for the business-intelligence practice at The Futures Group, a consulting company in Glastonbury, CT, the intelligence gatherers are the president and the salespeople. They are the ones constantly observing the external environment to sense if there is a threat or opportunity, talking to each other and acting on it. Once a company grows to more than fifty people, information gets mislaid and distorted as it slinks through management layers.

The streamlining of organizations, the increasing desire to rush to market and rapid globalization are energizing the demand for intelligence. At MassMutual, the halving of the product-development cycle has heightened the need for intelligence in the once highly-conservative insurance industry, according to Peter Daboul, the company's senior vice president of information systems services. Financial services is no longer an easy-buck industry.

[5]Baatz, E.B. "The Quest for Corporate Smarts," *CIO*. September 15, 1994. pp. 50-58.

To increase an organization's chances for success in deciphering the corporate intelligence code, the following are seven key insights from the experts.

Information is not Intelligence. Fuld advises that data, information, and intelligence may appear interchangeable, conceptually, but that's like saying you can use a kitchen knife to do neurosurgery.

For example, data is when one person is aware that his or her competitor had 100 employees last year and another person know that the competitor has 80 this year. Information is the distilled essence when these two people pool their data. Intelligence occurs in doing the analysis, noting that the competitor has a tight fiscal policy, and deciding that the company would make a good acquisition.

With regard to the CIO, these distinctions are critical. Computers and communication systems may ride roughshod over the information flow but intelligence gathering is essentially a tiring, human function.

Motorola Inc. may be the current intelligence gathering leader, but this was not always the case. Motorola had several failures before it achieved success. The story goes that chairman of the Executive Committee of the Board, Robert W. Galvin, fondly recalling his days as a member of the Presidential Foreign Intelligence Board, expended time and money assembling an extensive electronic library in the early 1980s. But according to Fuld, it became difficult to sell this expenditure with results and benefits being so intangible. To state succinctly, too much attention devoted to an information technology system can doom the intelligence system.

Intelligence Should be Action-Driven. For an intelligence system to operate smoothly and properly, the right information must be relayed to the right person in a timely manner. Observe once more, Motorola: Herring, who currently consults for The Futures Group, was the ex-CIA officer who helped Motorola implement a more-action-driven intelligence program in 1985. Describing his goal, Herring explained that it is important to note what intelligence is not at Motorola. It is not a library, not a collection of databases, and not trying to serve everyone.

Working with concrete strategic goals in mind, Motorola's intelligence group discovered that Japanese manufacturers were shifting their budgets from manufacturing to research and development. Motorola reacted by moving a portion of its R&D effort to Japan in order to take part in this new environment. Knowledge gathered that is related to business goals become a lot easier to justify, says Herring. The intelligence unit is still alive and well at Motorola, and it has grown considerably since Herring left.

The Vacuum-Cleaner Approach is Down, if not Quite Out. While this directly relates to the two intelligence insights previously detailed, it rates individual attention because of its direct involvement in creating technology tools to support an intelligence program. Describing the flaws of some current information systems, John E. Prescot, professor of business administration at the University of Pittsburgh and former president of the Society of Competitive Intelligence Professionals, says that a lot of information systems are not designed to provide the timely, accurate, reliable information that management needs. Companies relied upon the vacuum-cleaner approach, gathering everything anyone would want to know. But these systems suffered from data overload, delivering too much and too general information. What businesses systems need today are systems precisely tailored to meet specific needs; systems that help executives improve decision making on a project-by-project basis.

Meyer aides businesses in avoiding the vacuum cleaner approach by first creating an individualized intelligence profile for the company. A series of questions enables Meyer to sharpen the focus of the information-gathering process to allow it to yield intelligence that answers business questions.

Although intelligence-gathering experts like Meyer may dismiss the vacuum-cleaner approach, the vacuum cleaners themselves are becoming more refined, more powerful and easier to use. The CIA-type systems are slowly winding their way to the business community.

Ruthlessly Stamp Out information Distortion. At the April 1994 annual convention of The Society of Competitive Intelligence Professionals, Robert E. Flynn, chairman and CEO of The NutraSweet Co., stated that competitive intelligence must be a source of unfiltered, unbiased news.

Fuld advises organizations to install a coding system that evaluates expertly each piece of information on the network: a "one" has at least two supporting sources, a "two" has a single supporting source, and a "three" is a rumor. This coding system, according to Fuld, speeds information flow and allows people who are analyze the data to separate the wheat from the chaff.

Money Helps, But Support Must Go Deeper. Though a relative newcomer to the employment of systematized intelligence, MassMutual is not ignorant of the realization that top corporate support and employee buy-in mean more than money. The go-ahead for the set-up of an intelligence-gathering apparatus came with the arrival in 1988 of new CEO, Thomas B. Wheeler. Richard M. Dooley, a newcomer to business with no previous experience in competitive-intelligence gathering, was assigned the task.

In 1992, concurring with the internal systems department that purchasing an outside system to automate the information collection and dissemination function was the most sensible approach, Dooley perused the stock of the two or three vendors available at the time. After selecting Quest Management Systems' Incite program, he initiated a door-to-door sales pitch, gathering $15,000 from the strategic planning department and $5,000 each from three departments within the largest line of business. This funding plan, according to Dooley, was the soul of the program's success.

Once infused with cash and up and running, Dooley was able to ignite participation in the intelligence program, growing from 10 to 140 users in a single year.

Intelligence Is An Organic Process. Experts agree that it takes at least three years before a novice intelligence unit is firmly entrenched. As Southwestern Bell Telephone in St. Louis illustrates, this unit should always be dynamic. Since the formation of its intelligence-gathering system in 1989, the unit has evolved constantly—including a complete personnel overhaul.

At the onslaught, the intelligence function at Southwestern Bell was centralized, with a singular data collector and a singular analyst each dedicated to a separate functional area (marketing, technology, finance, government, regulations). When this set-up failed to provide actionable results, the team reorganized by merging the collector and analyst roles and refocusing on such key telecommunications technologies as wireless, cable telephony and the like.

According to Karen Wolters, a Southwestern Bell product-development manager who was a member of the of the original intelligence team, acquiring the information from the sales and line staff provided the most difficulty. The challenge was to get these employees to do a core dump of knowledge gotten from the day's tet-a-tetes and phone conversations.

Counterintelligence Counts, Too. As companies continue to be aware of the vulnerability of their faxes and e-mails to prying corporate eyes, their desire to counter attack with counterintelligence continues to grow. Preventing internal information leakage is becoming a major goal.

Intelligence gathering is a gray area but professionals in the field insist they do not cross the line...the information they obtain is publicly available.

Few companies that are seriously interested in being competitive are not using one or more of the techniques described in this chapter.

The techniques of info-marketing can assist the savvy organization in more areas than just marketing. Once these tools and techniques move into the mainline corporate environment, info-marketing just might become the pinnacle upon which corporate success relies upon most.

CHAPTER 3

The Productivity Paradox Births a New Form of (Re)engineering

by Jessica Keyes

The premise behind the "productivity paradox" is rather simple. Businesses that expected a big productivity payoff from investing in technology are, in many cases, still waiting to collect. According to research by Paul Attewell, then professor at the State University of New York at Stony Brook, there is an absence or paucity of productivity payoffs from information technology (IT), despite massive investment in IT over the last 25 years.

In an average year, U.S. companies invest some $51 billion in hardware, $20 billion in purchased software and over $44 billion in computer services, representing 25% of a firm's capital stock. This huge commitment is made at the expense of other kinds of investments while the U.S. placed its bet that IT investment would raise economic productivity.

During the 1980s U.S. businesses invested an awe-inspiring $1 trillion in information technology with little payback. Economists were the ones who invented the term "productivity paradox" as a way of explaining the anomaly of massive investments in technology which unexplainedly resulted in flat profits and stagnant productivity gains.

The most troubling piece of the puzzle is that the service sector, which alone accounted for $800 billion in information technology investments, experienced the most sluggish growth of all.

Since the recession bottomed out in 1991, some argue for the debunking of the productivity paradox. *BusinessWeek* for one has engendered yet a new phrase—*the Productivity Payoff*. The gist of the *BusinessWeek* premise is that a coupling of technology and sweeping changes in management and organizational structure (i.e., reengineering) have paved the way for the productivity payoff.

The statistics to support their view are compelling. In the two years since the end of the recession, productivity (measured in output per hour) has increased at an average annual rate of 2.3% while in 1992 that rate jumped to a long-awaited 3%.

Even more encouraging is that corporate profits are up. Although some economists, including the Federal Reserve's Alan Greenspan, believe that this profit surge can be attributed to higher productivity, not every economist is convinced that we are on the road to a productivity explosion.

Both sides of the debate provide studies to back up their position. On the productivity explosion side, Erik Brynjolfsson and Lorin Hitt of the Massachusetts Institute of Technology's Sloan School of Management, report that the return on investment in information technology averaged 54% for manufacturing and 68% for all businesses surveyed. In spite of these encouraging statistics, naysayers are pointing to stagnating wages, an unemployment rate persistently hovering at 7%, and lagging job growth.

One reason for the lag in employment statistics is the unprecedented rate at which corporations are downsizing their corporate hierarchies and a part of this, although economists cannot determine which part, is most likely due to the embrace of technology. Where dozens of workers were once needed to control the information flow now one solitary worker, armed with technology, can do the job. Case in point is Federal Express who rose to prominence on the back of its automated package tracking technology—Cosmos. Wal-Mart, the superstore king, uses technology to keep its stores stocked and its prices low.

There are always two sides to every story, however. Sometimes the use of technology adds more confusion to an already confused situation. Look at General Electric (GE). After aggressively spending a big part of their profits on technology in the 1980s, GE has begun reassessing its position to the point of taking automation out of its factories altogether. The reason? GE believes that in many cases technology impedes productivity.

Ultimately, this just might be a case of whose statistics are more correct. IT defenders insist that macroeconomic data produced by the U.S. Departments of Commerce and Labor is not reliable and does not accurately reflect the benefits being derived from IT. For example a shift to a service economy would be reflected as a decrease in labor-factor productivity. Additionally, service-sector work is much more complex resulting in an increased level of work output required to do the job. But macroeconomic statistics simply do not capture these vagaries, therefore it is impossible to measure productivity increases or decreases as a result of technology.

Perhaps the great leveler is a view proposed by Dr. Raymond Panko who is chair of the Department of Decision Sciences at the College of Business Administration at the University of Hawaii. During 20 years of researching the measurement of computers and productivity he has drawn the conclusion that investments in IT are not as massive as everyone assumes. Compared to the economy as a whole, they're actually quite small; only about one percent of Gross Domestic Product (GDP)—far too small to serve as a true test of whether or not they affect the economy.

The person with the most insight into this problem just may be the man who actually coined the term "productivity paradox." Morgan Stanley's Stephen Roach has been studying this problem for well over a decade. Economist Roach's position is that we have had a very serious problem for the past 15 years not because of any inherent deficiencies in the machines or in the software, but because of managerial ineptitude in applying technology to productive endeavors.

As Roach explains it, things have changed little despite the advent of faster processors, more powerful software and more programmers. Systems are still delivered late, with more than their share of attendant bugs and rarely deliver what they promise.

Steve Jobs, the creator of the Apple Computer and now president of NeXt, has his own theories on why software rollouts are almost always behind schedule. There are three factors to a successful implementation: features, schedule and resources. Two of the three can be controlled. To get a release out on time, features can be dropped. If all of the features are kept, a company might get behind schedule. To stay on schedule more people can be hired, and the features can be kept in, but it usually goes that the more people are involved, the worse it gets.

The Association for Computing Machinery (ACM), publishes one of the few worthy magazines in the industry, although it is a bit high-tech for most people's tastes. Probably the best column in *Communications of the ACM* is one written by Peter Neumann. Actually it's less a column and more a tale of woes. Neumann painstakingly collects system horror stories either told directly to him or written up in the various business and industry press.

What follows is a sampling of his collection. Those "war stories" with annotations were severe enough to warrant press consideration. The other war stories were culled from Internet messages by those involved in the untidy situations:

The state of Virginia acquired a new system for distributing child-support checks, but experienced massive delays, confusion, lost checks, delayed payments, and improper seizure of tax refunds. Operations costs were expected to be triple the original estimates (*Richmond Times-Dispatch*, June 8, 1987, p. B1).

The Bank of America spent $23M on an initial 5-year development of MasterNet, a new computerized trust accounting and reporting system. After abandoning the old system, it spent $60M more trying to make the new system work and finally gave up. Departed customer accounts may have exceeded billions of dollars (*BusinessWeek*, November 7, 1988).

Allstate Insurance began in 1982 to build an $8M computer to automate its business, with Electronic Data Systems providing software. The supposedly 5-year effort continued until at least 1993, with a cost approaching $100M.

Richmond, Virginia hired Arthur Young in 1984 to develop a $1.2M billing and information system for its water and gas utilities. After spending almost $1M, Richmond canceled the contract for nondelivery. Arthur Young retaliated with a $2M breach-of-contract suit.

Business Men's Assurance began a one-year project in 1985 to build a $.5M system to help minimize the risk of buying insurance policies held by major insurers. After spending $2M, the completion date slipped to 1990.

Oklahoma hired a major accounting firm in 1983 to design a $.5M system to handle its workers' compensation claims. Two years and more than $2M later, the system still did not work. It was finally finished in 1987 for nearly $4M.

Blue Cross and Blue Shield United of Wisconsin hired EDS in late 1983 to build a $200M computer system. It was delivery on time in 18 months, but did not work, issuing $60M in overpayments and duplicate checks. By the time it was finished in 1987, Blue Cross had lost 35,000 policy holders.

The U.S. Office of Surface Mining spent $15M on a computer system intended to prevent violators of strip-mine laws from getting new permits. The system could not keep identities straight, and the GAO called it a failure. (*The Washington Times*, February 15, 1989.)

Thousands of Los Angeles County homeowners were billed retroactively for up to $15,000 in additional property taxes, resulting from a 1988 glitch in an $18M computer system that was subsequently rewritten from scratch. The county was unable to collect $10M in taxes (*The Los Angeles Daily News*, February 25, 1991).

The B-1 bomber required an additional $1B to improve its ineffective air-defense software, but software problems prevented it from achieving its goals.

The software for the modernization of the Satellite Tracking Control Facility was about seven years behind schedule, about $300M over budget, and provided less capability than required.

The Airborne Self-Protection Jammer, an electronic air-defense system installed in over 2,000 Navy fighters and attack planes, was $1B over budget, 4 years behind schedule, and only marginally operationally effective and marginally operationally suitable.

General Bernard Randolph, commander of the Air Force Systems Command: "We have a perfect record on software schedules—we have a perfect record on software schedules—we have never made one yet and we are always making excuses."

The C-17 cargo plane being built by Douglas Aircraft had a $500M overrun because of problems in its avionics software. A GAO report noted that there were 19 on-board computers, 80 microprocessors, and six different programming languages. It stated that "The C-17 is a good example of how not to approach software development when procuring a major weapons system."

The computer is a machine. And like any machine there are appropriate uses of it and inappropriate uses of it. For example, a fork is an implement (i.e., a sort of machine): it can be used to scoop and pick up. While a fork works fine on one or two leaves of lettuce, you would not use a fork to pick up a head of lettuce. So too with computers. Just because they are here and available does not mean that it is the appropriate solution to every business problem.

Now this may seem obvious, but it is not—not from some of the stories in this chapter. When I got my M.B.A., the teaching methodology employed was to present case histories to make a point. The case histories in this chapter make a point, too. The point is, not all computerized solutions work. If we are going to go after the elusive goal of enhanced productivity, then not only do we have to "do systems right, we have to do the right systems."

From the daily tabloids:[1]

Thomas Rich was in an A&P supermarket in New York's Greenwich Village when he got into an altercation with a clerk about food stamps. When the clerk grabbed a five-pound sack of potatoes and flung it against his back, Rich went out and got a lawyer.

The defense lawyer retained by the A&P filed highly specific motions to the court. But the district attorney responded with a one-size fits all "boilerplate" reply spewed out of a computer. Unfortunately, it refers to demands for hearings that the A&P's defense attorney never made. It also ignores issues A&P raised.

A&P's defense attorney, Neil Checkman, had this to say about computer boilerplates, "Lazy. Sloppy. And I see this often. doesn't anyone bother to think anymore?"

State criminal court Judge Arlene Goldberg feels the same way. "The indiscriminate use of the word processor has resulted in the mass production of boilerplate responses which often bear little relevance to the motions they are ostensibly designed to answer."

The result of this practice is that some lawyers fill in the blanks on all-inclusive, formulaic motions. Thus a demand for autopsy photos might be made in a drug case, or a gun in Medicaid fraud. Some boilerplate motions by one side respond to boilerplate motions by the other. As Judge Goldberg puts it, "It's become the War of the Machines." Caught in the computer crossfire are overworked judges who have to read the whole dense mess.

The debate about boilerplating has gone all the way up to the supreme court. A decision by Justice Patricia A. Williams seeks to end this, "Judges should not be asked to wade through a paper swamp and identify those few items that stand on solid relevant ground; it appears to be assumed that this is quality legal representation. This court, for one, says it is not. The practice will not long go unsanctioned."

[1]Johnson, Randall. "Technology Use Must Match, Not Exceed, Human Needs," *Total Quality Newsletter*. November 1992. pp. 1-2.

IS PC SOFTWARE REALLY UP TO IT?

On some major roadways, PC stores are about as ubiquitous as gas stations. Go inside any of them and you will see sleek displays of even sleeker looking boxed PC software. And all these boxes are saying, "Buy Me."

The average number of software packages installed on an end-user's PC is four. The breakdown is usually a word processor, spreadsheet, database package and some presentation software.

From the start, the user knows he or she is in trouble when the installation program does not work. I had that problem recently when I installed a copy of Adobe's Photoshop. The installation program asked me for my serial number, which I dutifully entered. Unfortunately, it would not accept what I entered the way that I entered it. It turns out that Adobe had used a not-so-great printer to print serial numbers so what looked to me like PVV was actually PW. Chalk one up to lousy printers.

Perhaps my best war story is with the installation and running of my OCR software that came with my Microtek scanner. OCR (optical character recognition) is a wonderful boon to (my) productivity. Unless, of course, you have persistent and sometimes unexplained problems with your hardware/scanner.

I installed OmniPage Direct by Caere using what is referred to as the Twain driver. It installed and worked like a charm. Until two weeks later when, all of a sudden, OmniPage wouldn't "recognize" the typed page. The scanner was working. I knew this because I could scan pictures into my image editing software. I presumed OmniPage just got cranky.

I called Caere. Its service technicians were not interested in discussing what I considered to be an "Unsolved Mystery." They merely told me to download (at my own expense) an alternative driver and use the new one instead. I did. It works. But to this day I'm waiting for it to become temperamental and call it quits.

Then of course, there's that now famous bug in Microsoft's DOS 6. Some people found out the hard way that compression software is tricky business. Of course, Microsoft is not the only culprit here. I had much the same problem with Stacker.

I had used Stacker on my NEC computer. When I bought an ALR and began to run out of space, I decided that Stacker was just the thing to solve my problem. Unfortunately, I didn't read the readme.txt file. I should have. It would have told me that, for some reason, ALR and Stacker were mutually exclusive. I found that out the hard way when I wound up with an unusable 120 million-byte file. But at least I got more disk space.

I do not think there is a user out there without one or more of these war stories. The bone I would like to pick with software vendors is that there should not be any complaints. In spite of vendors' protestations to the contrary, PC-based software is neither carefully tested nor quality assured. At least, not in the same sense that mainframe software is. Again Computer Market Letter found that the user perception is that the error rate is 75% higher in PC-based software than in mainframe-based software.

There may be many reasons for this high error rate. Certainly, the quality of the programmer is one reason. And the number of software products that need to interact on a little PC is another. But the immaturity of the field and user expectations are another. Accept less than quality, and you get less than quality. The worst offender, however, is that the constant drive to "get it to market" pushes products out the door way before they are ready. In other words, a mainframe beta test is a PC product.

A commercial on TV for one wine brand or another stated, "we'll sell no wine before its time." Since we probably cannot rely on computer industry vendors to be this honorable, my recommendation is not to buy versions of software that offer radically altered functionality (of course, unless you really need to have it). Wait until the vendor shakes the bugs out. This is exactly what happened in the Microsoft DOS 6 case. After a few hundred organizations fumed that they had essentially killed their hard disks and lost hundreds of productive hours, Microsoft fixed the bugs. Those of us patient enough to wait got the new functionality without the trauma.

Even the Press Gets the Blues

Everyone in this field is created equally. The staff at *Infoworld* recently reported on their brush with PC-bugamania.

Apparently Apple computer sent a Macintosh Workgroup Server 95 to *Infoworld*'s Nicholas Petreley, Nancy Durlester and Laura Wonnacott. The first bad omen was that the floppy disk drive didn't line up exactly with the slot where you insert the floppy. The second omen was that Apple neglected to send along the color monitor. Then they opened the box that they presumed would hold the keyboard. Wrong. No keyboard. But it did have a few manuals, a CD-ROM, a cord, and a mouse. After stealing a keyboard and a monitor, the intrepid team put it all together and turned on the machine. After a few screen flickers they got—nothing.

Finding what they thought to be the answer to the problem, they reseated the SCSI card but had trouble getting a good fit. Once that was accomplished, they attempted to test out the misaligned floppy to great un-success since the floppy got stuck instead, and not even a paper clip would get it out.

But the team persevered and a screw driver did the trick. But an hour after it booted, the beautiful color screen develops a wee bit of an alignment problem.

Finally the team gave up, took the bait, and asked Apple for a new server. That delivered, the team tried again, got past the hardware intricacies, only to be foiled by the software installation. Although the instruction manual said they should be able to run

PING, a TCP/IP connection test facility, there is no PING anywhere to be found. So back to Apple tech support.

The first question the tech support person asks is whether or not the team was following what the manual says. They answered with a resounding "Yes." She said, "Well, there's your problem. The manual is wrong."

So, what's the answer to this dilemma. Apparently, PC software vendors, or at least the majority of them, are too busy getting the goods out to market to concentrate on quality. Their lack of quality is making users less productive then they think. The solution, I think, is to get producers on the quality bandwagon.

ISO9000 is a standards process that is rapidly gaining acceptance worldwide. What are the reasons? Internal quality improvement, market positioning and customer or regulatory requirements, particularly those connected with the European Community. Firms undergoing what is known as ISO9000 registration go through a grueling process of analyzing their quality shortfalls. And then they develop and implement the steps that lead to a better quality product.

If software firms would embrace this ideal many of the problems discussed in this chapter would simply disappear. The problem is few firms are making a commitment to ISO9000. Perhaps if we all clamor loud enough, this will indeed come to pass.

More Is Most Definitely More

A food company once generated reports of product sales and shipments, for the purposes of making marketing and promotions decisions, at the end of each month.[2] A few years ago, technology made it possible to turn them out weekly. Now they arrive daily.

The actual amount of information did not increase. The food company is still getting the same information, only faster, The problem in this case is not the technology but rather a belief shared by many organizations in the technology age: that more is necessarily better.

In the case of the food company, nothing was accomplished by crunching numbers daily that was not achieved when the statistics arrived monthly. These numbers were not so time sensitive as to create a competitive advantage for the company by being generated daily, nor does the company have the employee resources to study the data each day.

Yet, in a classic "climb the mountain because it's there" approach, the report is produced daily simply because it can be. This creates the illusion that it needs to be dealt with daily. Unable to accomplish this feat, employees are forced to choose between becoming hopelessly mired in reports or giving them only superficial attention. The

[2]Voboril, Mary. "Boilerplate Lawyering Draws Ire," *New York Newsday*. October 31, 1993.

technology defeats its purpose. Instead of making work faster, less tedious, and easier, valuable time is wasted and with no significant productivity or quality gains.

The Institute for the Future, a Californian think-tank predicts that by 1995, 90% of American white collar workers will own a PC. Given everything a computer can do today, and everything it is projected to do tomorrow, this should make us ponder the role the computer should really play in business.

The hardest problem posed by the computer is that it is so flexible. They can do all sorts of things. Increasingly the right question to ask about computers is not "What can the machines do?" but "What do we want them to do?"

THE (RE)ENGINEERING TREND

The preceding sections have given you a sense of what is wrong in the world of technology. From PCs that do not operate the way they should to software delivered with bugs. What's a business to do?

Reengineering is rapidly becoming the primary vehicle that businesses are using to stay on top of ever-evolving business requirements. Because reengineering is business-oriented rather than technology-oriented, organizations that venture forth are finding that not only are they benefiting from developing the right systems, but they are developing the systems right. What is interesting is that more and more organizations are finding that the methodologies employed during reengineering are quite useful for original development as well. Therefore I have slightly amended the name of this technique to (re)engineering for it emphasis the engineering aspects of the process, while pointing out that one can use it equally well for original development as for reengineering.

Although reengineering is really the art of renovating existing computer systems, the structural framework that surrounds the process is demanding when it comes to productivity and quality. According to Galler there are several components to (re)engineering.[3]

Planning

Sadly, it is not surprising to discover that few organizations invest any time at all in planning the (re)engineering effort. Development is not just a matter of using what is on hand. A successful product requires a thoughtful examination of the problem itself, the ramifications of solving that problem, those that the problem would be solved for, the methodologies to be used to solve that problem, and the techniques for implementing the solution.

[3]Galler, Craig. "The Reengineering Life Cycle," *Software Engineering Productivity Handbook.* McGraw-Hill. 1993.

Many companies are implementing Quality Groups. While the purpose of this is to promote quality within their organizations, I can see this extended to including a planning process as well. Software planning is one process that should definitely be done by a committee. Especially one that deals in quality.

The Objective

To define the (re)engineering objective, one must first recognize the level of (re)engineering needed. What is the purpose of the project? What needs to be accomplished—and when? How will one recognize success?

The Investigation

Surprisingly, most organizations dive into system development projects rather than just sticking a big toe in to test the waters. The first step in any (re)engineering effort is, therefore, to test those waters.

This is accomplished through a series of surveys that assess everything from user needs to data flow to work flow to ergonomics—and everything in between.

The first survey, in fact, should be survey of what needs to be surveyed. At this point in the (re)engineering life cycle, the methodology begins to look a lot like that practiced with great success by knowledge engineers.

Knowledge engineering is a tactic used to build knowledge-based systems. Based more on psychology than technology, its goal is to ferret out knowledge from those who possess it (i.e., experts). This knowledge is then turned into rules and/or objects depending upon the technology used to build the system.

By ferreting out, or knowledge engineering, the whys, whats, and wherefores of the system, you reduce the risk that the final deliverable will not be what the users wanted.

Thus far, the processes we have covered fall under the heading of "soft" processes. That is, planning, objectives, and investigation are all done manually without the assistance of any computers.

While not on the market yet, it should not take much longer for some enterprising company to realize that these processes are ripe for automated assistance. If we can write intelligent agent software to sift through information and ferret out items of interest, and we can use artificial intelligence to figure out when there is going to be a stock market crash, then we should have no trouble in developing software that helps us plan for software development.

The Tools

Much has been written about the proper selection of software tools. I believe that much bad code has been written simply because the tools were not a good fit for the problem at hand.

Many organizations standardize on a set of tools which are then used for all types of problems. While I do understand the rationale for doing that (i.e., costs, human resources and training), if one calculates the cost to the organization in terms of productivity loss (not to mention the errors that inevitably crop up in the code), one would realize that the best solution is to provide a selection of tools that one can pick and choose from. The tool workbench analogy is a good one here. Tim Allen from the Home Improvement comedy series aside, a tools workbench contains a plethora of tools. Does the handyperson only use a hammer? Or does he or she choose the best tool for the job? Development software are tools too. Pick the right one for the job.

(Re)engineering leaders are turning to flowchart and process-diagram tools to map corporate workflow and examine information systems.[4] This charting and mapping is especially helpful because certain (re)engineering projects are precursors to client/server implementation. Indeed, a number of companies have already reaped the expected productivity gains from client/server migration only after they first examined and then adapted their workflow processes.

Surveying 24 large users, most of which had already moved to client/server architecture for core business processes, the IBM Consulting Group in White Plains, NY, reached the same conclusion. Mike Sinneck, the group's VP of applications, stated that the in-depth survey revealed that the value of a client/server project is directly attributable to (re)engineering.

Business (re)engineers begin with a general question such as "What business am I in?" and build to increasingly detailed analysis, such as "Where are the bottlenecks in the product-return process? (Re)engineering tools help these workflow managers follow these methodologies.

The tools note in detail an organization's structure and business flow and present projected scenarios that copy idealized reengineered processes. Well aware of the tedious nature of designing process diagrams and flowcharts by hand, an increasing number of software vendors over the past two years have conceived programs that automate such documentation. Additionally they ease communication, thus ensuring a company-wide approach to documentation that speaks the same language.

Bernie Palowitch, director of business process reengineering, benchmarking, and quality, at Bristol-Myers Squibb Co.'s New York headquarters, has been using a tool called WorkFlow Analyzer from Meta Software Corp. of Cambridge, MA, to diagram present and future business processes in Bristol-Myers' human resources department.

[4]Wilson, Linda. "New Ways to Rebuild Business," *InformationWeek*. September 5, 1995. pp. 50-58.

Palowitch is assisting the HR department examine how Bristol-Meyers compensates its staff. Currently, the company's salespeople are paid based on the number of sales calls made to doctors' offices within a fixed period. With the current trend of large health maintenance organizations and employers purchasing may of the drugs patients will use, however, it made little sense to Palowitch to support a sales force that visits individual doctors.

Many (re)engineering tools can wrestle ever-increasing complex tasks. While many tools are basic flowcharting packages, others assist companies perform cost-benefit analysis, breakdown the time required to complete a given process, or even design animated, 3-D models of business processes. The majority of the tools operate on Windows or Macintosh PCs, though some run on the UNIX platform.

Priced as low as $250 to a high of $200,000, most (re)engineering tools can be used off the shelf. Some, such as Andersen Consulting's Value-Added Reengineering Workbench, are packaged with consulting services.

The ability to allow users to export data to computer-aided software engineering (CASE) products is just one of this technology's strong selling points. A case in point: Texas Instruments Inc.'s Business Design Facility provides data for the Dallas company's Information Engineering Facility CASE product, Application Development Workbench. The TI tool also provides users with a similar interface.

EDS Corp., a systems integrator in Dallas indexes off-the-shelf (re)engineering tools according to complexity—either unstructured or structured. According to Jason Grant, chief information officer of EDS's management consulting services, the company looks at various tools and decides what they are good at, not good at, and how they can guide their consultants.

Executives or (re)engineering groups that combine information systems and non-IS business professionals are generally drawn to the basic tools that sketch business processes with little detail. CSC Consulting, Inc. employs a similar package. According to Jim Ettwein, a CSC partner in Boston, most clients are educated in the uses of Micrografx in just one day.

An example of a structured business (re)engineering tool is Meta Software's Workflow Analyzer. This software includes a $3,995 flowcharting module entitled Design/IDEF that can export data to its $24,000 simulation module, called Design/CPN. Employers of the system can also run a cost-benefit analysis on the data with a desktop spreadsheet program. Additional complex (re)engineering tools include DECmodel from Digital Equipment Corp.; TI's Business Design Facility; and Metis from Metis Software in Horten, Norway, which is marketed by Digital in the U.S.

Overlooked potential business headaches can also be cured by (re)engineering tools. Using Metis to detail and analyze its sales forecasting process, Basic American Foods, a privately held 2,500-employee producer of dehydrated potatoes and beans in Walnut Creek, CA, found that this tool allowed it to develop a Mac-based system that could export sales forecasts prepared on a Microsoft Excel spreadsheet to a production planning system running on a Digital VAX minicomputer.

After a test of the prototype of the interface, however, Basic American decided to delay a full-scale rollout having discovered that the company needed to reevaluate the entire order-fulfillment process.

Improving customer service response to product inquiries was the goal of Intel Corp. in Santa Clara, CA. To achieve that end Intel employed the DECmodel. Headquartered in the company's Albuquerque, NM office, John Howe, program manager for business process (re)engineering, uses DECmodel to track how telephone operators respond to product support calls. It already has pointed the company in a new automated direction. Howe now believes an automated client/server system accessing a database of symptoms and resolutions would help operators to assist callers without the aid of a specialist.

The strong and growing interest in (re)engineering tools resulted from users' frustrations with manufacturing process-flow software and CASE tools. Creating a team five years ago to work on (re)engineering projects, Dow Corning Inc., in Midland, MI, is having the team devote time to evaluating KnowledgeWare's Maxim after concluding that traditional manufacturing software would not work.

Explaining company policy, John Torgerson, a Dow Corning senior information specialist stated that CASE tools were designed to develop computer systems; they do not do a good job at high-level mapping because they quickly get you into the bowels of the system. CASE products, unlike business process (re)engineering tools, do not allow users to deduce how much time and money a process costs, and they are unable to account for employees' whereabouts or functions.

Never forget this technological helpmate is just that—a *helpmate.* It is not a wonder drug—an easy cure-all for all business problems. It will not compensate for shaky methodology, ill-suited consultants, poorly trained reengineering team members, or a lack of support from top management. The biggest danger users face, according to Albert Case, VP of business process reengineering at the Gartner Group Inc. consultancy in Stamford, CT, is utilizing (re)engineering tools to help change processes without comprehending how one process may be connected to others.

(Re)engineering tools should offer workgroup capabilities; allowing all members of a project team to simultaneously enter data. Triune Software Inc., in Fairborn, OH, heard the call and announced one such product last May. The project, entitled Able Consensus, puts together group meeting and activity modeling in a package costing an average of $15,000 for 20 users.

In the end, organizations will find that the best tools for (re)engineering are those that they are familiar with today. While object-orientation (OO) is this year's savior (i.e., two years ago it was CASE, and before that one or another of the methodologies), it alone will not do the job. The problem is, that it is a programming tool. Therefore it is difficult to learn and less than useful if there are no procedures in place to create and use libraries of reusable objects.

Ultimately, salvation will come in the form of OO CASE tools. If the goal is to make systems development as painless and complete as possible, the tool to be used to CASE. If the goal is to make systems development productive and efficient, then OO is the way to goal. The marriage of these two technologies will enable organizations to quickly, yet methodically, develop systems from a series of reusable components.

The Workbench

We have already alluded to the tool workbench in the section above. It is a good metaphor since the hardware tools workbench is very much like the software tools workbench. Both need to be configured with a wide variety of tools to do the job.

This includes CASE tools, programming languages, and utilities. So far, I only know of one tool that combines the best aspects of all three of these requirements. Boston-based Hamilton Technologies markets 001 which, although termed a software development and systems engineering tool is really OO CASE.

The Prototype

Prototyping is very much the rage today. But how many of them are fully testing the full range of features of the proposed system? It seems that the purpose of most of today's prototypes is to "show" the user what the user interface will look like. While that is certainly part of the reason for the prototype, it is not the whole reason.

A good prototype needs to exercise the most rigorous components of the business problem. For example, if the system to be (re)engineered calculates loan amortization, then the team better find the most difficult case the users have ever seen and prototype that to their satisfaction.

Automatic testing tools, for this reason, will be popularized in the near future. As more and more organizations jump on the ISO 9000 quality bandwagon, they will be forced to quantity the quality of their organization. While technology is only one part of the ISO 9000 equation, "proving" software correct can only be done automatically.

The Final Design

Once the prototype is tested and accepted by the users, the final design is completed and documented for approval. While documentation is neither sexy nor fun to do, it is necessary.

Again, Hamilton Technologies' 001 comes to the rescue here. Not only does it generate code to any platform from a graphical design, it generates the documentation as well.

I have been following 001, as well as the rest of the software development industry, for several years now. Although it has not quite happened yet, I predict that where 001 goes, the rest of the industry will follow. It is not only that 001 is so radically different from anything else on the market, it is that the main mission of 001 is to promote unbelievable leaps in productivity. That is really the bottom line.

The People

Part of the problem just may be the people we are using to develop our systems. It is no secret that a good programmer runs rings around a not-so-good programmer.

One of the most intriguing of all trends is the trend towards hiring the best and the brightest—wherever they are located. More than one high-tech firm has outsourced some part of its development to high-tech countries like Israel, the former Soviet Union and India. Not only are these companies reaping the benefits of better code, but they are saving bucks as well.

One would like to be able to say that off-shore outsourcing is a temporary trend, but with the advent of virtual office there is really no reason why a programmer in Bombay should not be equally as accessible as a programmer in Brooklyn.

However, again using 001 as an example, this trend might just be offset by the trend towards programmerless programming. Yes, this is something that James Martin espoused years ago, but this time it just might happen.

001 permits a systems designer to sit at a workstation with one or more end-users. A system is jointed, graphically designed, and prototyped. After several iterations of this process, the final system is automatically generated as is the documentation. Is a programmer necessary for this? No. What is needed is a team of expert end-users and a systems analyst/knowledge engineering type who understands the business.

CRITICAL SUCCESS FACTORS IN (RE)ENGINEERING

In September 1994, CSC Index Inc., a Cambridge-based research consultancy, released a survey of over 600 senior managers on the critical success factors in the art and science of (re)engineering.[5]

What CSC found was that there are six key factors that facilitate (re)engineering success: project management, senior management commitment, corporate culture, case for action, early project success, and information systems in place.

[5]"Re-engineering: Critical Factors For Success," *InformationWeek.* September 12, 1994. pp. 25-32.

Project management is one of the keys to successful (re)engineering. Six factors are important: effective use of teams, commitment of time and resources by team members, clarity of scope and goals, establishment of strong communications, realistic planning and schedules, effective training, and education.

The 10 most difficult aspects of (re)engineering, in order or degree of difficulty are: dealing with fear and anxiety throughput the organization, getting the information systems and technology infrastructure in place, managing resistance by key managers, changing job functions and career paths, designing the new business process, having a clear vision of the new organization, achieving measurable results, communicating the changes and actions required, managing resistance by the work force, designing incentives and other management systems.

The nine corporate culture traits that appear to foretell successful (re)engineering initiatives: risk taking is encouraged within the organization, problems are openly discussed, people trust their managers, employees are rewarded for a job well done, employees' first priority is to please the customer, teamwork is encouraged and rewarded, employees are empowered to make their own decisions, employees are receptive to change, and information is shared freely.

Communications is probably the most important aspect of the (re)engineering process. While respondents used a wide variety of communications techniques from meetings to memos to newsletters to E-mail, a failure to communicate was cited as the biggest reason for (re)engineering failure. The ten aspects of communications rated by CSC were: getting across the magnitude of change, communicating frequently enough, responding to concerns about job restructuring and layoffs, using a common language and concepts understandable to all levels, responding to rumors, maintaining continuous visibility of the (re)engineering "champion," maintaining enthusiasm without being repetitious, maintaining a consistent message, devising a combination campaign, and maintaining communications among team members.

Finally, CSC rated the way in which technology issues affected the (re)engineering effort. The toughest technology issues are: ensuring that information systems are developed according to plan, understanding the scope and meeting the training needs of technology users, understanding the information system and technology implications of the (re)engineering process, ensuring that the IS organization has the skills necessary to support the task, ensuring that the technology called for in the initiative is viable, establishing technology standards and protocols, getting the IS organization to embrace the initiative, and involving the IS organization early enough in the initiative.

THE PRODUCTIVITY PAYOFF

No doubt about it, we have been suffering from a productivity paradox: we have invested billions in technology, but have not achieved commensurate benefits.

But things are beginning to change. Perhaps all the bugs we are seeing in the PC world have provided a catalyst for us to begin to understand just what we are doing wrong.

As we move relentlessly toward the year 2000, the depth of our insights coupled with (re)engineering techniques just might pave the way for us to reap exactly what we have sown for the past 25 years.

CHAPTER 4

Cultural Identity and Integration in the New Media World

by Red Burns

In the last fifty years there has been an overwhelming change in communications. This change has not yet been fully understood by any of us who are in the middle of it. It is of the magnitude of change occasioned by the introduction of writing and the printing press. Knowledge, power, and material relationships are changing radically as a result of the new forms of communications.

The electronics revolution in communications is multi-faceted. On the surface there are the startling convergences of computing, audio, video, graphics, and text. To these convergences we can add the expanding temporal and spatial reach of the new distribution networks.

As each new form of communications develops, it promises new opportunities and resulting benefits to society. Yet there is enormous potential for abuse. One of these abuses is an antiseptic technological world that cannot easily handle human paradox.

THE NEW MEDIA

These new communications technologies have almost irresistible allure, and their promise surrounds us. Extravagant claims are everywhere. In fact, there is much to be excited about. Yet we need to avoid becoming blindly seduced if we are to create desirable and equitable applications which justify our efforts and investment.

Soon, for individual users, a single wire will come into the home to provide access to local as well as international information. This wire will serve as the link for voice, data, and video transmission. This mode of communication and expression can be used by one voice or by many. Determining whose voice gets heard, and determining an equitable use of this new power is an important issue to consider.

The technology itself is as straightforward as a mathematical equation. The applications, however, are myriad. In the end it is people, not machines, who design applications.

My own engagement with communications technologies began a long time ago in documentary film at the National Film Board of Canada. At that time, a filmmaker went into a community, took notes, wrote a script, shot a film, edited it, and sent it off to be distributed. A most worthy occupation—telling other people's stories. We, the filmmakers, felt quite comfortable with our interpretation of others—their values, their lives, their aspirations. We were the professionals. We created a world about others. We selected the material we thought was important to represent them, rather than letting them represent themselves. In some cultures, it might be said that we robbed others of their identity. At the time, I believed we were performing a valuable communications service: telling people's stories. The professional framed the message; this was an accepted practice.

A current example of this kind of representation can be seen on CABLE NEWS NETWORK. CNN is broadcast to almost every major city in the world. During the Gulf War, CNN reached Israel, Saudi Arabia, and Iraq, as well as my home in New York City. Each viewer regardless of her/his culture received the same information at the same time. CNN is representing the world to us not only in America, but internationally. After the Army censors, the editors at CNN were essentially editing the war for all of us. That is very powerful.

As electronic media becomes the main mode of communications between peoples, the danger is that we may recreate the kind of colonialism found in classical literature where only one voice was represented. Interactive networks can provide a means for different voices and certainly more than one.

An example of a global interactive network was seen during the August 1991 activities in Russia. Computer conferencing provided an electronic link between Russia and the outside world. While it is estimated that there are only 1.5 million personal computers in the Soviet Union—and probably only 2% have modems—the personal and public communications they made possible didn't exist a few years ago. It has been reported that during the attempted coup many Moscovites received scanned images of the *Washington Post* and *The New York Times* describing the reaction of the West to the unfolding events in Russia.

In one instance NATO officials wanted to send a message of support to Boris Yeltsin and could not communicate directly from Brussels to Moscow. They sent their message through the Sovam teleport in San Francisco to Nikolai Kapranov, a top advisor to Yeltsin. Kapranov had an account on the computer mailbox. He sent a message back saying he had translated the NATO message and distributed it in the form of leaflets. He reported that the people inside the parliament building were grateful to know that they had a connection with democratic countries.

In the field of video, one of the first technologies I found that offered the possibility of this kind of self-representation was in 1970 when I witnessed a demonstration of a Sony Portapak—reel-to-reel 1/2 video. I was astounded! The skills required to operate the camera were not out of reach for non-professionals. The cost was not prohibitive, and for the first time, it was possible for ordinary people to make their own video documents. Interpreters became unnecessary. An option was available that had not existed before. The empowerment was thrilling for those who chose to use this new means of communications. By thinking about how they presented themselves, people learned about themselves and their community. Often that learning process was as important as the documentary ultimately produced. This modest advancement was quite revolutionary in its potential to allow people to tell their own stories and to control their information. Combined with a cable distribution system, people were able to present their own point of view in the dominant medium of American society: television.

This was an important time in the history of television in the United States. Broadcast television had always been a medium of scarcity, using a great deal of bandwidth and expensive equipment. Television production clearly was not available to ordinary people. The development of cable television changed that circumstance. Not only were over-the-air television signals transmitted but new origination opportunities arose.

The convergence of the low cost, portable video equipment and an equally low cost distribution network—cable television—created an unusual opportunity in the U.S. in the 1970s. At the time, cable television reached about 20% of the American homes—(now it reaches an estimated 60%). While the initial purpose of cable was to bring television reception to areas blocked by physical obstructions, the cable carried with it much more bandwidth than was then available to broadcast stations. With very little investment, small communities and cable systems could originate programming. No spectrum space scarcity here! You could originate channels as well as re-transmit over-the-air television.

EMPOWERMENT THROUGH MEDIA

My research group, The Alternate Media Center at New York University, was set up in 1971 to examine a possible relationship between communities, video, and cable. We set out to train people to develop and distribute their original documents locally on new cable channels. We set up models of Public Access Centers in partnership with cable systems in several U.S. cities from New York to California. We provided equipment and training for community members to shoot and edit video tapes of local concern.

People who were not media professionals, but who had community interests, were offered training and time on the cable systems on a first come, first served basis. City council meetings and protests were videotaped, interviews were conducted, conditions in schools and neighborhoods were recorded. Discussions of community problems were raised. It was an extraordinary political opportunity. Ultimately these access centers were absorbed by the cable systems and mandated by the Federal Communications Commission. Public Access in the United States is now established in communities across the country.

While this is an enticing story of empowerment through technology, important lessons can be learned from it. Many of the programs on these channels, were ignored by the communities. Producers, anxious to imitate the broadcast style of commercial television with none of the resources or professional know-how, lost their audiences quickly.

When communities stopped trying to mimic broadcast television, and found their own voice, their own appropriate use for the video technology, and the distribution channel, their audiences returned. Authenticity—a real and original voice—made community cable appealing. When people were themselves, not what they thought they should be, they were far more interesting than they realized. It was an alternative. It clearly pointed out the difference between access to a communications channel and access to an audience.

The flip side of the access issue is CNN. In the late 1970s CNN exploited cable and combined it with satellite technology to distribute news worldwide from Atlanta, GA. While this service is available worldwide, it is not distributed to many individual homes in many countries because the requisite satellite dishes are still very expensive. So in Europe, for example, CNN can usually be found in hotels and government offices and in newsrooms outside of the United States. Its influence is even more powerful under these circumstances since many of those who have access to it are members of government and the local press—in other words, the world's most influential people.

INTERACTIVITY

Marshall McLuhan's global village may be too simple a model for today's complex interactions of culture, politics, technology, and media.

One way to better navigate the course of the future global technologies might be to examine these complex issues and international situations in terms of smaller, local ones.

With technological change comes the widening opportunity for interactivity. One of the attributes of interactivity is two-way communication. Television broadcasting is an example of one-way communication. One way media as we know it originates from one source and goes to many. Interactive media creates very different opportunities. It allows exchange and origination from many different sources. Two-way communications and one-way communications are as different as a conversation and a lecture.

Interactivity has the potential to provide a new perspective on the issues of access to an audience, access to technology and access to information. It can give a voice where one was not heard before. How we use it is a question of some significance.

As described above, projects developed by the Alternate Media Center demonstrated to me the importance of allowing users to design their own systems and of allowing the content to create the application. In the mid-1970's, we responded to a National Science Foundation question about whether two-way television could deliver social services in a cost effective manner. We chose to work with old people because we knew they were large consumers of social services. We believed that the key was to design a communications system that would be used by senior citizens and programmed by them for their use. The technology had to be transparent and unthreatening. The real job was to find a way to have the users make the system their own. There are many ways to communicate information. One way is to push the information out, and the other is to pull the information out. We chose to organize a system that would have people pull the information out.

We began in 1975. One of our first efforts was to set up a governing board of community members to help us design and implement a two-way, interactive television system for senior citizens in Reading, PA. Using a return band of the local cable system, we originated from three neighborhood communication centers in existing senior citizen facilities. These were our permanent origination points. We also set up several mobile origination points from different city agencies such as City Hall, the social security office, schools, and the county commissioner's offices. These locations were introduced once a week.

Four separate origination sources were sent to the head end of the cable system. Only one channel was transmitted to cable subscribers. This was done by combining two locations on one split screen with a simple switcher. Switching was determined by who was talking. Because the cameras were placed directly beside the receiving signal, people were able to look at each other while talking from different locations. Those at home could participate by way of a telephone patched in at the head end and heard by everyone on the cable system.

While I have described briefly the technical grid, this project was not about technology. Communications technology does not exist apart from the people who create and use it. This was not broadcast television, although initially everyone wanted to emulate what they saw on television. Our analogy was the telephone. We said, "You don't make a program when you call someone on the telephone, you begin with the reason for the call and the content is the dialogue which follows."

The important aspect of the programming was exchange and interaction.

A few examples: We found a woman in her eighties who had been active in local politics. She was a natural to "host" the weekly "live" exchange between the mayor and city council members each week. Other people hosted discussion programs that took place in four different locations at the same time. Those who hosted were in fact, producers. We avoided the jargon of television, such as the title of producer, and instead used the metaphor of a telephone to suggest conversation, rather than one-way television. Presenters determined the topic of the day, invited the participants, and stimulated discussion in four different locations.

Another woman who was interested in nursing homes took a camera person to various homes in the area and brought back a "taped" report for others to share. Her questions to the nursing home staff were far more relevant to the needs of older people than any questions we might have designed. The videotape and the organizer's accompanying live commentary provided a lively exchange of questions and answers about the nursing home from people who did not have the opportunity to visit.

Research and planning of programs was required and many of the old folks became completely engaged in the process. I suspect the benefit of the activity kept many a mind agile. Voluntary committees who were responsible for the programming developed original ideas. The programming they created could never have been developed by a program director who did not share their interests and problems. As a result of the broadcasts, people who had not seen each other in years discovered each other on television. When they waved to each other on camera, some thought it would trivialize the system; it turned out to be an unexpected benefit because our intent was to encourage communication.

As familiarity and comfort were the order of the day, we needed to have open microphones at each of the centers. People had to be able to jump up and get involved in the discussion easily. The engineers advised that we could not have open audio with fourteen microphones because of feedback. We knew that if the system were not spontaneous and easy to use, people would reject it. So we found an engineer who listened to our needs. He designed a feedback suppressor to solve the problem. The difficulties we encountered strengthened our recognition that users ought to drive the technology—not the other way around. It is a point I feel very strongly about. Too often technology drives an application because users are intimidated by the technology and do not participate in designing it. This project demonstrated that if people are given the tools, the environment, and the encouragement, they can create something that reflects them.

I remember vividly the day I was driving in Reading, PA, and on my car radio a local news announcer referred to a comment that someone had made on one of the programs. I knew then that the channel had been accepted as a permanent fixture in the community. The board of elders that we left in place in 1977 still operates to this day. They found the funds to sustain the system and managed the governance of what is now a model in the United States. They do not remember that we were ever there, and we consider that our greatest success.

WHAT GLOBAL NETWORKS CAN LEARN

The global reach of television has made globalization a current buzzword in media. And multinational companies are rushing to serve new markets with the promise of profit not far behind.

There are many options. Nowhere will issues of design and transparency be more important than in the interactive global networks now being formed. These networks are facing multinational issues of control, ownership, intellectual property, transborder dataflow. But if these issues overshadow user issues, such as accessibility and ease of use as these networks develop we will lose an extraordinary opportunity to create interesting and exciting new forms.

The same factors that proved to be crucial at Reading apply to global networks. Transparency and clear and simple design are essential to demystify technology and make it accessible to its users.

The telephone is a wonderful example of interactive applications designed by users. The telephone company provides an empty pipe and a universal interface. It is easy to use. Business people discuss business, teenagers discuss whatever teenagers discuss, people use it to combat loneliness, others use it to order groceries, some to make appointments, still others to make love. It is truly a user-generated medium.

The June 27, 1991 issue of *The Washington Post* featured a story about the telephone. A sociologist with Bellcore, the research arm of the Baby Bells, was quoted:

> "When the Eskimos in remote northern Canada got direct-satellite telephone communications for the first time about 20 years ago, the Canadian government figured they would just use it to talk to each other. Instead, the Eskimos started calling Seattle to check the market for seal meat and started cutting their own deals. That kind of thing has happened all over the Third World."

Historically the landscape is littered with examples of technologies that were developed for one purpose but used for vastly different ones. If flexibility is included in the design, users are able to adapt technology to their own needs.

Initially, when Bell developed the telephone, Western Union refused to invest in the new company. Western Union was convinced that people would not conduct business on the telephone because they needed information in writing. But the convenience of use, the transparency of the technology (we do not know whether a phone call is carried on terrestrial lines or through a satellite), ease of use (you do not need a manual), and most importantly, the opportunity to use it for a variety of communications, has made the telephone an indispensable part of our lives.

Tomorrow's high-speed networks will have the capacity to carry multimedia in real time. The same need of transparency and ease of use applies to these networks. If a doctor is operating on a critically ill patient and needs consultation with doctors elsewhere, whether the distance is small or great, a CAT scan can be sent down the network quickly. Museums and libraries can increase their distribution and availability to many more people in distant locations on these same networks. Physical distances can be cut dramatically. But like the early days of the telephone, we can only "imagine" applications. But once the empty pipe is laid and accessible, imagination and need will drive new uses.

THE PEOPLE, NOT THE TECHNOLOGY

It is startling how quickly virtually any discussion of the future of computing, entertainment, publishing, broadcasting, office technology and education evolves into speculation about the technology. But the technology is straightforward. What is not clear is how we can benefit from these new and powerful developments.

Now that users and artists can use these new tools because of lower costs and less obtuse interfaces, something new has been added to the mix. I use the term "artists" in a general sense. Art has traditionally shown us new ways to look at the world. The sensitivity and sensibility of the artist can temper a singular technological approach.

There is no question that we will have the ability to move large amounts of digital information across space and time. Fiber optics, digital switching, compressed video, and direct satellite transmission will take us to this electronic nirvana. Broadband digital switched networks will be accessible almost anywhere in the industrialized world, television receivers will be computers, imaging technology will get cheaper, smaller, and improved displays will incorporate graphics chips capable of decompressing full-motion video. High resolution video will merge with computer displays. This new digital technology translates images, graphics, text, video, and sound to a common language.

Still people lead the development of applications. People bring humanity to the machines, people are the ones who will speak this common language. How we train people, how we encourage people to recognize a moral as well as an aesthetic responsibility, are serious considerations.

Training the New Technology Designers

In 1979, the Alternate Media Center developed a graduate program in the Tisch School of the Arts at New York University—the Interactive Telecommunications Program. Its focus is telecommunications and the production of multimedia applications. The fact that the program is in the School of the Arts is important. We are not housed in schools such as Computer Science, Social Science, Education, Business or Journalism. We are in the School of the Arts because we are primarily interested in creative communications.

The focus on interactivity requires that we look to the new technologies as the engine, that we look at technology as the verb, not the noun, that our commitment is to people not technology; that there is a new community of disciplines. Today, fields that were once worlds apart are converging.

Our students and faculty come from a wide range of backgrounds. Computer scientists work with students from the fine arts, sound engineers, videomakers, musicians, writers, journalists, architects, are all in the mix. What can they teach each other? A great deal, and perhaps, most importantly, that there is no one way to create, rather there are many different ways. Skills and approaches are exchanged. This is an environment with no fixed circumference, fluid enough to move with the acceleration of technological change.

The field is not yet defined. The ability to communicate must be combined with creative vision. This is not an inherent attribute of the technology, however, it comes about when the technology is accessible to talented and imaginative people.

We believe that a new kind of communications professional is needed, one who can adapt to changes, who is aware of issues of power, control, representation, interface, aesthetics—one who is able to demystify technology and yet has enough knowledge of the attributes of technology to put it all in perspective.

This new kind of professional is comfortable with both creative and analytical modes of learning. Our curriculum is set up for the student who wants to "learn" as different from the student who wants a "skill." This is a new breed of person. One who understands the value of pictures, words, critical thinking, judgment, understanding, technology, and aesthetics. A student who has the ability to manipulate tools and the critical capacity to sort out information. In short, the student who reflects the knowledge shift from static knowledge to a dynamic, searching paradigm.

We believe that technology is not value-free. It takes on the values of the designers. How can designers provide an environment where diverse, talented people can harness these technologies so that they add to the human spirit?

We encourage collaboration, rather than competition. The emphasis on collaboration creates a kind of collegiality that is exciting in any learning environment. Group projects allow students to add their particular skills to a project while they learn about other disciplines. Students are encouraged to fail—to try ideas and concepts that may or may not work and to learn from mistakes. We focus on asking questions, rather than on looking for solutions.

Along with classes, students work on projects—experimenting, pushing the technology to the edge of the possible, combining tools and ideas in new ways. The tools are personal computers, video, graphics, sound, animation, and text. The medium is transmission. These new convergences of technologies are providing capabilities to individual users which were once reserved only for highly skilled technicians operating very expensive equipment.

Labs are the heart of our graduate program, a crucible in which a mix of ideas and talents create new forms and applications. Students in the Interactive Telecommunications Program developed and produce a program called WINDOW which is cablecast each evening over Manhattan Cable, the local cable television company. The content ranges from homelessness, racism, and sexism to other community issues.

Interactive videodiscs and telephone call-ins allow viewers to choose portions of a videodisc produced by students. The telephone interface was designed by one of our students. This student is not an engineer who was interested in experimenting with interactivity in a mass medium. The opportunity to control a television program is appealing to people who watch WINDOW. We know that the audience is involved because the call-in lines are constantly in use during the cablecasts.

The tools to mix and edit audio and video and electronic text are now accessible to a broad range of creative professionals—producers and artists—who want to control the process directly. Computer technology was once the domain of the computer scientist. Now because of miniaturization and less obtuse interfaces, artists are working with computers. This may be one of the most exciting developments because artists will bring a new dimension and creative investigation of the possibilities of communications technology. The computer can be a creative partner but it cannot do anything by itself. A talented and imaginative creator can make it sing.

We are now able to move beyond traditional, linear exposition to explore interactivity, a non-linear presentation of information. Parallel, even concurrent, story lines are possible in this new medium.

This is an entirely new art form. We are freeing the music, film, video broadcast, and publishing industries from the limitations and practices imposed by previous generations of technology.

Our students are joining the ranks of the experimenters and the "imaginers." We expect them to raise the ethical questions, the aesthetic requirements, the empowerment issues, and to address the design needs we emphasize when they go out and create the technology applications of our future.

Our graduates have set up their own companies while others work in many different areas for both profit and nonprofit corporations such as the NYNEX Media Lab, Apple

Computer's Advanced Technology Group, the American Museum of Natural History, Chase Manhattan Bank, the Educational Testing Service, the Telecommunications Office of the City of New York, Philip Morris, IBM, Optical Data, HBO, and ABC News Interactive, to name only a few.

In our graduate program we try to maintain a balance between the excitement and allure of potential new and enhanced forms of communication and the responsibilities and pitfalls inherent in technological advances. We walk a tight rope, learning as we go.

Sometimes the use of a technology is not obvious, and it needs to be introduced into a community of users who can design it themselves. Sometimes a technology can be adapted by a creative team or event. Often a need creates an application of the technology.

Whatever way a communications technology comes to life, it takes on the values of the creators and the users. It is important to remember that there is no one way—there are many different ways. In a time when a big, popular, centralized media can control how a person, a political point of view, or a cultural group is represented, it is critical to consider ways to provide alternate means for people to create their own media and, as a result, their own way of representing themselves.

> A Zen monk once asked a Zen teacher, "What is the way?"
>
> The teacher answered, "Ordinary mind is the way."
>
> The monk asked, "Then should we direct ourselves toward it or not?" The teacher replied, "If you direct yourself toward it, you go away from it. The monk said, "If I don't direct myself toward it, how can I know the way?"
>
> The teacher said, "The way is beyond knowing or not knowing."

ACKNOWLEDGMENT

Presented and published originally with the "Cultural Identify and Integration in the New Media World" Conference at the University of Industrial Arts, Helsinki, Finland November 1991.

AUTHOR BIOGRAPHY

Red Burns is chair and a professor of the Tisch School of the Arts' Interactive Telecommunications Program at New York University. The focus of this graduate department is the production of electronic multimedia and telecommunications. She is a currently co-principal investigator of an interactive cable/telephone project. She is also the Director of the Alternate Media Center, a research and implementation center for new technologies that she co-founded in 1971. During the 1970s and 1980s she was the Director of Implementation of a series of projects such as two-way television for senior citizens, developing telecommunications uses to serve the Developmentally Disabled, a field trial of Teletext as well as the designer of media programs for drug abusers, an electronic community service on cable television, and an experimental videotex database. Professor Burns teaches in the graduate program, and has served on many committees, attended numbers of working meetings, and spoken publicly on new communications technologies.

CHAPTER 5

The New Face of Data

by Joe Celko

"It's not your father's Oldsmobile"

—*Advertising slogan, Oldsmobile Motors, 1992.*

It is also not your father's data. The quantity is greater, the quality is higher, there are more kinds of data today, and we do not use it in the same way as we did before. In his book *The Third Wave*, Eric Toffler referred to three major changes in human society. The first wave was the invention of agriculture, the second wave was industrialization, and the third wave is the information revolution.

Agriculture feeds more people than was possible when humans were hunting and gathering. But there were more changes than simply a full belly. Land became valuable and could be owned the way a spear had been valuable and could be owned. Religions and customs based on harvests appeared. A surplus of food created leisure time.

Likewise, industrialization did more than provide cheap consumer goods to the population. Capital became valuable and could be abstracted as shares in the way that land ownership was valuable and abstracted. Man's time horizon extended beyond a harvest season, and the range from which materials were obtained extended. Agriculture now employs fewer people than at any time in history and the major problems are distribution and surpluses, not famine.

The information revolution is going to do more than simply tell us things faster. Data, in and of itself, is becoming valuable and more abstracted. People are not thinking in terms of three dimensional space, but in terms of cyberspace. Industry will employ fewer people than at any time in history, and the major problems will be distribution and surpluses, not shortages.

While speculation about the ultimate fate of humankind is fun, the focus of this chapter is on current events and trends.

THE QUANTITY OF DATA

Everyone is aware of the explosion in the amount of data kept in the world today. Every year, a newspaper filler piece informs us that the average American is recorded in so many hundred databases. Every year that number grows.

Why so much data? The major reason is that it is cheaper to collect, store, and process than ever before. The cost per bit of data is still dropping, while the speed of the computers that process it continue to increase (Moore's law, a rule of thumb in the semiconductor industry, says that processor speed doubles every 18 months). Likewise, the density of storage media increases at an exponential rate. This is part of the reason that Americans have more food in their house than Europeans or Third World countries. Until someone invented good, cheap refrigerators, people had to shop for perishables every day or every week.

A second reason for the increase in data available is that much of it is free. Public agencies are making data available on tapes, diskettes, and CD-ROM in greater volumes than before. Several states now have computer access to legislative information, such as bills, laws, and voting records.

Even private sector data is more available than ever before. In 1992, the Feist decision (Feist Publications vs. Rural Telephone) ruled that a compilation of facts (a telephone directory in this case) is not subject to copyright protection. In order for a compilation of data to be protected, it has to have some creative work in it. For example, a database of baseball statistics in which the number of men on base was calculated was ruled to be protected since that fact did not exist in the original data published by the league.

Another reason for the increase in the data we keep is that anybody can buy it wholesale. For example, a friend of mine was investigating opening a gift store in Hawaii, and she wanted to find a source of good beach towels. In one evening, I logged onto CompuServe, went to the Dun & Bradstreet information service, and downloaded information on over 140 wholesalers and manufacturers of towels in North America. After a manual inspection of the results, I could generate letters to the selected companies to get prices—or I could have even avoided paper altogether and faxed them directly from the computer, if we had the letterhead scanned in.

The total time required was less than one work day, and the total cost was under $200, including stamps and envelopes. If I had known more about the database and had a SIC (Standard Industrial Classification) code book, it would have cost under $150 and run faster.

To do this work before the advent of online data services would have required a trip to a large library, a manual inspection of volumes of printed industrial listings and then manual typing of letters. It would have taken days to complete and the listings, would not have been as up to date as the database.

THE QUALITY OF DATA

Charles Babbage, father of the modern computer, observed that an inseparable part of the cost of obtaining data is the cost of verification. His remarks were made in regards to mathematical tables used by the British Navy at a time when even the errata in the books contained errata.

Quality is a fairly recent concern in data processing. It is not that people were happy to have bad data in their systems and suddenly they woke up one day. Manufacturing has considered quality control as a separate discipline which has to be part of normal factory operations for over a century. The industrial quality control people, and Dr.Demming, the father of Total Duality Control, had some solid proof of results on a scale that we software types cannot yet match—the post war recovery of Japan.

Data processing has lagged behind. Perhaps it is because a new technology must follow a growth pattern. At first, data was kept by hand for particular purposes, such as accounting. Then the physical storage of the records was centralized so that the records could be easily maintained. Logical integration of all data in an enterprise is a recent concept and this concept will not work if the data is in error. The data warehouse is the step after the enterprise wide database. The data warehouse stores historical data for later use—a use which we do not yet know about. Data is being seen as something which is valuable in and of itself.

Verification and validation became easier when data started to be put into databases. Instead of trying to validate data in thousands of programs, hoping that all the programs are doing exactly what they are supposed to do, we simply check it at the database before it goes in. This also centralizes the rules as to how we do business.

Suddenly things that we could not check before became easy to verify. When Returns and Orders were handled in two different file systems, the salesmen could think that customer X was a big spender because all they see is his orders. They did not talk to the people in the Returns department who knew that customer X was returning most of his merchandise.

THE TYPES OF DATA

The types of data are more numerous than before. The term "data processing" implies record-oriented databases to most people even in the trade. Such things as video or graphics were viewed as interesting toys, used only for amusement or presentations. This is not true any more; they can be the principle product delivery method of a company.

There is a talent agency in Los Angeles that started sending out a CD-ROM of the people it represents in 1994. The casting company gets a search program on the CD which allows them to see headshots, hear voices, view video clips, find union affiliations, and search by other attributes ("Send me a Marilyn Monroe type").

This is not an advertising stunt; it is the basic product of a talent agency. Remember that most casting work does not deal with famous stars, it provides extras. The agency that gets the photo on the desk first has an advantage.

The original method of doing business was to send lots of headshots with one-page resumes to the casting directors. Physically carrying paper is slow, so mail was replaced by messengers. Messengers were replaced by Fax. But notice that we are still dealing with headshots with one-page resumes as the product. The casting people had to see the actors and weed out those they did not want. Who wants a little girl with a voice like James Earl Jones? How did we know that the guy was only four feet tall?

The filtering process meant telephone calls, appointments, auditions, and lots of time just to do a first filter. The news stories about casting hundreds of people for a single part in a major picture are not just publicity stunts. The CD is quicker than turning pages and quicker than looking for SAG (Screen Actors Guild) membership in a hundred different places on the back of a thousand headshots. With a video clip, casting personnel can see a preview of the person.

If you go to airports and large shopping malls in North America, you will see touch screen maps that locate points of interest. This is providing graphic information.

ATM machines that give you money away from home are turning pure information they got off of your teller card into money for their owners. In the near future, that data will not be just record-oriented keyed inputs; it could include fingerprint or body geometry scans as well.

THE USES OF DATA

If we wanted to sell something, how would we do it? First, we need to find people who want to buy the product. If we do not do that, we cannot sell anything. The next question is how, how do we find them?

Proctor & Gamble did a successful newspaper coupon promotion for a cleaner that removes rust stains. Proctor & Gamble has done a lot of successful promotions, but the unique part of this one was how they found their customers. They worked from simple, obvious facts, and collected data.

Sinks and tubs get rust stains from hard water. So we take a U.S. Geological Survey map, and put that in our marketing database. This is two-dimensional graphic data.

People live in houses. So we get a household listing from the U.S. Census Bureau and overlay the Geological Survey map. This tells us who probably has rust stains in their household sinks. This is a mix of two-dimensional and record-oriented data.

Newspapers and magazines have subscriber files with address information. By joining the map data with the subscriber files, we can find which publications give us the best coverage of the target audience for the money. This is pure record-oriented data.

It is time to print coupons and get them into circulation. The coupons are returned to the company via grocery stores. If we happen to be dealing with POS (Point of Sale) systems to redeem the coupons, there can be other bonuses. Unlike the old cash register, the POS units scan bar codes, handle credit cards, ATM cards, food stamps, and other data. And they keep that data for us.

We can now use the POS database to ask questions. What else do people who use our coupons buy at the same time as our rust removing cleaner? Steel wool? A competitor's soap?

You expect some obvious correlations, like peanut butter and jelly, to show up, but the good stuff is the unexpected. The sales of certain ethnic items goes up around ethnic holidays, so it would be nice to have a table of all the holidays to correlate with the demographic data and the delivery schedule. Did you know that a sale on beer at convenience stores in Chicago increases the sale of disposable diapers? Working class fathers buy disposable diapers on their way home, and cheap beer is a good way to get them into the store.

By looking at how the purchase was made (credit card, ATM card, food stamps, cash), we have an idea of the income level to which the product appeals. By looking at what magazines they buy, we have an idea of where to advertise. By looking at buying patterns caused by where and when we issue the coupons, we can plan delivery schedules better.

THE NEW DATA

Industrialization created demand for secondary services, such as packaging and advertising, which in turn create a demand for more specialized services, such as marketing research. The nature of a fundamental revolution is that you can never see how things will cascade.

The information revolution is doing the same thing. As data becomes more public, there is a demand for privacy. That creates encryption products and security services—things that apply only to pure data. As the volume of data increases, there is a demand for ways to convert it into usable information as well as ways to physically compress it. As the volume of outside data increases, there is a demand for programs that will do the searching and trading.

Cyberspace, the stuff of science fiction stories, is becoming more real every day, business is conducted on network services among people who have never physically met. Partnerships are negotiated and executed without the help of lawyers. The day of the fully electronic corporation is just around the corner. Today, at least one network service puts investors and entrepreneurs together.

One start-up company has already attempted to make a stock offering over Internet in 1994. They were blocked for legal reasons, but you can see the day coming when this will be the way to do business. The government could not determine what laws applied to the offering. You see, they have this idea about a physical location for a company, and they cannot relinquish it. You might compare this to a Neolithic hunter trying to understand how a stock certificate can be valuable.

AUTHOR BIOGRAPHY

Joe Celko has had eight regular monthly or biweekly columns in the past ten years in the computer trade and academic press. His current columns are: "SQL Explorer" in DBMS (Miller-Freeman Publishing) and "Celko on Software" in COMPUTING in the UK.

Celko has been a member of the ANSI X3H2 Database Standards Committee since 1987, and he is a regular speaker for Digital Consulting Inc., Norm DiNardi Enterprises, and Miller-Freeman Seminars. His first book on advanced SQL programming is due in 1995 from Morgan-Kaufmann publishers.

CHAPTER 6

Time is Money: Increasing Velocities, Decreasing Costs through EDI

by Bert Moore

Despite what bookkeeping says, everyone knows that raw materials and finished goods inventories are liabilities, not assets. Inventories represent money spent (or owed), material handling costs, facilities space expenses, and inventory labor costs—much of which is unnecessary.

One major manufacturing company estimates that if it can cut its raw materials inventories by 30%, the company can save more than $2 million per day, just in the cost of money.

Yet reduced inventories have typically meant increased risks, and few materials mangers have been willing to face those risks. They know that there are fixed amounts of time required to generate and send purchase orders, for suppliers to process and pick materials, and the inevitable delays in transportation, receiving, and handling.

Cut inventories? You might as well cut your own throat. The truth is, however, the use of bar codes and electronic data interchange (EDI) can allow businesses to reduce inventories significantly without increasing risks. There are two ways to reduce inventories. The first is deceptively simple: to know exactly what you have, where it is, and when it is used.

But gains through internal controls alone, while significant, are merely incremental. The second way is to build partnerships with suppliers and customers, based on the efficient collection and exchange of critical business data. These partnerships allow significant inventory reductions at the same time you decrease risks. Data partnerships cut the "fixed amount of time" required between order generation and receiving, increasing the velocity of information, and therefore goods. Gains here apply to all trading partners, allowing them to realize significant cost savings, increased responsiveness, and reduced waste. And that's only the beginning.

Accurate Inventories

One materials manager insisted that his inventory was 99.8% accurate. When pressed about what he meant by "accurate," he admitted that he knew what he had—but he was not so sure where it all was. Another company that felt it had a good inventory system finally admitted that up to 20% of its inventory might be obsolete parts—but its managers were not sure.

A third company, which had invested millions of dollars in hardware and software to implement a sophisticated MPR II system, found that it could not rely on the reports generated by the system because the data going into it was so inaccurate.

How much time and money are these companies wasting looking for materials, moving obsolete parts, providing storage space, utilities, insurance, and labor costs to count and recount all these inventories—and to what end? We know what we have, but we cannot find it. We know what we have but we don't know if we need it. We are not sure what we have or where it is.

These are not accurate inventories. These are liabilities. What is an accurate inventory? One example is a major manufacturer in the electrical/electronics industry: the automated inventory system in its distribution warehouse is more than 99.9% accurate. Everything is inventoried. Who says so? The company's Big-8 auditing firm. So, convinced of the accuracy of the system, the auditors eliminated the requirements for physical inventory counts.

That meant that the warehouse did not have to be shut down for a week, which meant that droves of employees armed with clipboards and pencils did not have to spend their time counting and recounting. That, in turn, meant that an army of data entry clerks did not have to laboriously key enter the data and that supervisors did not need to validate every entry. That is an accurate inventory.

Increasing Velocities

How long does it take for an order to be processed?

One day for it to be written up by the materials manager. One day for it to be key entered into the materials and order databases. Three days (if you're lucky) for it to be printed out and mailed to the supplier. One day for the order to be key entered at the supplier's facility. One day for picking and validation. Then, who knows how many days for it to get from the shipping dock to your receiving dock. At least a week—not counting transportation—and five of these days, an entire work week, are wasted time. Even if you fax your order, at best one day is eliminated from this example.

How long should it take? Two days, plus transportation.

ENABLING TECHNOLOGIES

Bar codes are by far the most commonly-used form of automatic data collection (ADC) technologies used for inventory tracking. A bar code is nothing more than a series of light and dark bars arranged in a specific pattern to represent numbers, letters, punctuation, and other characters.

Bar codes, which can easily be printed on-site or bought from outside vendors, are used to enter data directly into a computer system, bypassing manual recording and key entry. Not only do bar codes provide quick entry of data, they eliminate errors along the way.

A government study of key data entry shows one character out of every 300 entered is in error. This study did not investigate the number of errors made in reading and recording information from an item, nor did it study the errors made when key entry personnel tried to read handwriting.

Compare this to typical bar code accuracy of one in one million (or better). Then consider that a single swipe of a pen-style bar code reader takes, at most, one-quarter of a second to enter a 12-digit number—versus approximately 3-5 seconds to record and later key enter the same 12-digit number manually. Also consider that bar code readers are available to read bar codes on shipping containers moving at 300 feet per minute on a conveyor belt and require no human intervention.

Bar codes are typically used to identify products and product packaging, storage locations, and other pertinent data. How does this improve accuracy?

First, bar code eliminate all the time and errors associated with manual data collection and entry. Second, they making it so easy for employees to record material moves that employees will actually do it. Third, these codes provide all the information required for accurate inventories: item number, quantity, and location—information that is often not cost-effective to gather manually.

Recording every item and every move provides the tools to do better planning. Materials that do not move can be quickly flagged for further investigation—is this material obsolete, is it simply excess inventory, is it stored too far away from where it is needed?

Perhaps the biggest benefit is that, at the same time that you are collecting more data, and more accurate data, your employees can concentrate on doing what they are paid to do (receive, make, or move products), not on doing clerical work. Another benefit is that given adequate computer capacities and software, a bar code inventory system has a payback time of between nine and eighteen months.

Electronic Data Interchange (EDI)

EDI truly began during the days of the Berlin Airlift when planes had to land and take off every two to three minutes. Manual methods were out of the question. So a means to allow computers to communicate directly was devised.

Today, electronic data interchange allows companies to maintain the same type of tight coordination of their own logistics programs. Consider the following example, using Company A and Company B.

Using standard formats developed during the 1980s, Company A's computer can generate a purchase or release order, based on the data generated by the bar code inventory system, and send it to Company B's "mail box." (This "mail box" provides a holding area, a buffer, between the two computers used by the two companies, so that there is never any danger of a supplier gaining access to a customer's database, or vice versa.) Company B's computer picks up and processes the order the same day. No key entry, no delays.

Company B, which has an accurate inventory itself, notifies the transportation company that it will have a shipment ready on a specific date and time, destined for Company A. This allows the transportation company (carrier) to plan its activities more accurately and to pick up the order on time.

As Company B picks the order, using bar codes to verify each pick, it updates its own inventory and builds a shipping manifest. As the order is loaded for shipment, a unique bar code serial number on each carton or pallet is scanned. This serial number provides a key to a record in the computer about the container's contents, quantity, customer, purchase order number, and other relevant information. Once the shipment is loaded, this data is sent both to the carrier (for billing and routing) and to Company A as a Ship Notification.

The carrier can track the shipment accurately as it is handled, consolidated, and delivered. Knowing the identify of every part of the shipment can also help the carrier expedite the shipment and tell Company A exactly when it will arrive.

Company A, receiving the Ship Notification from Company B knows exactly what is coming—and the carrier can tell Company A when it will arrive.

In the most sophisticated companies, Company A keeps Company B updated constantly on the inventory levels in order to allow Company B to accurately forecast orders. Once the system is operating smoothly, not only can Company A reduce inventories, Company B will do so as well, providing savings throughout the pipeline.

These savings are only possible when data partnerships are established and inventories are accurate.

A CASE STUDY

If the examples above sound far-fetched, they are not. In fact, the textile and apparel industry in the U.S. is a classic example of the kind of partnerships that can be formed for mutual survival.

In the early 1980s, textile and apparel manufacturers in the U.S. were under intense pressure from low-cost offshore competition. Not a few insiders began to wonder whether textiles and apparel might go the way of other sectors in the U.S. manufacturing economy that simply packed up and moved to countries with lower labor costs—or buckled under competitive pressures. One leading U.S.-based textile manufacturer recognized that the only way it could continue its strong market position, and preserve jobs in the U.S., would be to provide its customers something offshore manufacturers could not. That "something" was responsiveness.

Starting in the early 1980s, this company spearheaded a coalition of retailers, textile, and apparel manufacturers to find ways to eliminate waste and increase responsiveness at every level of manufacturing and distribution. In truth, the prospects looked bleak.

For a clothing retailer, each fashion season is eight weeks long. That meant that many products had only an eight week life. Because of production lead times, retailers could reorder goods only once per "season"—if at all. That forced retailers to buy everything for a season at one time. It also meant that apparel and textile manufacturers had to produce everything based on their estimates of what buyers wanted.

This lead to tremendous waste at every level of manufacturing and sales. Retailers would sell out of some items (angering customers and losing sales) and, at the same time, have excess inventories of other items (which had to be sold in "bargain basements," often at a loss).

At times, up to 15% of a retailer's floor space could be termed "unproductive" because of stock-outs or the wrong product mix. With a typical operating margin of 2%, the financial implications are obvious.

For manufacturers as well, sales were missed because of out-of-stock situations and unsold goods had to be sold at or below cost, resulting in more waste, more inefficiencies, and more missed profits. The problem is that accurately forecasting buying trends for products as volatile and unpredictable as clothing fashions is not possible. The answer, then, appeared to be for manufacturers to somehow react more quickly once buying trends became apparent. But where to squeeze out wasted time; how to increase speed, efficiency, and responsiveness?

The answer was deceptively simple. One-quarter to one-third of the time in the reorder process was required to generate, process, and mail purchase orders up the production pipeline manually. That is, it could take one week for the purchase order to go from retailer to apparel manufacturer, one week from apparel manufacturer to textile manufacturer, and possibly one week from textile manufacturer to raw materials suppliers. Two to three weeks of the eight-week window of opportunity—wasted.

Starting with a cross-industry coalition of companies, and eventually working under the auspices of the Uniform Code Council, these companies developed electronic data interchange (EDI) standards for purchase orders. Using third-party networks, orders are now transmitted directly from the customer's computer to a supplier's "electronic mailbox"—from the retailer all the way to raw materials supplier in only a few days.

EDI transactions were also developed for shipment information. Information on materials shipped to customers is now transmitted downstream as shipments are loaded, providing downstream manufacturers the data they need to accurately schedule production runs. These standards also included transmissions to the supplier's common carriers (trucking companies) so that they, too, can develop more accurate pickup and delivery schedules.

With EDI alone, not only does the velocity of goods and information in the pipeline increase, internal scheduling improvements are also realized. But it's not enough. Exchanging order information quickly is only half of the equation. Identifying and tracking materials and products during manufacturing and shipping is another important part of the equation. Standards were also developed for bar code identification of goods and materials. Product information received from the customer through EDI, combined with data already in the manufacturer's database, is used to create bar code labels to identify and track goods and materials as they move through the system.

On the receiving dock, these bar code labels are read (scanned), identifying the supplier and item serial number to the computer system. This information, in turn, is used to access database information, transmitted from the supplier through EDI, about each item in the shipment.

Instead of workers manually checking shipments, it is now done by the computer and discrepancies are immediately identified. The final part of the equation is the use of bar codes at the retail level. Retailers now scan each item as it is sold, providing accurate, up-to-the-minute sales data. Their purchasing departments can make restocking decisions based on actual sales trends, not estimates. With the possibility of quick replenishment, retailers are now able to restock two, and even three times, during a selling season.

BOTTOM LINE BENEFITS

In addition to becoming more responsive and productive, members of the textile/apparel/retail coalition have gained real benefits for their companies. Because they have the information they need to work smarter, their operations can shrink without affecting output—or they can increase productivity with the same personnel—and all without increasing risks. Whichever way they calculate, profits go up because of:

- Reduced inventories and significant savings in the cost of money
- Reduced manufacturing, warehousing, distribution, and data entry costs (salaries, benefits, and paperwork)
- Improved utilization of existing production equipment or sales area
- Reduced facilities (space) requirements due to reduced inventories
- Reduced facilities overhead (insurance, lighting, heating)
- Reduced equipment maintenance and operating costs due to optimization of use
- Reduced obsolescence of inventories
- Elimination of shipping errors and associated costs
- Improved inventory accuracy; reduced cycle counting costs
- Improved competitive position against offshore and less efficient competitors

CONCLUSION

Excess inventories are the result of fear, uncertainty, and doubt. Bar code and EDI technologies, properly applied, can increase the velocity of information and goods through the pipeline. By reducing lead times, increasing control, and driving down costs, bar code and EDI can put materials managers in positions of strength, knowledge, and certainty.

RESOURCES

For information on bar code and other automatic data collection technologies, applications, and standards, contact:

AIM USA
634 Alpha Dr.
Pittsburgh, PA 15238
Tel: 412-963-8588
Fax: 412-963-8753
Online BB: 412-963-9047 8-n-1

For information on the U.P.C. bar code and UCC standards, contact:

Uniform Code Council (UCC)
Suite J, 8163 Old Yankee Rd.
Dayton, OH 45459
Tel: 513-435-3870
Fax: 1-513-435-4749

For information on EDI standards and education, contact:

Data Interchange Standards Association (DISA)
Suite 200, 1800 Diagonal Rd.
Alexandria, VA 22314
Tel: 1-703-683-6814
Fax: 1-703-548-5738

AUTHOR BIOGRAPHY

Bert Moore is the Director of IDAT Consulting & Education, a Pittsburgh, PA-based firm that helps companies to understand, evaluate, select, and implement a wide range of "automatic data collection" technologies including bar code.

Moore is a 1992 recipient of the AIM USA Service Award for his contributions to the Automatic Data Collection industry. He is currently the secretary of ANSI MH10 SBC-8, an American National Standards Institute committee, and maintains liaison with a number of national and international standards-setting organizations. He is the former Director of Technical Communications for AIM USA, the national trade association for vendors of ADC technologies, and former Executive Director of FACT, the major organization for users of ADC technologies.

Moore is a Contributing Editor for *Automatic ID News* and *ID Systems Magazine*, and he is a columnist for *Materials Handling Engineering* magazine. He served as a bar code consultant for the Time/Life Book's *The Computerized Society* volume of the *Understanding Computers* series.

A frequent speaker and writer on the technologies, Moore has been involved in the ADC industry since 1984 and is an acknowledged expert on the technologies, standards, and uses.

CHAPTER 7

PCMCIA and PC Card Technology

by Steven R. Magidson

As we peer across the sweeping landscape of the computing market, we cannot help but notice, sometimes with amazement and wonder, the dramatic changes that have occurred, and the trends that continue to reshape it. One area which has seen rapid changes is the mobile computing market. With a growth rate outstripping that of the desktop market, mobile computing has had a profound impact on the way we compute and the ways in which we work.

Within the broad spectrum of mobile computing, one technology stands out which has been a key element in the rapid adoption of mobile PCs. PC Card technology, based upon the standards developed and published by PCMCIA (Personal Computer Memory Card International Association), has brought forth a myriad of diminutive credit card-size products, from hundreds of vendors, that have expanded the functionality, usability, and convenience of the portable workhorses we increasingly grow to depend upon for our daily work (see Figure 7.1).

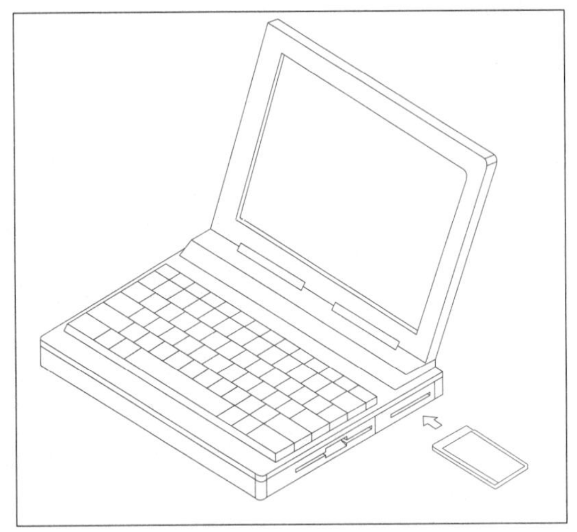

FIGURE 7.1. Workstation and PCMCIA card

But what is this PC Card technology, why are these credit card-size PC Cards so important, what are their applications, and what impact will the technology have upon us both now and in the future? We examine these questions, and hopefully provide the answers necessary to better understand both the technology and the products, and to help you become better equipped to gain the maximum benefit from them.

THE CHALLENGE OF MOBILITY

It should come as no surprise that today's business environment is a highly mobile one. What with the increasing number of business trips, sales calls, visits to remote offices, as well as the trend towards telecommuting, more business professionals are spending more time on the road, away from their home office, than ever before. This is what prompted the development of portable PCs.

Even a few years ago (eons in this business) portable PCs did not have the power or flexibility of their desktop brethren, and they were treated mostly as "second" machines, relegated to fairly mundane tasks. In recent times, however, that has changed. Portable PCs turned into laptop PCs, then into notebook PCs. Now there is wide range of mobile PCs all the way down to battery-powered handheld and PDAs (Personal Digital Assistants). They have not only gotten smaller and lighter, they have gotten more powerful as well. Today, many mobile PCs are the equal of desktops in performance, and they are being used more and more as the sole, or primary PC.

However, until the advent of PCMCIA's PC Card technology, the mobile PCs lacked one significant capability often taken for granted in desktops—the ability to expand the

PC's functionality by adding cards into the desktop's internal expansion slots. There had been no industry-standard way to extend the capabilities of a mobile PC for such things as data/fax modems, LAN (local area network) adapters, sound cards, storage extension devices and more. Users were relegated to external boxes (and their assortment of associated cables) that connected to the PC via the serial or parallel ports.

While these external box solutions were acceptable in the absence of a better alternative, they did not offer the convenience, flexibility, or performance that mobile users demanded. In order for mobile computing to continue to flourish, a better way was necessary, one that would satisfy users' needs, and one that would offer the necessary compatibility that comes from a set of industry standards. Proprietary solutions that varied from manufacturer to manufacturer simply would not be acceptable. It was from these seeds that PCMCIA was born, and with it PC Card technology as shown in Figure 7.2.

FIGURE 7.2. The PCMCIA card

PC Cards and What They Do For Us

PC Cards are small, credit card-size adapters designed to fit inside a mobile PC rather than hanging outside it. They offer both convenience and flexibility, along with performance that makes them suitable for virtually any task. Products designed in accordance with the PCMCIA PC Card Standard also offer compatibility and interoperability not attainable with vendor-specific proprietary solutions. They provide a universal expansion facility for properly-equipped mobile PCs, allowing users to expand the basic capabilities of the PC to whatever their unique needs are.

In the original 1990 release, PC Cards (then known as PCMCIA cards) were designed exclusively for memory applications. In 1991, the standard was expanded to Release 2.0 to include full I/O capabilities, and the potential for PC Card technology exploded. By allowing the same type of I/O functions that were available with a traditional ISA bus, almost anything that could be done on a desktop with an internal card could be done on a mobile PC with PC Cards.

Today, in addition to basic memory cards, there is a wide range of PC Cards designed for an equally large range of applications. PC Card data/fax modems can now provide support for V.34 at 28.8Kbps and V.17 fax at 14.4 Kbps. PC Card LAN adapters are available from many vendors, supporting both Ethernet and Token Ring across all of the major network operating systems. For powerful presentations and entertainment, sound cards and SCSI (Small Computer Systems Interface) cards in the PC Card form factor that add true multimedia and CD ROM capability to mobile PCs. Video capture cards with TV tuners can even provide full-motion video.

Wireless PC Cards give users the capability to communicate from their mobile PCs without requiring any wired connections. Pager cards that operate both inside and outside of the mobile PC give users another alternative for staying in touch—anytime, anywhere. There are even GPS (Global Positioning System) cards designed for transportation delivery applications.

PC Cards can also be used as mass storage devices expanding on those in the mobile PC itself. Flash memory cards are available with a capacity of up to 80 Mbytes. Many of these offer an ATA (AT Attachment) interface making them look just like any other mass storage device. However, since they are really memory, they are a lot faster. There are actual rotating storage devices in the PC Card form factor, just like the hard drives in a PC. Some have a capacity of up to 170 Mbytes, which can even be doubled to 340 Mbytes with commercially available disk compression software. Imagine having an additional 340 Mbytes of storage that can be carried in a shirt pocket. The applications for PC Card technology are almost limitless.

The benefits of PC Cards, goes far beyond their diminutive size. Since they are designed for mobile PCs, many of which are battery-operated, PC Cards are far more energy-efficient than traditional internal cards, drawing significantly less power. This goes a long way towards extending the battery life of the mobile PC.

The real benefit of PC Cards may well be the fact that they can be dynamically inserted and removed at will by the user without having to open up the PC and screw in a circuit board. PC Cards contain no switches or jumpers, and they are configured automatically by the appropriate software on the host PC. On a mobile PC that has been properly configured for PC Cards, the user simply has to insert the card and use it. In fact, cards can be removed and others inserted without having to power down the PC or reboot it.

This facility, called Hot Swapping, is one of the significant convenience benefits of PC Cards. Consider the following example: A user inserts a Flash card (or a rotating disk card) to copy some files to the PC's internal hard disk. Once done, the user removes the Flash card, inserts a PC Card data/fax modem, and then calls an outside service to get or send some electronic mail. Later, the user wants to browse some information from a CD ROM. By removing the data/fax modem and inserting a PC Card SCSI adapter, full access to the CD ROM is possible. Finally, the user removes the SCSI card and inserts a wireless pager card to download the messages that it has been collecting while outside of the PC. Keep in mind, that at no time did the user have to reboot the PC, turn off its power, or open its covers. Also, the mobile PC did not need to be large enough (and heavy enough) to carry all of these cards at one time. Even a sub-notebook at four pounds, can carry all of the power of a fully-equipped desktop.

You can carry the example one step further and see even more of the power of PC Card technology. Suppose I want to transfer several graphic data files, totaling 15 Mbytes, from my mobile PC to yours. Instead of doing the "floppy shuffle" with a dozen or more diskettes, or using a cable-based file transfer program, I simply copy the files onto a Flash card, using it like a very large, very fast floppy disk. I then insert the card into your PC (which temporarily creates an additional "drive") and copy the files directly to your PC's hard drive. Look at your own mobile PC usage and applications—I am sure you will find many ways in which the ability to easily and quickly move peripherals and storage devices from one PC to another, without time-consuming reconfiguration, would come in quite handy.

PC Cards on the Desktop

While PC Card technology was originally developed for the mobile computing market, it is now being used in desktop PCs as well—and for good reason. The small size of the cards and slots will allow the development of smaller footprint desktops. Their low power utilization makes the PC far more energy-efficient than older internal cards. Users will not need to "open the covers" and get inside the PC just to add a card. The configuration and setup will be greatly simplified making these products usable by those with little or no technical experience.

The ability to dynamically insert and remove cards will be an added benefit for those with both a PC Card-equipped desktop and a mobile PC. Flash cards or PC Card disk drives make it easier for users to transfer information between their machines than other alternatives. Where appropriate, a single peripheral such as a data/fax modem or SCSI card could be shared among PCs as needed, reducing the cost of outfitting both PCs individually.

Some manufacturers are already delivering desktop PCs that contain PC Card slots, and more are expected to do so in the future. But even those with existing desktops are not left "out in the cold." There are many PC Card "reader/writer" kits available that can add up to four, or more, PC Card slots to virtually any current desktop. It is expected that as the growth in the desktop arena continues, even more applications for PC Card technology will appear. We have seen how we might put PC Card products to use: now we will look at the technology itself.

THE PC CARD TECHNOLOGY

PC Card technology is made up of three essential components—cards, slots, and software—all designed to operate together offering an industry-standard method of expanding the functionality of mobile and desktop PCs alike. Products that conform to the PC Card Standard as published by PCMCIA offer users a wide range of choices with a maximum level of compatibility. PC Cards also offer unique features such as Hot Swapping, which make them far more flexible and convenient than traditional "static" internal cards.

PC Cards

The PC Cards themselves are the most visible of the components, since they are the ones that actually do the work. About the size of a credit card, all PC Cards are the same size in length and width—85.6 mm X 54.6 mm (or roughly 3.37 X 2.15 inches)—and all use a 68-pin connector to provide the electrical and physical interface to the host PC (see Figure 7.3).

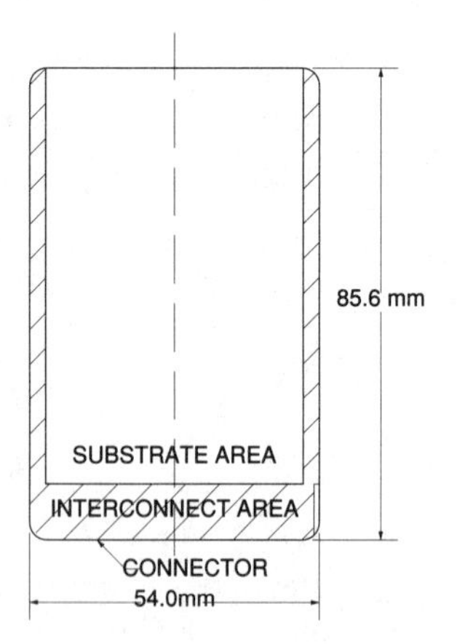

FIGURE 7.3. PCMCIA interfaces to the physical PC

PC Cards do differ in their height, referred to as the card's "*type*." Type I cards are 3.3 mm tall, and are most often used for memory applications such as SRAM and Flash memory. Type II cards are a bit taller, measuring in at 5 mm. Type II cards are the most predominant for traditional I/O applications such as data/fax modems, LAN adapters, sound cards, and the like. These applications require chip components that do not fit in the Type I form factor, even with current surface mount technology. Type III cards are the tallest at a height of 10.5 mm. Type III cards are most often used for rotating storage devices. These are actually mini-hard drives and contain rotating platters and moving read/write heads just like the hard drives in PCs, only much smaller. The physical size of the platter and head assemblies in these devices, however, currently requires more space that a Type II card can provide.

One of the real convenience features of PC Cards is the absence of the troublesome and error-prone switches and jumpers often found on traditional internal cards. Those with experience in installing and configuring such cards will welcome a solution that eliminates this nightmare. With PC Cards, the configuration and setup is handled through software. Each PC Card contains a CIS (Card Information Structure), which can include all of the configuration information necessary for the host PC to properly incorporate it into the system without conflicts. This information is provided on the card by the card vendor, and is transparent to the user. The CIS is an important element in providing compatibility and interoperability across a wide range of cards and systems.

Cards Must Have Slots

PC Card slots, or "*sockets*" as they are sometimes called in PC Card parlance, are built into the PC and accept the PC Cards via the 68-pin interface. They are essentially internal slots designed to house the PC Card inside the PC rather than outside. Like PC Cards, slots also come in three Types—Type I, II, and III—as shown in Figure 7.4.

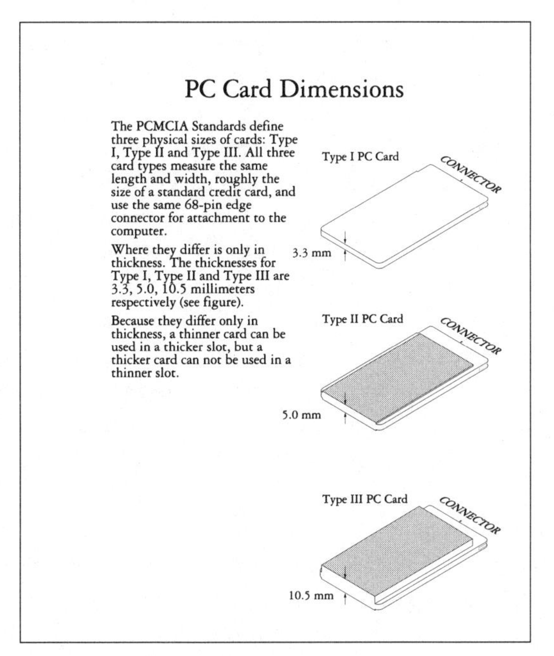

FIGURE 7.4. PCMCIA card dimensions

There does not have to be a one-to-one correlation between the "Type" of the card and the "Type" of the slot. Since the 68-pin connector is located in the same physical position on a PC Card, whatever its type, a Type I card (3.3 mm tall) can fit into a Type I, II, or III slot. Correspondingly, a Type II card (5 mm) can fit into either a Type II or a Type III slot. A Type II card, however, cannot fit into a Type I slot, nor can a Type III card fit into either a Type II or a Type I slot. When selecting PC Card products it is a good idea to make sure that the PC being considered is capable of holding the cards being considered—so no surprises will occur later on.

Many notebook PCs contain two Type II PC Card slots that are stacked vertically. In this case, either two Type I/II cards can be inserted at the same time, or a single Type III card can be inserted which will use the connector of the bottom slot. Stacking the slots in this manner provides a good deal of flexibility for the user, particularly considering that cards can be removed and inserted at will (Hot Swapping) without having to restart the PC. You need to be careful, however. Some sub-notebooks may only contain a single PC Card slot, and it may well be Type II. Unless you are planning to use a PC Card Type III rotating storage device (or some other Type III card) you will probably be fine, but itís a good idea to check first and make sure.

Software Ties It All Together

The final component of the PC Card technology is the software, which is composed of Card and Socket Services and client device drivers. The Card and Socket Services Specification is contained in the PC Card Standard, and it is most often distributed either by the PC vendor or the card vendor (sometimes both).

Socket Services is the host PC software layer that interfaces directly with the PC's socket controller hardware. There are several controller chipsets available, and, while they all perform essentially the same tasks, their implementations may be different. Socket Services' responsibility is to mask these differences from the upper layer software modules; it does so by providing a standardized software interface to them. Among other things, it is responsible for managing the socket resources and notifying the upper software layers when a card has been inserted or removed. In earlier versions of the Standard, Socket Services was considered optional. Now, because it is so important in providing compatibility across a range of systems, Socket Services is a required PC Card component.

Sitting above Socket Services in the software hierarchy is Card Services. While Socket Services is hardware dependent and not concerned with the PC's operating system, Card Services is software dependent, acting essentially as an extension to the operating system itself. That is to say, Card Services for DOS/Windows is different from Card Services for OS/2. It is Card Services that is responsible, for example, for proving such PC Card unique services as Hot Swapping. It manages the cards, their resources, and the interaction between cards when more than one is in use. It utilizes the standardized interface provided by Socket Services, and in fact requires Socket Services to be present.

Typically Card Services software comes together with the Socket Services software, and it is most often provided from the PC vendor. While there are different vendors of Card Services software, the standardized interface to the software layers above it allow compatibility across the various vendors products. Also, like Socket Services, Card Services is transparent to the user once it is installed.

The final software component includes the client device drivers that come from the PC Card vendor with the card. These client device drivers interface with Card Services on one side, and provide an interface to applications in the PC that are "PC Card-aware." One typical application for such drivers are PC Card LAN adapters which interface with a variety of network operating systems. Such drivers work in concert with Card Services to bring the full range of PC Card capabilities to the user.

ON THE HORIZON

Although the Standard has been around since 1990, it has been significantly enhanced over time, and it has really only been within the last 12-18 months that PC Card technology has been widely adopted. The Standard is still a "living" document, and PCMCIA continues to enhance it in response to the growing user needs and the dynamic development of the technologies that surround it. In October of 1994, PCMCIA announced a major new revision of the PC Card Standard specifically to address both short and long term technology trends. It includes several new features that should be incorporated into products in 1995, and which will greatly expand the applications for PC Cards.

As PC systems become more energy efficient, they will rely more and more on 3.3-volt technology as a means of lowering power utilization. 3.3-volt systems draw significantly less power than to 5-volt systems, making them ideal for battery-powered subnotebooks, handhelds and PDAs. The new PC Card Standard includes optional support for such 3.3-volt systems and cards in addition to the current 5-volt specification. And it allows for PC Cards to support both 3.3-volt and 5-volt operation in the same card to take advantage of new low voltage systems, while maintaining compatibility with the large number of existing 5-volt systems.

Somewhat related to the low voltage issue is that of power management. Many mobile PCs now incorporate extensive power management facilities, such as APM (Advanced Power Management), that are designed to reduce the PC's power consumption while in an idle or suspended mode. The PC Card Standard now incorporates an interface to APM allowing card vendors to easily avail themselves of these power management facilities. Power management-aware cards will be able to draw less power while idle, thus increasing the PC's battery life.

Another significant enhancement in the recent release is support for Multifunction cards. Virtually all current PC Cards offer a single function, be it a data/fax modem, LAN adapter, or SCSI interface, to name a few. As the number of applications for PC Cards has increased, users have wanted more of them on a single card. With advances in hardware technology that allow fewer integrated circuit chips to provide more functionality, it became possible to have PC Cards perform multiple functions, such as incorporating both a modem and a LAN adapter, or a sound card and a SCSI adapter onto a single PC Card. Under the new Standard, Multifunction PC Cards that can integrate several related technologies in a single product can be built. Having fewer but more powerful cards is not only more convenient for users; but should also reduce overall costs.

One trend that has been with us continuously since computers first appeared is the need for increased performance. As PCs got faster and had greater capacity, the application demands placed on them increased accordingly. Can you imagine trying to run Microsoft Windows and its applications efficiently on an 80286 PC that was current technology only three years ago? And the power trend continues. On the horizon we see 100 Mbps Ethernet, ISDN, and other high-performance networking technologies. The demands of multimedia and full-motion video also place heavy demands on today's technologies. Higher speed mass storage devices will demand throughput that may outstrip the capabilities of today's solutions.

PCMCIA has recognized this trend and has incorporated a higher speed interface into the Standard. Called "CardBus," it provides a 32-bit bus mastering interface capable of a 132 Mbytes/sec. maximum throughput rate at 33 Mhz. The current PC Card technology operates at 16 bits and roughly 20 Mbytes/sec. While this is more than adequate for today's applications, PCMCIA felt it was important to provide standards today for tomorrow's applications. In that way, as these applications appear, vendors will still have an industry-standard method of addressing them and providing compatibility for their users.

These and other enhancements included in the October 1994 release of the PC Card Standard, will provide a powerful base upon which developers can rely for their future products. These products will bring even more functionality, flexibility, convenience, and performance, and they will further expand the opportunities for PC Cards.

Looking further ahead, the markets and opportunities for PC Card technology will continue to broaden due to the PC card's unique attributes of size, convenience, and power consumption. Already, digital cameras are being developed that will use PC Cards to store digital "pictures" which can then be viewed and edited on any number of devices. Printer manufacturers are planning to incorporate the technology into tomorrow's printers to store fonts and to provide additional plug-in processing intelligence.

PC Card technology is also expected to play a large role in a variety of "consumer-oriented" products ranging from sophisticated entertainment systems to interactive exercise equipment where PC Cards can store individuals' profiles, exercise programs, and their progress. The automotive industry will be incorporating PC Card technology into cars as early as 1995, coupling it with the global positioning system (GPS). PC Cards will hold detailed city and street maps (and related information) to give drivers their precise location and to provide exact directions to their destination. Combined with wireless radio technology, such a system could warn the driver of traffic jams or delays and offer suggestions as to alternate routes. Other consumer applications for PC Card technology, such as banking and purchasing, are also being investigated.

As mentioned earlier, the applications for PC Card technology are almost limitless. Even those of us close to the technology believe that we have only seen the tip of the iceberg. The creative minds that are responsible for so much of the innovation we have seen already will have one more powerful tool to add to their arsenal. While mobile computing is certainly where PC Card technology got its start, clearly it will not end there. With so many technology trends advancing at an ever-accelerated pace, PC Cards will continue to play a vital role today, tomorrow, and in the future.

AUTHOR BIOGRAPHY

Steven R. Magidson is an independent consultant and author specializing in the areas of mobile computing and networking. In addition, Magidson serves on the board of directors for PCMCIA, the international standards body and trade association for PC Card technology. Previously, Magidson was Vice President of Marketing at Xircom, a leading manufacturer of networking products for mobile computers.

Magidson is a frequent author and has been published in *Computer Reseller News*, *Network Computing*, *LAN Times*, *ComputerWorld*, *LAN Technology*, *Stacks*, *Portable Computing*, *HP Palmtop*, *USA Today*, and others. He has been a panelist and panel moderator at several industry trade shows and technical symposia including Comdex, IC Card Expo, LAN Vision, and Innovate. He received a B.A. in Journalism from Rutgers University and an M.S. in Computer Science from West Coast University.

CHAPTER 8

Do-it-Yourself Software

by Dr. Joachim Schurmann

"Today I look around and see whether we can change our business practices to the way our software works," Pete Walker, Accounting Manager at the Superconducting Super Collider facility in Waxahachie, TX tells me. "We used to rewrite accounting systems in the past, because we felt we had to, and, also, because we could afford it," Walker continues. "People were cheaper than systems, but that has changed in a big way."

The person speaking is a system user, practice-hardened and goal-oriented. "And even if we had all the programmers that we could use, changing an accounting package would cause us endless grief with the next release."

Managing your business after your software: that is the concept of the 1990s. If you are unconvinced, take a look around—it's really happening. *Federal Computer Week* of September 5, 1994, carried the following headline: "DOD Builds Command System with Off-the-Shelf Software." No kidding, go look. If the Department of Defense can do it, you can do it too.

Indeed, there is enough software available for just about any application. Pick the right software, and build your business around it. This makes sense, because it saves you time and worries, and it is also a great way to acquire knowledge without going to school. Take advantage of the competence of countless experts who have contributed their skills. Consider this: if a package is good enough to keep its designers from starving, there must be a few thousand users out there who think that it works all right. Their edge can be your edge too.

The Stars, Our Destination is a specialized Science Fiction bookstore in Chicago, IL. Recently I noticed that they were about to install an Information System. Since the store seemed to be doing just fine without such a system, this intrigued me quite a bit. I finally got hold of one of the owners, Alice Bentley, who confirmed that they were planning to use BookWizard on a Macintosh Quadra 610. "You see," she confided, "up to this point we have simply stashed our books on shelves and restocked when the shelves got empty. We would like to do better than that. We need to know exactly which books do sell and which do not so that customers will find what they want." When I asked whether she had a good idea of how to go about doing it, she pointed out that was what she bought the software for. "We thought about it a lot before spending the money. But the software will help us to do things right, and that is worth the expense."

WHERE WE ARE COMING FROM

Software used to be developed on an individual basis, created for the needs of each individual customer. The philosophy was to write the software to fit the way business was done in each individual case: automate the process without disrupting the workflow. Early software was as individual as the tools created by a medieval artisan.

As everybody automated processes, marginal ROI decreased. Soon automation would not generate enough benefits to justify individual solutions. The market responded by inventing the concept of packaged software. The most common applications were targeted first. After all, there are only so many ways to do, for example, payroll, accounting, warehouse management, and what have you. Still, people wanted personalized solutions. Hence personalizing packages became big business.

Unsurprisingly, interfering with a program logic that had cost several designer-decades to put together proved cumbersome and expensive. Much of this expense, though, did not show as the cost of a software package. The software itself was relatively cheap. But the man years required to make the modifications kept countless consultants and jumbo sized DP departments in business for years. The bulk of the changes was as thrilling as: change the report layouts because we have always done things this way, expand the title fields because our company name does not fit, and add the oddball feature that is the buyer's pet idea. I know, because I was there. It was like a leasing plan: you got in for cheap, but in the end you paid a lot more. The industry was in need of a serious attitude adjustment.

THE DECLINE OF THE BEHEMOTHS

In the 1960s, and well into the 1970s, computer logic was the exclusive domain of a segregated priesthood that did magic things to the great black box and saved the corporation a lot of money. Only it didn't. And it didn't stay that way either. The great black box was a million dollar item and the priesthood perpetuated job security by keeping data processing procedures as arcane and obscure as humanly possible.

Along came fellows like Ed Roberts, Les Solomon, Steve Wozniak, Bill Fernandez, and Steve Jobs who emancipated the idea of data processing into a totally new way of life: power to the people. From a breadboard, to a small box, to a personal computer: the development was swift and fateful, heavily undermining the existing DP establishment and introducing totally new paradigms into the industry.

Don't get me wrong, DP was still as arcane an art as ever, but, with the wider availability of hardware, it found a large number of new practitioners, people who actually thought computers were fun. They coaxed sound out of the box and invented games, graphics, and images. Personal computers were cheap and fun, the mainframe monsters were dull, expensive, and hard to manage. Now it was gaming programs vs. accounting packages. And the computer action moved out of the hallowed halls into people's living rooms. What fertile ground for experimentation! The sling was poised to bring down the Goliaths of the industry. Leverage was the keyword. People could structure their ideas as a logical sequence of computer commands, and, if they so desired, they could reproduce leverage thousands of times with very little cost. A new medium was born, and with it came new ideas.

I have fond memories of running VISICALC on a battered Apple computer. VISICALC is a spreadsheet. Spreadsheets were a truly new idea, conceivable only in an era with personal computing power. They were a major ingredient to the first recipe for the instant programmer. Get a hold of a piece of hardware, load a spreadsheet onto it, and thumb your nose at corporate America. No pleading, no waiting, no red tape: you need a new formula, you write a new formula. You could actually test and experiment for yourself. That his how Johnny Application-User became a programmer, even without being aware of it.

To program a computer one does not need to sling COBOL, or whatever other language may be fashionable at any given time. Programming means to convey one's original ideas to an electronic device in order to enable autonomous action. This goal can be achieved by more means than one, some more painless than others. The availability of desktop computing power allowed more realistic interfaces to be established for many day to day activities. Soon the typewriter would become a back-office curiosity, and the slide-rule would go the way of the Dodo bird.

WHY WE ARE HERE

Soon a new productivity plateau was reached. There is a point beyond which even cheap and available hardware ceases to be cheap and available. That point is reached when the office clerk spends more time coaxing productivity out of a computer than being productive for himself.

ROI started declining again. A desperate search for the silver bullet ensued. Pundits prophesied an era where computers would program themselves, relieving humanity of this last technological burden. But this vision proved to be as ephemeral as an Elvis sighting.

Other technologists had more realistic goals in mind. One of these goals was dubbed business process reengineering. While this expression carries significant rhetorical punch, there is really not very much to it, other than that now we are modifying business processes to accommodate automation. Because it is too expensive to adapt computers to traditional human ways, we change the ways humans do things to better capitalize on the strength of computers.

Another goal that is being pursued is improving the ways computers work together. The industry tries to tackle the interoperability issues of their systems. But since each vendor puts forward its very own brand of interoperability, most computer programs end up interoperating only with themselves. Over time, however, a few de-facto standards do emerge. They came into being because a few big kids on the block grabbed the initiative and bullied a few other kids along. The MAC platform, MS-Windows, X-Windows, and a few other standards ended up prevailing. The secret of interoperability is that if you write your program to meet certain standards, those standards will be able to work together with other programs that meet the same standards. There are two important factors in place now. Firstly, people have learned to harness the fabulous, albeit moronic, intelligence of computers. Secondly, different pieces of logic, written by total strangers, for totally different purposes, can work together. The era of functional specialization has begun.

THE SUM IS GREATER THAN THE PARTS

History is repeating itself. What used to be self-contained products, individually crafted by high skilled artisans, are now becoming an industrial age commodity. Specialization engenders quality, which causes more widespread use, which improves the cash flow of the software producers. This permits investments into bettering ergonomics, which, again, makes more users willing to use computers. This again, improves the balance sheet of software developers. All in all, through this combination of factors, a software industry spirals into existence, which has resources to spare to create its own tools and rationalize the software construction process.

At this point no single vendor needs to provide cradle-to-grave, complete and comprehensive user satisfaction. Users can pick and choose the desired logic from different software factories and combine as they see fit. Disparate logic elements can work together like a single family of products. This, of course, spells the doom of the large providers, and brings on an entirely new breed of professionals.

The prevailing DP professional is now the Information Systems yenta, a matchmaker, fitting existing solutions to perceived requirements. He or she does not do programs but knows enough about existing products to put a comfortably integrated system on the user's desk. He or she does not work for any specific hardware and software vendor but knows enough about all of them to understand their idiosyncrasies and make their products work together. He or she touches upon, and successfully bridges, a lot of skills that used to be separate: he or she can do a hardware repair, a software patch, or even train the user. Where do you find such a professional? Your local small business association, college bulletin board, and, not to forget, the yellow pages.

The profession of a programmer has not gone away, although it has changed a lot. The traditional image of the pizza gobbling, freckle faced college kid is passé. Today, software is too complicated an affair to be conceived by a single genius. The complexity of recent operating systems alone rival the Space Shuttle project. Hence further specialization ensued. There are those programmers that forge the tools for programming; and those who use the tools to build final products.

The tool builders are highly specialized, come in large groups, and are narrowly focused on interacting with hardware and the operating systems. The product designers are industry-specific knowledge engineers who packetize procedures and processes. They focus on user interaction, human interface, and industry-specific application logic. Both types of programmers are equally important to the current picture. They have evolved from common ancestry, but they are now playing widely differing roles. Let us sum up the current situation: Processing logic is becoming an off-the-shelf commodity. Users are increasingly able to put together their own software from existing building blocks. The world of Information Systems professionals is changing: standardization allows for specialization.

WHERE WE ARE GOING

Data processing is becoming ubiquitous. We are already at the beck and call of our electronic minions wherever we go. We will have more of the same: from voice communication, to e-mail, to data preparation and display. Raw processing power is modifying the way we see the world. Processing power requires software to be effective.

Where there are many similar pieces of hardware, a lot of money is to be made by selling software, since significant economies of scale can be achieved. This means that by its sheer size, the market will be able to support a large number of diversified products.

The end-user of future data processing products will find it even more challenging to identify the right product for his or her needs. He or she may have to retain the services of expensive professional intermediaries in order to make sense of the market. On the other hand, those software producers who can bypass this middleman will corner the market and command top prices. The challenge is to figure out a way to do that. The companies best able to provide highly specific products for individual users and yet exploit the leverage of a large market are going to be the winners.

I see only one way to come out on top: increase the granularity of the logical program elements until they are only a small part of the whole, and then allow a mechanism to fit them together like Tinker Toys. Each logical element can be reused under all sorts of circumstances, but the combination thereof might be highly user specific. Establishing a logical process could be as easy as going shopping at the mall. The difficulty is to invent the glue to hold these pieces together.

At this point, purists introduce the concept of Object Oriented Programming (OOP). OOP is a fancy name for just such a modular environment. Scientists have revisited the cherished von Neuman logic and changed the foundations of data processing based on radically new assumptions. Unfortunately, the tangible results of this new technique have so far been rather underwhelming.

When explained in the Queen's English, most assumptions driving OOP make a lot of sense. It is an ecology of self-contained pieces of logic that have the intelligence to know their own limitations and practical uses. They place certain expectations on the environment and provide specific services in return. Each of the modules has a life of its own. It initiates processes and independently interacts with the world. Theoretically this technique is extraordinarily well suited for an intuitive information processing environment that mimics reality as we experience it. Unfortunately actual implementation is very slow, albeit quite steady.

Object Standards Accelerate is the headline in the September 5th issue of *ComputerWorld*. This sounds upbeat, but realistically it will probably be a decade or more before the end-user will reap the direct benefits of this new technology. Users have been holding their collective breath for a long time, hoping for a more practical approach to data processing. As it is, they are probably not going to wait for OOP to be fully matured. There is plenty of room for compromise.

On the one hand operating systems are growing increasingly sophisticated, providing more common services, and allowing for more granularity in the program logic than ever before. NeXt is a case in point and, in a different way, Windows 4.0 should prove to be on the same track.

On the other hand, OOP thinking is heavily affecting the software development industry. New ways of manufacturing programs are coming to benefit the end-user, albeit indirectly. By using large amounts of precompiled code and advanced toolkits to let the programmers manage computer logic more efficiently, program development time is shortened and programs are becoming more reliable.

Visual Basic is an example of such a hybrid development tool. Some of its design features are borrowed from the OOP world. Other characteristics betray a rather mixed parentage but are all pragmatically implemented to take advantage of current technology and techniques.

All factors converge to provide a plethora of user-friendly, sophisticated, highly specific, yet flexible software. Suitable operating systems and increasingly better standards make it easier than ever to pick and chose from a wide array of software which will work seamlessly together.

This should squeeze a number of DP professionals out of the market. In particular the providers of advice, the cherished matchmakers of the computer industry, will feel the pinch as end-users become increasingly self-sufficient. At the same time a larger number of people than ever before will be needed in the software manufacturing process. Employment in the DP area will not go away anytime soon, it will just be significantly restructured.

WHAT TO DO TO STAY ON TOP

Reading, 'riting, 'ritmetic are the three traditional skills cherished by the Anglo-Saxon educator. I am afraid this will not do in the future. We will need to add computer skills, or at least the ability to use computers in a meaningful way, to make people functionally literate.

A medieval merchant could afford to hire a scribe to read correspondence or take dictation. Today's merchant is doing the reading him- or herself. Today's professional can hire a computer specialist to manage the computer. Tomorrow's professional will need to structure his or her business around computers to best take advantage of their strategic edge.

The biggest mistake we have made in the past was to isolate computers from the real world, to relegate them into an aseptic environment that would remain uncontaminated by common sense and the concerns of the balance sheet. The Luddite in us has kept this newfangled contraption out of sight as much as possible. Economics dictate that this will need to change. The computer is coming out of the closet, and we must be ready for it.

AUTHOR BIOGRAPHY

Dr. Joachim Schurmann has been a data processing professional since 1981 with several large organizations. His career encompasses the full spectrum of data processing activities: He started as a programmer and has since operated in various positions, including as Project Manager on large private and government projects. He is also an experienced EDP auditor. Schurmann has practiced his profession in the United States and in various European countries and has achieved a global view of Information Systems.

At this point in time Schurmann is Director of Application Development at Academic Advantages Inc., and he is a consultant with Crossbridge Connections Ltd. Contact him on CompuServe at 72122,144; on the INTERNET as joachim@Advantages.mystery.com; or drop a note to 7223 S. Rte. 83, Suite 301, Willowbrook, IL 60521-7561.

CHAPTER 9

Pool the Risks, Share the Benefits: Partnerships in IT Innovation

by Sharon S. Dawes and Mark R. Nelson

The California Department of Motor Vehicles recently experienced the failure of a major IT undertaking. A project to move nearly 70 million vehicle, license, and identification records from an antiquated system to a new relational database was both behind schedule and over budget. When California's lawmakers finally decided to end the agency's IT project, over $44 million had already been spent and no end was in sight. One of the reasons the DMV project failed, says California Assemblyman Phillip Isenberg, is because the agency personnel "were over their heads with a technology they did not understand." The project also lacked a clear link between agency operations and the goals of the selected technology. Due to procurement restrictions, the agency was married to a specific hardware platform before available IT solutions could be explored. As a result of the failure, California's technology procurement process faces even greater control and oversight (Miller, 1994).

Despite these problems, California has an annual IT budget of over $1 billion, and more big projects are on the horizon as they are in every state in the nation. Federal, state, and local government constitute the world's largest consumer of information technology. Government serves a vast array of public needs, manages billions in annual operating expenses, and has well-recognized service delivery problems. It cannot operate effectively without using advanced information technologies. However, as the California DMV crisis amply demonstrates, the risks of IT innovation in government are daunting.

RISKS OF IT INNOVATION

Companies have wasted money on IT because they made one or two fatal mistakes. Either they bought the wrong technology for the job, or they bought the right technology and installed it but did not implement it. Far too often, people in business environments do not take the time to learn what computer technology can do, or they do not change their business processes to capitalize on technology. So they buy the wrong tools for the job, or they never learn about how to use the tool properly (Currid, 1994:7).

IT innovation is risky business in every organization. Repeatedly, organizations abandon IT projects because they fail to accomplish the objectives they were intended to meet. In both the public and private sectors, a well-documented set of risks attends IT innovation:

Lack of Organizational Support and Acceptance

Adoption of a new technology is unlikely to succeed if it does not have widespread organizational support and acceptance (Alter, 1992). Organizational perceptions of new technology are critical factors for generating that support. Realistic and positive expectations lead to success; negative ones lead to failure (Abreu and Conrath, 1993). Much has been written about the critical importance of top management support. But support and acceptance throughout the organization, especially among the people who will use the technology, is equally important, and it is often more difficult to achieve.

Failure to Evaluate and Redesign Business Processes

Systems may not meet performance expectations because "companies tend to use technology to mechanize old ways of doing business. They leave the existing processes intact and use computers simply to speed them up" (Hammer, 1990). Meeting the needs of customers, employees, and decision makers means carefully studying and evaluating business processes. In most organizations, new processes are added as needed, but old processes are rarely evaluated to determine whether they still make sense. When new technology is thrown into the picture, a cumbersome and inefficient process is often automated "as is." The results are systems that do not serve business needs, systems that are too expensive for the small productivity gains they provide, and systems that are not flexible enough to meet changing demands.

Lack of Alignment Between Organizational Goals and Project

Another risk factor involves the alignment (or lack of alignment) between organizational and project objectives. The goal of IT adoption should be to enhance or improve an organization's ability to carry out its main mission or business objectives. It should improve customer service, reduce inventories, speed production, increase sales, and prevent errors. Technology cannot be adopted for its own sake. An IT organization that becomes enamored of a database or office automation project without understanding how real people use information to accomplish real work is setting itself up for failure.

Failure to Understand the Strengths and Limitations of New Technology

Information technology is constantly changing and improving. No one is able to keep up with the details of new developments or to understand comprehensively how each new technical tool works. Add to this the fact that most new technologies must work in tandem with others, or must be incorporated into existing older systems, and the potential for trouble mounts rapidly. Since most organizations are not in the IT evaluation business, they tend to rely on word of mouth, vendor claims, and trade publications for the bulk of their knowledge about which tools may be right for which jobs. There is little opportunity to learn first-hand before making irrevocable decisions.

PUBLIC SECTOR RISKS

Government seems to have even more trouble than the private sector in successfully applying new technology. The public policy choices and public management processes that are part of government make it an especially difficult environment for IT managers (Andersen and Dawes, 1991). Some contend that bid protests, low government wages, and legislative interference lead large government information technology projects into trouble (Miller, 1994). Others recognize that the structure of government decision making, public finance, and public accountability complicate the government-manager's job and limit the array of management choices (Kelman, 1990). Risks unique to the public sector include:

Divided Authority Over Decisions

Executive agency managers do not have a clear line of authority over agency operations. Their decisions are circumscribed by existing law, the limits of current appropriations, a civil service system, and a variety of procedures mandated by both the legislature and the courts. These restrictions do not blend well with the complexities of managing a multi-million dollar IT project in a rapidly changing technical environment. Worse, as can be seen in the California situation, when IT projects fail, the first legislative response is to place more restrictions and more controls over the IT management process.

One Year Budgets

Uncertainty about the size and availability of future resources weakens the ability of government agencies to successfully adopt new IT innovations. Most government budgets are handled on an annual cycle. As a result, funds promised for a project in the first year may not be continued during the second or subsequent years. While many agencies have developed planning mechanisms to cover a three-to-five year period, annual appropriations (influenced heavily by changing government-wide priorities) tend to negate long-term planning.

Highly-Regulated Procurement

Most decisions to adopt emerging technologies are made through the traditional competitive bidding process, a one-shot technique that is ill suited to the experimentation and learning that should appropriately accompany such large investments. While the goals of competitive procurement are goals of integrity and fairness, the processes are often a source of problems and delays. Agencies write RFPs using the information they have been able to gain from limited research. Vendors spend large sums of money trying to develop a winning response. Lengthy, arms-length reviews and negotiations ensue. Losers often take advantage of bid protest procedures which can delay contract awards for months. Due to insufficient project support or understanding by top management, IT procurement requests may receive low priority (USGSA: 1988). The resulting delays can mean time and cost overruns which in turn yield negative publicity, decreased support from top management, and negative perceptions of the technology overall.

Lack of Government-wide Information and Information Technology Policies

The absence of a government-wide information policy in most states adds additional risks and problems. Many government officials do not appreciate the importance of information as a management resource for sound decision-making, for program accountability, and for effective and efficient service delivery. Without a strategic focus on technology management and a high level overview of how information and information technology can support government operations, a wide range of potential new services, public-private relationships, and operational efficiencies cannot be realized. Instead, each project is an isolated initiative of a single agency, unrelated to other activities of the government which serve the same people or use the same information (Dawes, 1991).

A Partnership Model for Reducing Risks: New York's Center for Technology in Government

Throughout the public sector, there are competing demands to both reduce the cost of government and increase the quality of public services. In order to achieve both goals, public managers attempt to redesign processes, increase productivity, improve coordination, reduce costs, and improve quality. To do any of these effectively, they need to experiment with innovative technology. Large businesses often meet this challenge by creating research and development (R&D) offices, but government agencies almost never have funds or people to devote to this kind of work. As a result, the public's business tends to be conducted with older, "safer," less efficient technology.

New York State is the first government to establish a program that pools the resources of three sectors to develop or demonstrate advanced information technologies and thus reduce the risks of IT innovation. The Center for Technology in Government (CTG) was established by the Governor and the Legislature in 1993 at the State University of New York at Albany. Its $1 million budget supports competitively selected prototype projects which address critical public problems such as land use, emergency psychiatric services, small business development, and services to mentally retarded citizens.

Projects are structured to give full attention to management and policy issues so that the prototypes are designed and evaluated in recognition of the needs and limitations of the real organizations that will eventually implement them. The development of prototypes gives agencies first-hand knowledge of how technology can improve their operations, how it can be integrated with existing systems, and how it can solve their problems.

A particularly valuable element of each project is the ability to demonstrate the prototype to the people who will actually work with any new system and to involve them directly in helping to redesign their own jobs, processes, and support systems. The results of all Center projects are documented and disseminated widely to public and nonprofit organizations.

Through a program of research, publication, and education, these agencies have ready access to an expanding pool of information about applications of technology to public problems.

Partnerships Underlie the Center's Program

As a neutral site for experimentation, the Center solicits project proposals from public agencies and conducts them with the on-site involvement of experienced government managers. Through the University, the Center draws upon the research interests of a multidisciplinary faculty. The Center offers opportunities for interns, fellows, and graduate students to work on applied problems as part of their academic or professional training. Corporate partners support the Center and its projects with donations and loans of equipment, software, and expert consulting services. They also help equip the Center's network-based computing and communications laboratory which in turn supports many areas of investigation. Professional CTG project managers coordinate operations on behalf of the participating partners.

Partners' Contributions

Government organizations in New York and around the nation are laboratories for learning, experimentation and problem-solving. Public agencies bring to the Center a set of mission-critical service delivery problems embedded in practical organizational, political, and community contexts. These are not theoretical questions; they are issues that significantly affect the lives of citizens, families, businesses, and organizations throughout the state. Moreover, each Center project involves experienced public managers with knowledge of their agency's programs, history, resources, and clients.

Corporate partners contribute the latest technology tools as well as their own experts, who work directly on project teams with agency and university participants. Their participation reflects both interest in the practical applications of technology and a long history of successful industrial research and development programs. Since information technology and service companies have both business and government customers, they offer a broader context for problem-solving than either sector can achieve on its own.

The University community offers a wide range of expertise in a variety of fields including information science, public affairs, business administration, and computer science. Faculty and students serve on project teams in many capacities ranging from project management, to software development, to research and evaluation. Thanks to the participation of trained research scientists, CTG projects produce well-documented publicly available evaluation reports to assist both current and future decision makers in making sound choices about IT-based innovations.

The Benefits of Partnership

All three partners gain significant benefits from the CTG partnership model.

Government Benefits. Government agencies learn how technology can improve their operations and how it can be integrated with existing systems. Agencies become more knowledgeable about particular technologies and are able to write more informed and accurate requests for proposals. They become better able to manage implementation and the organizational adaptation process that inevitably accompanies IT innovation.

CTG provides agencies the ability to experiment and to learn what works and what does not work in a low-cost, low-risk pilot project before high-stakes, high-cost systems are developed or purchased. Because projects are usually conducted away from the agency site, participants are able to focus on the project free from the day-to-day operations at the agency. Working in the Center, they assess the policy, management, and technical dimensions of their problems, work with academic and business partners to apply technologies as tools for problem-solving, build and test prototype solutions, and create new ways to use and share information. All of this experimentation and analysis takes place before the agency makes potentially extensive and costly capital and human resource investments in full scale systems.

Corporate Benefits. Private corporations contribute state-of-the-art technologies and services to projects in exchange for better understanding of practical applications in government. Corporate partners have a unique opportunity to work with both practicing government managers and academic research teams. As a result, they achieve a better understanding of the capabilities and constraints which government agencies face. This knowledge helps them to better define the government market, which in turn allows them to provide that market with better products and services.

In New York, like many other states, there are hundreds of government entities, and no single, central information management or technology agency is responsible for all purchasing. Since each state agency or local government makes its own decisions, technology vendors must establish relationships with each of them as individual customers. By contrast, the Center organizes the interests of the NYS government market in one place. Involvement in the Center gives corporate participants a unique picture of the needs, capabilities, and constraints of state and local governments and an opportunity to demonstrate their products and capabilities to a wide government audience in the context of a practical application.

University Benefits. The Center's research agenda is organized around the contemporary and emerging problems of government in the electronic information environment—providing a rich resource of research topics, field work, and teaching material. Each project brings applied research questions which are present at every level of government. Problems of citizen contact, service integration, policy analysis, program evaluation, and information sharing are just a few examples of areas where practical research can yield knowledge that will improve services. Moreover, the Center's research directly enriches the information science, government, business, and related curricula of the University.

As research faculty gain knowledge, they expose their students to a wider array of issues and practical applications. The students themselves become involved not only in project teams, but also in field projects, master's thesis, and dissertation research based on CTG projects, all of which help them make greater contributions to their future employers in both the public and private sectors.

PARTNERSHIP IN ACTION: NEW YORK STATE DEPARTMENT OF MOTOR VEHICLES PROJECT

CTG's first project began when The New York State Department of Motor Vehicles proposed to develop a prototype imaging system for a portion of its vehicle title operation. Agency managers expected that an automated image storage and retrieval system would improve operations by increasing the speed of title processing, improving data accuracy, and increasing employee productivity and customer satisfaction.

The agency had been criticized for long delays in issuing new and duplicate vehicle titles, and was under pressure from the public and the legislature to significantly improve its operation.

Following standard practice, the DMV had requested information about imaging technology from all major vendors expecting to issue a request for proposals to implement an imaging system. The resulting vendor suggestions varied greatly in technology, scope, and application. Rather than write specifications in a Request for Proposal for a technology with which they had little experience and no clear guidance, the agency managers sought instead to explore these technologies in the Solutions Laboratory at the Center for Technology in Government (CTG).

The DMV's major goal at CTG was an evaluation of how imaging, automated workflow management, networking, and other information management technologies may improve quality or lower the cost of customer service. The development of a prototype application provided experience with and accurate information about state-of-the-art technologies and educated decision makers about possibilities and limitations of IT for achieving these goals. By developing a prototype application for a portion of their Title Bureau processing, agency staff gained hands-on experience with the technology and the effect it would have on agency operations and customer service. This knowledge better equipped them to make informed decisions regarding procurement and deployment of imaging technology. Managers and users learned what the new technology could do and began to reengineer their processes to take maximum advantage of its features.

Project Goals

Understanding the technology. The Title Imaging project at CTG gave the DMV and University participants an opportunity to learn how a basic imaging and work flow management system works. The project provided first-hand experience with the capabilities and limitations of these technologies. The prototype demonstrated how government services that rely on documents could change with imaging. Demonstrating the technology in a prototype application was educational—especially for the agency staff whose work would depend on it.

Organizational acceptance and support. More than 300 DMV staff, including managers and users from all levels of the agency, viewed and discussed a demonstration of the prototype system. Virtually all supervisors and employees working in the Title Bureau saw and critiqued the application. These demonstrations and discussions allowed staff to see how their jobs would be affected—and improved—by the technology. This knowledge helped build organizational support and prepared staff to participate meaningfully in the eventual design and refinement of a full system.

Linking agency goals and project goals. Agency program staff saw first-hand how imaging and work flow technologies could be merged to reengineer a process and greatly improve its efficiency. Workers from all levels of the organization, from the Commissioner to staff in the mail room, had an opportunity to review the current processing steps and see how they might be performed in an imaging environment. For many staff members, this was a first opportunity to see modern automated work flow including a "windows" environment. A key strength of the prototype was the use of images of actual title documents to illustrate how new technologies can simplify the processing of requests. At the end of the project, DMV decided not to issue an RFP solely for the Title Bureau, but to step back and think more broadly about how imaging could improve a number of other processes as well. As a result, its eventual procurement will be designed to take advantage of the technologies used in the prototype to improve operations agency wide.

Opportunity to reengineer business processes. DMV experienced significant productivity gains in its Title Bureau simply by seeing how the chosen technology could improve the process. In the months following the demonstration, the agency began the task of process reengineering that involved the entire organization. The Center project brought focus on the current processing steps and brought about a reexamination of the current process. By looking at processes closely in relation to IT, the agency learned to redesign the processes before a new IT is adopted. As a direct result, the risks of adopting a new imaging system were reduced. The agency was more prepared to face the new technology with a process that would fit, rather than automating the inefficient existing process.

Without partnerships in the Center, this project could not have taken place. By taking time to experiment with the technology in a neutral, low-risk environment, DMV learned a great deal about how imaging and related technologies could improve its services. This learning took place in the context of a prototype application which incorporated, but did not directly affect, actual data and technology resources in the agency's existing offices and applications systems. It was built in full partnership with private sector experts and with assistance from computer science faculty and students. Agency staff participated in every aspect of the project, from definition, to system development, to reengineering studies, to widespread evaluation. In the context of a CTG project, the agency has no obligation to buy from its corporate partners, thus removing the project temporarily from the procurement arena. When the agency is ready to buy or build a system, all the usual procurement procedures do apply. The difference is in the quality of the RFP they are able to write and the sophistication with which they are able to evaluate responses.

According to New York Department of Motor Vehicles Commissioner, Patricia Adduci, the project was "a win/win situation using a quality team approach with vendors providing technology; the University providing academic support, knowledge, and student involvement; and DMV providing program expertise...CTG gave us 20-20 foresight in developing the system."

Corporate partners in the project (primarily AT&T Global Information Solutions and NewVision Systems) had a unique opportunity to work with both practicing government managers and academic research teams. As a result, they achieved a better understanding of the capabilities and constraints which government agencies face when adopting new IT in general, and work flow and imaging technologies in particular. Their successful participation in the project spoke for itself during the demonstrations which included not only DMV staff, but nearly 200 representatives from more than thirty other government agencies. Since completion of the project, the corporate partners have been involved with full implementation systems in other branches of the agency and are well-prepared to respond to the growing interest in these tools in many other state government organizations.

The DMV project provided faculty and students an opportunity to apply new technologies to public service problems. Students learned how to solve real problems using state-of-the-art technologies in an out-of-classroom setting. Both groups gained experience with applied problems of information technology in government operations.

By documenting the project results, the University initiated a permanently available body of knowledge for use by others. In addition, the CTG experience provided the basis for Albany's first doctoral dissertation in information science. It also led directly to a professional position for one of the master's students upon graduation. Finally, it cemented a relationship with the information science doctoral program which has resulted in the involvement of additional faculty and students in the Center's ongoing program and new projects.

CONCLUSION

Technology can be used effectively and creatively to increase productivity, enhance quality, and reduce the costs of government services. The risks of IT innovation are great, but the benefits can be substantial for taxpayers, clients, and government itself. The CTG partnership model enables government to substantially reduce the technological, financial, and organizational risks of IT innovation by providing a forum for government, business, and the university to pool the risks, while sharing the benefits. In his 1994 State of the State Message, Governor Mario Cuomo, summed up the idea in these words:

> The Center gives us the opportunity to test new applications on "real-life" problems before we try them in the field. We are able to stretch our scarce State dollars by tapping into the wealth of knowledge in the academic community and drawing on pro bono assistance from businesses in the form of cutting-edge technology and expertise.

AUTHOR BIOGRAPHIES

Sharon Dawes is the Director of the Center for Technology in Government at the University at Albany, State University of New York. Dr. Dawes works with a variety of government and corporate partners to design projects which increase productivity, reduce costs, increase coordination, and enhance the quality of government operations and public services. Before coming to the Center, she was Executive Director of the New York State Forum for Information Resource Managanization. Repeatedly, organizations abandon IT projects because they fail to accomplish the objectives they were intended to meet. In both the public and private sectors, a well-documented set of risks attends IT innovation.

Dr. Dawes is co-author of *Government Information Management: A Primer & Casebook*, published by Prentice Hall, and she has written articles, research reports, and case studies which take a variety of perspectives on the "politics of information." She holds a Ph.D. in Public Administration from the Rockefeller College of Public Affairs and Policy at the State University of New York at Albany.

Mark R. Nelson is a staff associate at the Center for Technology in Government. As an associate, he works on various innovation projects within the Center. He is currently working on a project with the NYS Office of Mental Health involving the creation and evaluation of a prototype decision model to assist hospital emergency room personnel in making effective psychiatric assessments and admission and discharge decisions. Mr. Nelson has also worked with state agencies and corporate partners on Center projects involving groupware, workflow management, document imaging, and system implementation issues.

Nelson is a student in the Information Science Doctoral Program at the Rockefeller College of Public Affairs and Policy at the State University of New York at Albany. He holds a B.S. degree in Computer Science from Saint Michael's College in Vermont. His research interests include database marketing and decision support systems.

CHAPTER 10

The Information Network: Broadband and Multimedia Services

by Rick Faletti and Mike Frame

By now, most of us in business and industry along with many segments of the general population have become familiar with the concept of the "Information Age," with its Superhighway and associated range of services. Companies such as Pacific Bell, New Brunswick Tel, and MCI, among many others, are spending billions of dollars on broadband networks, each of which will become linked through a honeycomb of electronic on- and off-ramps to a world of on-demand voice, data and video services. This network is already taking shape, and we are well on our way to what some industry analysts have estimated to be a $3.5 trillion, global market by the year 2001.

But what's in it for business and industry? How can they benefit from the breakthrough technologies that will make the information superhighway a reality? Historically, businesses have been the early adopters of new telecommunications technologies. Centrex services, such as call forwarding and conference calling, were first embraced by the business community and eventually migrated into the residential marketplace. To a great extent, broadband multimedia products and services will take the same path. But just as the initial success of Centrex depended upon developing meaningful applications for strategic deployment, so too must broadband services address real needs and deliver measurable and immediate benefits. With a plethora of alternate technologies vying for our attention—ISDN, ATM, and frame relay, just to list a few—the need for an overall vision and clear direction has never been greater.

This chapter examines some of the ways in which business and industry are beginning to use broadband multimedia technologies to create immediate benefits and solve real problems. It also provides a glimpse into the areas that will contribute to the creation of broadband networks, and discusses their merits.

WHY BROADBAND MULTIMEDIA?

It has often been said that information is power. Certainly, information is critical, but it must be accessed, assimilated, interpreted, and shared if it is to be of benefit. To a brokerage house, immediate access to trading activity is important, but interpreting that knowledge and sharing it with others is also important. To any corporation, large of small, condensing decision-making to its smallest increment through fast communications translates into market leadership, increased efficiencies, a more productive workforce, and greater profitability.

In today's business world, then, power lies not only in information but in the degree to which bandwidth—the width of the "pipe" through which information is carried—is deployed and utilized. From copper telephone wiring, the traditional channel for delivering telephone services, to the massive bandwidths created by fiber, coax, and wireless technologies, the availability and ubiquity of high-capacity channels will be a key determining factor in the spread of broadband services.

Today, organizations that possess broadband capabilities have the means to allow employees to fully explore and exploit creative ideas through every medium at their disposal—voice, data, image and video. Desktop videoconferencing allows engineers to collaborate on a project from the same screen at the same moment, whether they are on different floors, across the street, or halfway around the world. Telecommuting gives employees at home access to the services to which they are accustomed in their corporate offices, increasing productivity, lowering overhead, and benefiting the environment by reducing auto emissions. Information retrieval technologies place an entire world of information directly into the hands of those who need it most—and who know how to use it. Those businesses who recognize and introduce similar broadband capabilities into their workforce will reap these benefits, and more: improved employee productivity, increased network efficiencies, and significant cost savings.

Early adopters of broadband technology can enjoy another advantage—that of competitive superiority. Timely access to information—and the ability to disseminate it effectively—is critical to the operation of any business, in any industry. The speeds with which vast quantities of information may be accessed, retrieved, processed, and acted upon can give those who possess such capabilities a distinct advantage over others in their markets—until the competition catches up. This broadening usage, however, will create the need for new competitive advantages, which will drive greater service innovation, and ultimately benefit businesses all the more.

ISDN—THE ON-RAMP TO BROADBAND SERVICES

In the 1980s, the development of the integrated services digital network (ISDN) enabled simultaneous delivery of digital voice, data, image and video over an existing copper telephone line. ISDN extends the advantages of fully digital telecommunications—principally, faster speeds and increased reliability—all the way to the end user. Basic Rate ISDN provides two 64 kilobits per second (kbps) channels that can be used for voice, data, or video, and a 16-kbps signaling channel, which can be used for network signaling or data transmission. Basic rate is used for such applications as telecommuting, point of sale, and small businesses multimedia services.

Primary Rate ISDN offers 24 64-kbps channels of information (one of which is used for signaling). Primary Rate is used for applications such as private branch exchange (PBX) to network connectivity and mainframe-to-computer connections. Multi-Rate ISDN, developed more recently, enables the combining of any number of the 24 channels in Primary Rate ISDN for applications such as local area network (LAN) bridging and high quality video conferencing.

Today, approximately 950,000 lines of Basic Rate ISDN installed are in the North American public network, and that figure continues to rise. Four of the Regional Bell Operating Companies provide 100% coverage—meaning that all subscribers can order the service—and other Bell companies, independent telephone companies, and major long distance carriers are deploying ISDN aggressively. In addition, an increasing number of telecommunications equipment manufacturers are producing ISDN equipment, such as ISDN telephones and terminal adapters, resulting in a proliferation of options and lower costs for consumers.

The key to ISDN lies in its ability to support a wide range of applications over existing telephone lines. Telecommuting is more efficient when a simultaneous data transmission and voice call can be conducted rapidly and reliably over a single ISDN line. Integrated voice and data applications, such as those used by Domino's pizza, use ISDN to access and deliver customer database records—instantaneously displaying a customer's buying history on a screen when a clerk answers a call—and increasing the security of its drivers by identifying crank callers. Point of sale, desktop videoconferencing, Internet access, and small business multimedia services also bring a heightened level of efficiency to business users.

The ISDN infrastructure is growing rapidly. Pacific Bell is making five Basic Rate ISDN lines available to every school in California. Ameritech is selling 1,500 lines each month to end-users. By 1995, the majority of North American residents will be able to order an ISDN line for a reasonable charge, and purchase affordable off-the-shelf ISDN terminal products.

Because ISDN is based on an international standard for end-to-end digital connectivity, investments in the technology are protected against obsolescence. In fact, as ISDN becomes more ubiquitous and new applications are developed, it is quite possible that Basic, Primary, and Multi-Rate ISDN connectivity will be provided over coax or fiber, in addition to copper. Whatever the carrying medium, the network intelligence and 64 kbps building blocks of ISDN will be fundamental attributes of the emerging broadband network, and today's growing number of ISDN connections will be the digital on-ramps to the nation's Information Superhighway.

MULTIMEDIA ON THE DESKTOP

The ability to send and retrieve data, voice, images, and video from one desktop workstation to another is one of the greatest productivity enhancements yet offered to the business community. Editing a document with a colleague who is miles away and sharing information at the press of a key enables faster and more efficient operations. Speaking to a co-worker whose image appears in one corner of a computer screen brings a sense of immediacy and collaboration to a task. Work is completed more quickly, decisions are made faster, and money is saved by reducing travel costs, faxes, and time spent on projects.

Desktop multimedia combines videoconferencing, (which was once available only for groups through expensive private line connections), and data and voice transmission (which once required separate telephone lines). Traditional group videoconferencing systems cost anywhere between $15,000 to $80,000; desktop versions range from $2,000 to $5,000, and offer a range of additional advantages. They are less expensive to operate, they can be accessed at any time without reserving facilities, they do not need dedicated personnel for operation and maintenance, and they can be used by individuals. Productivity rises as costs go down. The low cost also makes videoconferencing affordable to smaller businesses that otherwise could not justify the more expensive systems.

Perhaps the greatest advantage of desktop systems is their integration of videoconferencing, screen-sharing, and file transferring capabilities. Not only can users see and talk to each other, but they can also simultaneously view and manipulate documents, data, or graphics on their workstations in real time. Communication is instantaneous. The cumbersome and expensive practice of shipping data through fax, mail, or overnight courier is lessened and often eliminated. Workflow is smoother, interaction is more efficient, and tasks can be completed in far less time—often within the space of a single conversation. In instances where decisions need to be made quickly, the access and convenience of desktop conferencing facilitates timely results.

Another area in which desktop multimedia revolutionizes employee productivity is that of multimedia messaging systems, which integrates voice, fax, and E-mail messages onto desktop computers. This combination results in an economy of time and effort by enabling users to manage all messages from one location. A list of messages may be viewed quickly rather than listened to sequentially; important messages may be identified and accessed before less important ones. It also allows users to streamline and simplify office systems. Instead of maintaining separate and independent systems for E-mail, fax, voice mail, and graphics, multimedia mailboxes unify these functions on a user's desktop. Having the ability to send and receive faxes directly from the desktop is a tremendous cost savings and efficiency improving development over conventional fax methods.

The technology exists today in almost every office in North America to implement desktop multimedia services. They can be provided over private and public networks, generally requiring either two digital telephone lines—one for voice, the other for data and video—a digital line for data and an analog line for voice, or one Basic Rate ISDN line. Current desktop multimedia systems represent the cutting edge of what is becoming a standard business tool. The real benefits derived from these multimedia applications will grow as additional bandwidth is allocated.

With such a network in place, desktop multimedia communications become even more powerful. Desktop conferencing can allow entire "classrooms" of students—or executives—to watch, listen to, and interact with a teacher or a seminar leader from around the world. Full motion video enables lectures, presentations, and product demonstrations to be stored for on-demand access by customers, suppliers or in-house use. Instant access to stock market activity, massive research databases, and a myriad of other online business services can improve performance to a degree that cannot even be measured today. Downloading television news reports or video executive seminars directly to your desktop will be as common as acquiring information through online services.

TELECOMMUTING

Telecommuting is a fact of life for nearly 30 million full- and part-time corporate Americans—in addition to the millions who are self employed. With growth estimated at 15% per year, working at home has become not only the latest labor trend, but a tremendous business opportunity for communications service providers. The additional pressure of the federal Clean Air Act, along with various state equivalents, ensure that corporate telecommuting will continue to gain in popularity as a viable working alternative. (After the recent earthquake which destroyed freeways in Los Angeles, for example, ISDN lines were provided to stranded workers, allowing them to remain viable and productive.)

Telecommuting increases employee productivity, saves office overhead, gives businesses a flexibility in structuring their organizations, and benefits the environment. For those employees whose presence on-site is not required for the performance of their duties—at least full-time—it is a viable and cost-effective option. Attaching modems to computers and linking them to the home office through the public telephone networks give work-at-home employees access to many corporate facilities. Adding a fax machine provides an additional link to the outside world. Installing a desktop multimedia system over a Switched 56 (a digital 56 kbps service provided by local telephone companies) or ISDN line links an employee to fellow workers as effectively as if the telecommuter were sitting in a corporate office.

However, a former office worker turned telecommuter will not be as satisfied (or as efficient) with computers, faxes, and telephones that do not provide the same level of service as those at the head office. In order to fully realize the potential of telecommuting, businesses must have access to the technologies that will provide a comparable level of service to that available in the corporate offices. Cable providers are exploring ways to provide such services through their existing coax network, and telephone companies are looking at the widespread deployment of ISDN to bridge the gap. It is ultimately the building of national broadband networks, however, that will give telecommuters the full benefits of broadband service.

BROADBAND IN THE PRIVATE ARENA

It is the rule and not the exception that large businesses operate a wide variety of networks. In fact, the average Fortune 1000 corporation maintains six or more different networks. They are used for routing voice traffic, Systems Network Architecture (SNA) computing, and LAN routing, among others. They are functionally unique, must be individually maintained, and are likely supplied by different equipment manufacturers. And while they serve the needs of the corporation, they do so at a premium: multiple sets of hardware, software, and associated personnel result in high costs for the company.

The invention of a dual architecture design allows one system to perform multiple switching functions, collapsing the intricate mosaic of local area networks into a single, manageable entity. This new switch—the enterprise network switch—addresses the entire spectrum of an organization's needs, from video, voice, and various styles of LANs to frame relay and Asynchronous Transfer Mode (ATM)—a high-speed, multimedia technology. The ability to manage these different types of traffic means that communications characteristics can be tailored to suit the specific needs of businesses and their applications.

With enterprise network switching, businesses have the ability to access unused bandwidth for reallocation to other applications. As a result, they can realize tremendous cost savings above and beyond the economies of a single system. The total investment in an enterprise network switch can have a payback of as little as twelve months, at the end of which the network will not only be cost-effective, but poised to receive every new innovation in the broadband multimedia arena.

Deciding which technologies work on which level in the network within the enterprise network switch environment, however, will depend upon the individual needs of a given organization. No single technology will solve all problems. This point is clearly illustrated by comparing ATM with frame relay. ATM, which uses fixed-length cells and high capacity trunks, is optimized for high-speed multimedia applications, such as voice and video. Frame relay, on the other hand, is optimized for high-quality data transfer and is best suited for data applications such as accommodating data traffic between local- and wide-area networks. Clearly, both ATM and frame relay have a place in the broadband network. The best solution for any business network operator is one which provides the option to use ATM, frame relay, or any other service which optimizes performance.

Private broadband capabilities are also well within reach of small- to mid-sized businesses. One product that caters to this segment, the key system, is literally being transformed by its emerging multimedia capabilities. Northern Telecom's Meridian Norstar, a voice and data system for smaller businesses, has superseded its original product mandate—and even its original name. Now known as a multimedia communications system, this product provides interfaces for desktop multimedia functionality, in-building wireless systems, and Basic Rate and Primary Rate ISDN for digital connectivity to the public switched network. It will ultimately migrate to provide interconnections to the broadband network. With such compact yet versatile communications systems, smaller businesses—and branch offices of larger corporations—have cost-effective access to broadband multimedia services through existing equipment.

THE PUBLIC BROADBAND MULTIMEDIA NETWORK

Public telephone service providers, along with the dozens of alternate carrier network providers, will evolve their networks in much the same way as will business network administrators. Typically, a network will retain some existing data capabilities, along with its switched voice services, but an overlay of broadband technologies will augment services provided by private networks. Offices equipped with desktop video conferencing systems that are integrated into existing personal computers can use public networks to interact with each other and to access external databases or information resources. Wireless handsets will have similar functionality to their wired office telephones—and computers. Telecommuting employees will work from home with the technological services to which they are accustomed in their offices.

All of these capabilities are available today in one form or another. Some, such as desktop video conferencing over ISDN or switched 56 digital telephone lines, are well-established. Others, such as multimedia switching through ATM, are just beginning to be deployed. Many hotels provide data lines through the public telephone network for accessing home databases. Cellular phone service, which has become a way of life for millions, continues to evolve toward truly ubiquitous service.

In fact, the keyword in the growth of broadband multimedia services is "evolution." In order to make it worthwhile for a business to begin the process of evolution, a network must be able to integrate existing applications and services while simultaneously providing access to emerging multimedia technologies. This calls for a focus on an architectural strategy rather than a specific technology; in other words, it is more important to maintain the flexibility to adapt to multimedia environments and to provide a migration path from today's heterogeneous mix of services and technologies, than it is to choose a single winning technology. For this reason, broadband services will not appear overnight; instead they will prove themselves over time in both public and business environments, regardless of technological availability.

INTEGRATED COMMUNITY NETWORKS™

Integrated Community Networks (ICNs) are shared broadband resources that interconnect local entities such as businesses, schools, health care facilities, and government offices. ICNs use video, voice, data, and image technologies for interactive applications such as Electronic Data Interchange (EDI), distance learning, and telemedicine—applications that benefit society and business by bringing people together across geographic, cultural, and economic boundaries. ICNs are already bringing broadband multimedia capabilities to individuals and groups all across North America. As they proliferate and begin to interconnect, they will form a powerful infrastructure for the further evolution of the broadband network.

As for business applications, the central decision factors regarding ICNs lie in the financial benefits of cost and labor savings, and in productivity and efficiency issues. When measurable improvements can be demonstrated, there is a clear rationale for participation. For example, business sites can be linked directly into universities for executive education, high schools for basic skills training, or community colleges for specialized course-work, enabling employees to participate in classes without leaving their offices. Technical and product training can be conducted between business sites, or between a company and its customers. Stores of data residing in local library systems or government databases can be immediately accessed through desktop computers.

The intangible benefits of being tied (in a literal sense) to the communities in which businesses reside can only prove beneficial to both the community and the business. Providing community access to corporate resources—online search services, marketing and business libraries, and mentoring programs, to name a few—can bring innovative services to schools, supplier businesses, customers, and others. High-speed video transport capabilities tie local TV stations, special effects houses, production studios, remote source locations, and editing sites together. These benefits, similar to those realized in the medical, government and education industries, can serve the interests of both business and industry while providing a strong tie to the communities in which they operate.

ICNs have already been created all over the country. The Mississippi FiberNet 2000 network, a well-known distance learning network, has been delivering broadband, fiber-based educational services to the citizens of Mississippi for several years. The State of Iowa runs one of the largest broadband networks in the world, providing greater access to educational opportunities and government services for state residents. The Maryland educational ICN links public schools and state government. In North Carolina, an emerging broadband information highway is beginning to link hundreds of sites, including senior citizen groups, corporate offices, health care groups, and state educational systems.

Each of these ICNs—and there are many more—were developed to meet specific needs: state-mandated equalization of educational opportunities or sharing of information between specific interest groups. In Western New York, for example, four network providers and a host of communities are building an ICN with the express purpose of extending educational opportunities, spurring economic development and attracting new industry to the area. In fact, the Integrated Community Network is recognized by state and community leaders as a critical economic development tool precisely because it creates a high-powered backbone network that new businesses can access and use to their competitive advantage.

THE WIRELESS WORLD

Cordless phones in the home have long been commonplace, pagers are used by everyone from executives to teenagers, and cellular phones are in the hands of 19 million subscribers (and counting). Each of these wireless technologies have drawbacks, however; cordless is limited to a relatively small area and is not secure; pagers are one-way and thus do not allow dialogue; and cellular, while valuable between work and home, is benched in favor of wirelined systems when not in the car or walking down the street. In addition, none of these systems allows access to the true digital capabilities of an office, including access to video services. While cellular networks can accommodate fax and data exchanges, they are generally provided over analog cellular facilities which are not as reliable and efficient as digital. It is this next generation of cellular telephone that will truly provide the additional advantages sought by business and industry.

Two things must occur before this new technology can take hold. First, the existing analog cellular networks must be converted to digital before broadband services can be accommodated. The improved voice quality, better hand-off, longer battery life, and clean data transmission of digital cellular are essential to the implementation (and success) of wireless services.

Second, the personal communication services (PCS) spectrum, defined at 1.85 Gigahertz (GHz) to 2.0 GHz, must be placed in the hands of providers: more than 2,000 licenses are currently up for auction. PCS will provide low-power hand units that will communicate through numerous call areas connected to the public wired network and, eventually, to international satellite networks. The promise of PCS is best described by the scenario in which a telephone number is associated with a person rather than a location, since the portable sets will be carried and used from the home to the office, and everywhere in between. PCS trials are underway all over North America. As the services become commercially available, and the concept of "reaching" someone at a desk becomes outmoded, benefits to business will be enormous.

Future handsets, prototypes of which already exist (BellSouth's "Simon," for example), are called personal digital assistants (PDAs). These portable multimedia handsets will have liquid crystal display (LCD) screens with a dialpad for sending and receiving faxes, data, and ultimately, video signals. They will also have the capability to function as personal organizers. Instead of merely being able to make and receive calls when away from the desk, employees will be able to carry the complete multimedia functionality of their desktop with them.

Until that time, available wireless technologies abound in the business and industrial environments. Unlicensed offerings using low-power, local networks offer distinct advantages in a variety of applications, including manufacturing, healthcare, and business.

One metal-components manufacturing company has greatly increased the efficiencies across its campus by testing an in-building wireless communications system that provides service to its highly mobile workforce. With the system, employees can repair machinery, solve engineering problems, and speak with customers and suppliers directly from the manufacturing floor. Functioning as an adjunct to the resident private branch exchange (PBX), the wireless handsets share the phone number of the desk telephone, allowing either to be used. The system has other advantages: no airtime charges, no license required to operate the system, and no monthly fees to service providers. The digital sound quality equals that of wired telephones. The savings, both in terms of revenues and functionality, have been enormous. Such systems will play a key role in providing on-site wireless service to businesses and industries.

Cable TV Companies

Cable companies are gearing up for direct competition with telephone companies in the voice arena, a market which is currently off limits, but which by all indications will open through regulatory action very soon. For example, CableLabs, representing several cable companies, has already issued an RFP for telephony-over-cable equipment, and Time/Warner has "agreed to compete" with Rochester Tel in the voice services market. There is a nationwide cable TV movement toward consolidating broadcast offices (called head ends) which, aside from the obvious economies of scale, will further strengthen their broadband capabilities. Most significantly, cable companies have purchased (or created) competitive access providers (CAPs) in metropolitan areas, through which they are offering a variety of services, including voice and high-speed data, on fiber networks deployed throughout large cities.

Cable TV companies and telephone companies are each preparing for entry into the other's market. Both have a number of distinct advantages they can exploit to provide broadband services over large areas. Cable's chief advantage is its existing coax infrastructure. Because the broadband nature of coax will make two-way—or interactive—services possible, installing a digital voice switch at a head end will do much more than enable voice services. On the other hand, telephone companies have a national (and international) switched network already in place, decades of experience behind them, and a deployable technology in the form of ISDN. The ensuing battle for market share will greatly benefit business users, since the competitive climate created between the cable TV providers, the telephone companies, and other third-party service providers will drive down prices, cut business operating costs, and foster the development of new and innovative services.

The developments in cable TV are of interest to business and industry because they offer an alternative broadband network to businesses located in metropolitan areas-networks which will ultimately be linked on a national level. The move to consolidate head ends and implement digital technology will not only create new options for residential consumers but will provide new opportunities for local businesses through community-wide networks. In order to achieve this, however, cable TV companies must recognize and capitalize upon the business needs of its subscribers, enlarge their networks to include businesses, and form alliances with (or purchase/create) competitive access providers to offer services on a national scale.

With a strong presence in the residential market, the cable TV industry is likely to have an enormous impact on telecommuting. As businesses and industries move increasingly toward work-at-home scenarios, a cost-effective link to the office—with the full functionality provided through broadband channels—will become both economically and functionally plausible. When cable companies are able to connect subscribers through Ethernet WANs, and link these networks to the fiber rings provisioned and administered by the CAPs, businesses will have a powerful incentive to allow select employees to work from home, confident that technology will provide them with the capability to function efficiently.

CONCLUSION—THE BROADBAND, MULTIMEDIA NETWORK

Our underlying premise throughout this chapter has been that only a focused and strategic implementation of broadband technology will assure both profitable and productive returns to business and industry. To that end, the broadband multimedia network will be an amalgam of old and new technologies, incorporating whatever existing solutions remain feasible with an assortment of new products. Network simplification will lead to cost-effectiveness and innovative feature development. Strategic applications will drive deployment.

The overriding questions for business and industry will continue to be when, and in what manner, they will incorporate broadband multimedia into their day-to-day lives. With the various approaches available in the marketplace today, the wise network planner should consider which mix of technologies will provide the most flexible infrastructure for future innovation, and which broadband multimedia applications will bring the highest value for the dollar invested. What is clear is that no one solution will bring broadband services to our desktops (or doorsteps) overnight. There will be an ongoing need to carefully assess, implement and administer various pieces of the broadband puzzle in order to optimize the evolving multimedia network—and realize its greatest benefits.

AUTHOR BIOGRAPHIES

As president of Northern Telecom Limited's Multimedia Communication Systems global product organization, Rick Faletti is responsible for the company's design and manufacturing of voice, data, and video digital communications products, systems, and related application solutions to public carriers, businesses, institutions, and other organizations around the world. He is responsible for fifteen major product lines for Northern Telecom including: PBX, Key Systems, Multimedia Terminals, Data Communication Products, ATM, and Broadband Multimedia Switching Systems in addition to Multimedia Business Application Systems. He is also responsible for marketing sales and service of the MCS product portfolio in North America including Global Accounts Marketing and Network Integration projects.

Faletti, who joined Northern Telecom in 1976, has held various senior level management positions with Northern Telecom and BNR, the company's research and development subsidiary.

Prior to assuming his current responsibilities Rick Faletti was President of Private Networks global product(s) division, which was responsible for PBX, Key Systems, Terminals, and Data Communications product lines, with nine manufacturing locations and six R&D labs located around the world. Northern Telecom achieved Global market share leadership in CPE products in 1992 and 1993, and PBX Market Share Leadership in the U.S. in 1993.

Faletti is located in Richardson, TX and reports to John Roth, President of Nortel North America of Northern Telecom Ltd.

Mike Frame is Vice President, Broadband Networks, Northern Telecom Inc. The Broadband Networks global line of business addresses the emerging range of multimedia services for the home and for community networks.

Prior to this appointment, Frame was Regional Marketing Vice President responsible for the MCI Account. He was previously Vice President Marketing, Northern Telecom Inc., Southern Region. In this capacity, he was responsible for account management and new business development in the South. Earlier, he was Vice President, Systems Engineering, Transmission, Northern Telecom Inc. Since joining Northern Telecom in 1969, Frame has held numerous senior management positions in technology, engineering and marketing.

Frame holds a Bachelors degree and a Doctoral degree in electrical engineering. He is married, has three children and lives in Atlanta, GA.

CHAPTER 11

Middleware: Laying the Foundation for Open, Distributed Computing

by John Senor

Prologue: Like Waking from a Bad Dream

Picture it: after a year of planning and construction, your dream home is finally completed. Everything is ready, the landscaping is done, even the furniture has arrived. It's time to cut the ribbon and move in.

As you round the corner and pull into the long, tree-lined driveway, the house breaks into view. It's an architectural gem, perfect in every respect. You approach the hand-carved front door, slip the key into the lock, and walk inside.

But something is amiss. The structure is complete, the decorating is right, but all the fixtures and appliances are missing. There are no sinks or faucets, no dishwasher or stove, not even any electrical outlets. Instead, you are confronted with exposed pipes and bare wires at all the places where there are supposed to be functioning gadgets and utilities.

Now, imagine the installation of a new corporate information system. All the host computers are in place, the networking gear has been installed, even the application software has been configured for the individual desktop PCs. But when the users arrive and sit down to do their jobs, they are confronted with cryptic program calls and confusing data access statements.

Sure, all the equipment is there, but nothing is integrated. None of the systems work together, and users in one department cannot access the data and resources in another. The only way the Information Services professionals can remedy the situation is by wrestling with tedious, low-level communications protocols and database languages, the bare wires and messy pipes of software technology.

A new software technology is appearing to make life easier for corporate developers, a technology that brings true plug-and-play compatibility to dissimilar database applications on various kinds of networks. Wake up, shake off the bad dream, and read on: middleware has arrived.

INTRODUCTION

For some time now, the computer industry has been striving towards the lofty goal of Open Systems. It has been an up-hill climb. Open systems' promised benefit of easy communication among many different types of systems is still heavily dependent upon home-grown solutions. Today's corporate programmers still have to work with the "bare wires" of the underlying technology. They must know the physical location of data and wrestle with a host of networking details and languages to get applications speaking across interconnected systems.

Yet the vision remains: a vision of an enterprise network that offers seamless access to programs and data, regardless of hardware platform, operating system, database type, or networking protocol.

Open systems rely on hardware, software, and networking standards to help programmers fit the pieces of a distributed computing environment together smoothly and easily. Many of the most important standards for database and network communication are embodied in today's emerging middleware products.

Think of how important standards are in the construction industry. Builders can combine doors from one vendor with door hardware from another vendor, without concern for whether they are compatible. But what if carpenters had to tailor each door to each lock, or if electrical fixtures had to be specially engineered to handle different currents and voltages.

Before middleware came along, this type of custom work was the norm in the software world. Despite all the talk about distributed computing, standards for interprocess communications are still being hashed out, meaning it takes a lot of work to get the various components of a computer network talking to each other.

Unfortunately, in our rush to embrace more open, cost-effective multi-vendor systems, we lose the consistency that comes with relying on a single "environment provider." A decade or two ago, most companies relied on large, host-based computer systems with all the components supplied by a single vendor.

These monolithic computing architectures have since fallen out of favor due to their tremendous expense and proprietary designs. But at least they offered programmers a known, consistent way to connect programs and access the resources of the network. Developers could write business applications for a single type of homogeneous system.

They could concentrate on the problem that the application was addressing, rather than being bogged down with the low-level details of connecting all the pieces.

THE CHANGING FACE OF SOFTWARE DEVELOPMENT

Programming itself has changed dramatically to account for this market diversification. In the early days of computing, programs were procedural blocks of code that specified every bit of the processing that would take place, including the way data was stored on the disk drives. Each program had its own interface to the data, including data definition statements and data access statements.

Once databases became separate entities, as they are today, a split occurred between applications and the data those applications act upon. Databases began to include maps of the data, allowing programmers to devise standard data access routines. But programmers could still devise applications based on a single source of data, stored in a central, protected location.

Then diversity began to creep in. At first it was just a trickle, a few special purpose mini computers, maybe a high-powered workstation or two for engineering activities. Later, with the rise of personal computers in the early 1980s, the trickle became a torrent. PCs and workstations began to proliferate, and as their types and numbers grew, so did the potential sources of data available to users. Networks sprang up, workgroups were formed, LANs linked up with WANs, and the people in the glass house began to lose control of their most important resource: corporate information.

In many cases, the primary computers and applications that kept the company running were still within the glass house, still under the watchful eye of the Information Services (IS) staff. But those applications gradually became separated from the rest of the computing world springing up around it on all sides. And something else began happening to confuse the situation even further: the rise of client/server computing.

THE CLIENT/SERVER DELUGE

Client/server gets its name from the way computers are connected in a network. Typically, desktop clients (personal computers) share the resources of a common server. A server is a larger computer designed especially for storing lots of information. It serves files to the clients and runs central applications that all the clients can access.

Client/server architectures are attractive because of their flexibility: users can share the resources of the entire computer network, yet each user has his or her own personal computer to perform unique tasks.

The real challenge is developing the applications that can run on these client/server networks. Older computer applications are designed to process information and store data centrally, on one big computer. Client/server applications, on the other hand, divide up the processing load among many different computers. Intensive database processing is handled by powerful servers. Report writing and spreadsheet analysis is done right on the clients, giving users maximum control. Data can be stored on both clients and servers, depending upon the computing task at hand.

Also, because the overall information system surrounding each application is constantly in flux, these applications may need access to new data sources that were not considered by the initial developers. The networking software may change, or the application might get ported to a different type of computer.

Which brings us to the current nightmare, the bad dream which middleware was designed to solve. Today, there are many types of computers, operating systems, and networks online at an average company. The potential number of data sources continues to grow, and there are also the networking complications to consider. This is the tricky part, defined by communications protocols such as SNA, TCP/IP, DECnet, and X.25. Like Alice lost in the forest, confused by a hodge-podge of signs pointing in all directions, getting from point A to point B on today's client/server networks is like navigating through a maze. Specialized programmers are required to sort out these cryptic networking details, programmers who are hard to come by and not readily interchangeable.

Herein lies the true challenge of client/server computing: providing users with transparent access to system resources, while sparing corporate developers the anguish that comes from coding at the network level.

MIDDLEWARE TO THE RESCUE

Middleware is client/server software that helps developers connect front-end applications running on PCs and workstations to back-end database servers on mainframes, minicomputers, or local area networks.

Middleware shields developers from complex network topologies by providing a software bridge between the application and data that takes into account the differences between operating systems, network protocols and operating environments. It helps them to sort out today's tremendous diversity of protocols, platforms, and programming languages so they can spend more time focusing on what they do best: building applications.

In a client/server environment, middleware represents the software which translates user requests on the client into a form which the server can respond to, wherever the server might be, whatever type of data it holds, and whatever network path is required to get there.

What makes middleware unusual is not that it can sort out these low-level programming details; it's that it can do it so easily. Links that might take months to program by hand can be achieved with middleware in a matter of days, or even hours. Middleware even resolves differences between dissimilar standards.

You don't think about where your electricity is coming from. You just plug in your appliances and turn them on. Whether the electricity comes from a hydro-electric plant or a nuclear generating station doesn't matter. The plug is the same and the cycles are the same from one socket to another and one building to another.

Extending this analogy to computing means that a particular application can be dropped into whatever computing environment is available, regardless of the details of the underlying system. The middleware layer understands and sorts out the complexities—finding a clear path through the forest without programmers having to trouble themselves with all those confusing signs along the way.

Middleware products disguise system resources by neutralizing the details of each system element behind an interface that is generic to all such elements. Thus, rather than expecting programmers to master the intricacies of several different network interfaces, those interfaces are hidden behind a single, generic interface to the network. A programmer using that generic interface can effectively use all the networks in the system, because his or her expertise applies to all of them.

This is the ideal world for the corporate application developer, equivalent to the homogeneous computing environments that we left behind more than a decade ago: an environment in which developers can write a program one time, and that program—the actual source code—does not have to change as it moves from one database to another, one operating system to another—one hardware platform to another.

TYPES OF MIDDLEWARE

There are many types of middleware products, but essentially, they all work by providing developers with three standard components:

- A "front-end" application programming interface (API);
- "back-end" database gateway technology, which enables requests sent by the client API to be processed against a specific database or file;
- A networking technology which enables the front and back-ends to communicate;

Developers rely on these standard components to connect client applications to the diverse resources of the surrounding network. This allows them to query many different types of database management systems in the same way, and it gets those queries from point A to point B without the developer needing to know the specific types of networks in between.

Most middleware products employ an API that supports four key functions: a connect function, a send function, a receive function and a disconnect function. These functions are embedded in each application in the same way, regardless of where the applications are ultimately run.

As Carl Potter says in his definitive *Client/Server Development Tools, An Evaluation and Comparison* (published by ButlerBloor, Ltd. in Milton Keynes, U.K.), the usefulness of middleware ultimately depends on three things:

- The number of interfaces it supports at the calling end (e.g., APIs)
- The number of interfaces it supports at the data storage end (e.g., gateways)
- The number of different environments it is able to span (e.g., networks)

An effective middleware product can translate among a maximum number of network operating systems and database access languages simultaneously, just as the translation teams used at meetings of the United Nations can simultaneously translate many spoken languages to a diverse group of listeners.

The well-established middleware vendors hold the high ground when it comes to satisfying these three requirements, partly because their middleware products have been around long enough to make the necessary ports and work out all the kinks.

All middleware products work by establishing a common ground, a single point of contact to and from many different types of systems in a heterogeneous network. There are three basic types of middleware, Potter tells us, which correspond to the three points at which dissimilar types of systems find their common ground:

- Common interface
- Common protocol
- Common gateway

A common interface means all client applications can be written to a single user interface API, irrespective of current or future server types. Client applications talk to this generic API, and a database gateway or driver, which can be located on either the client or the server, translates the client calls into requests that the server can understand.

Common interfaces for databases typically center around Structured Query Language (SQL). Examples are Microsoft's Open Database Connectivity (ODBC), Information Builders' API/SQL, and X-Open's Call Level Interface.

Common protocol middleware products achieve a similar effect by translating messages from both clients and servers into a common protocol, such as IBM's DRDA, for transmission across the network.

Common gateways act as interpreters between local client/server applications and any foreign data sources those applications need to access. Gateways make remote database tables appear local. A Sybase table named CUSTOMER might actually be an IMS segment on the mainframe. A gateway allows a client-based application on the LAN to retrieve that IMS data from the mainframe with an SQL statement, then goes across the gateway to get it, makes any necessary conversions to put it in a form the client can understand, and returns the data to the client for processing.

MIDDLEWARE AS SHARED APPLICATION SERVICES

Client/server has proven itself effective for today's diverse information networks, but the first-generation of client/server applications are starting to show their limitations. Many of these applications are based on a simple, two-tiered architectural model.

Within the two-tiered model, the client handles nearly all processing tasks, partly because there is no mechanism for dividing application logic between client and server nodes. The presentation layer of an application is partitioned to the client processor. The database is partitioned to a server, usually under the control of a database management system.

As the user base for a client/server application grows, the application often needs to be scaled to new servers or multiple servers. This has given rise to a new generation of application development tools that allow developers to have much more control over how application logic can be divided among all the different processors in the network.

The client/server applications created from these new tools are comprised of three tiers, instead of two: clients, data, and shared services. The client tier resides on the user's desktop as before. The data tier represents one or more sources of data spread among the available servers in the network and can also come from an online source, such as a machine process controller or bulletin board service (BBS).

The third tier is a set of shared components that can "service" both client applications as well as their data sources, across interconnected systems. These middle-tier services perform all types of tasks. For example, in one location a single, shared distribution service might transparently dispatch client application requests for data to various remote databases or files, in much the same way that telephone calls are routed between interconnected long distance phone systems. In another location, a single, shared stored procedure service might be automatically reordering inventory from several suppliers whenever client applications delete specific items from the inventory database.

Other types of shared services can also be distributed throughout the network, such as copy management, data replication, governing, and data access services. In each case, the use of the service only occurs when needed by the application and is completely transparent. Just as we do not worry about what the phone company does to get our telephone calls from "here to there" across multivendor phone systems, these shared middleware services enable companies to build their own transparent networks across multi-tier computer systems.

Today's leading middleware products are evolving in tandem with this new client/server model, allowing shared services to be distributed throughout the network in a modular fashion. This allows companies to configure systems which meet their exact application needs, purchasing middleware as a series of independent modules, and mixing and matching combinations of services to meet the needs of their evolving client/server environment. Each service fulfills a specific need within the overall information network and can be added to or subtracted from as application needs change.

TOWARDS AN OBJECT-ORIENTED FUTURE

On another front, client/server applications are becoming more object-oriented, and, once again, middleware is helping to lay the foundation for network connectivity and communication.

Object-oriented applications are divided into a set of logical domains that can be spread around the network—not just one or two domains, but as many as makes sense to solve the business problem at hand. Middleware is the ever-present thread that ties these application domains together. The goal is a complete neutralization of the operating system, network, and database differences, making the resources of the network accessible to application programmers in a common, consistent way. Each client application can seek services from many different servers, but developers only have to contend with a single API.

This type of architecture can only work effectively if the client application itself does not have to worry about the location or format of the data. It makes a request of the middleware layer, which automatically sorts out differences in data types and locations, networking protocols, and hardware platforms.

Most of today's middleware vendors are gravitating towards industry standards, such as the request model used by the Object Management Group's Common Object Request Broker Architecture (CORBA), and the Open Software Foundation's Distributed Computing Environment (DCE).

These standards specify interfaces for object messaging, such as how client interfaces should be used by applications to make requests of objects. As various products converge around these standards, the real payoff will come over time: an infrastructure in which objects can message each other without any knowledge of the structure of the network or the kind of communication involved.

UNDERSTANDING YOUR MIDDLEWARE NEEDS

Most companies have made a good start at building intra-network infrastructures among mainframes, mini-computers, and PC LANs. What they lack is an enabling layer of software between the operating systems and the applications they use. They lack middleware.

To understand your own middleware needs, start by dissecting your overall information system into its application subsystems. Then you can start to see what type of middleware is required at each point in order to connect the three critical layers—presentation, application logic, and data management—particularly when those layers reside on different platforms.

The following are five primary guidelines you can use to determine whether or not middleware is essential to your operations:

- Your information system is comprised of multiple applications communicating with each other over more than one type of network
- The applications run on different platforms with different types of operating systems
- Several different types of databases are involved, particularly both mainframe databases and client/server relational databases
- The applications are being developed independently of one another at different times
- When several applications or tools need to access the same (or common) data

Remember, middleware works by supplying programs with a common interface or protocol. Where data exchange is concerned, it must manage access from clients to data resources, must assist with copying data between resources, and, possibly, with warehousing data at certain locations.

When selecting middleware, you must not only ensure that it can mask differences among the systems you use. You must make sure it can offer acceptable levels of performance for the type of data exchange you will be doing. Performance considerations center around the type of processing, whether it is ad hoc querying and reporting to support the decision support activities of knowledge workers or online transaction processing (OLTP) for mission critical applications.

Decision support activities are often best served by a data warehouse. Data warehouses represent local copies of the data used by production applications. For example, a mainframe IMS/DB database, which is used for performance-critical transaction processing in a CICS environment, might be periodically copied to a departmental server on a local area network where it can be warehoused for ad hoc querying and reporting. This way, knowledge workers can access the data locally for decision support purposes, without affecting the performance of the mission critical CICS applications running on the mainframe.

Of course, your typical knowledge worker is not a programmer, and thus does not have the skill or patience to contend with the technical nuances of IMS data access mechanisms, segments, and so forth. Thus middleware lowers the mainframe intimidation factor by performing two essential functions. A middleware gateway between the mainframe and the LAN not only copies the data to the departmental server, it also translates it into a more accessible format, typically a relational one.

Decision support activities at Rogers Cantel Mobile Communications Inc., Canada's largest national cellular phone company, represent a good example of this type of middleware. Cantel relies heavily on an IBM mainframe for intensive online processing, but the company recently transferred most of its end-user reporting activities to a pair of IBM RS/6000 midrange computers. These high-end IBM midrange systems, running the UNIX operating system, operate on a TCP/IP local area network. One is RS/6000 services Sybase Inc.'s SQL Server relational database management system. The other one is used to handle communication with the mainframe with Information Builders' EDA/SQL middleware product.

Each day, after Cantel's corporate billing procedure has been performed on the mainframe, a remote execution procedure makes a copy of the IMS billing and subscription data, creating a data warehouse environment in Sybase for end-user access.

In this case, the EDA/SQL middleware product does all the dirty work: establishes the network connection between the mainframe and the UNIX processor; schedules the copying of data each night; and handles all necessary conversions and translations to put the data in the new relational format.

Other companies use middleware to perform data exchange and datatype conversion in high-performance OLTP environments. In these cases, speed is essential and response times of a second or less are required.

Some IS managers worry that bringing in a middleware package will mean one more layer of overhead and one more set of products to learn. But most middleware products require learning a relatively small number of commands. The payoff is substantial: learning one new environment so that you can win many. In the case of EDA/SQL, for example, by learning how to program to a single API, programmers gain access to 60 different data formats on 35 different platforms, regardless of the intervening network configuration. EDA/SQL also provides API resolution layers between other programming it and other APIs, such as ODBC, DBLIB, and DB2.

Value Added Resellers (VARs) and software houses can especially benefit from middleware technology, since it enables the software products they create to work in a diverse array of operating environments. They can write an application one time for many different platforms, broadening their market opportunities.

A prime example is Encompass, Inc. in Cary, NC, a vendor of logistics management information systems. Encompass helps companies with global shipping requirements establish electronic links with trading partners and service suppliers. Its Encompass product combines a set of logistics applications with an Electronic Data Interchange (EDI) service for scheduling, document transfer and other functions. A middleware layer makes it all possible, shielding Encompass customers from the complexities of the various systems involved.

The middleware also enables Encompass to meet market demands more flexibly, since the logistics management system can be offered to customers with many different types of computer environments. Encompass users can send or receive transactions from multiple trading partners around the globe, despite variations in each partner's hardware, software, and networking technologies. Various circuits, routing information, networking protocols, and database access languages are defined homogeneously through the middleware piece.

PAVING THE ROAD TO THE FUTURE

As companies such as Rogers Cantel and Encompass have learned, middleware presents solutions to some of the toughest problems facing the software industry. Companies throughout the world are facing the problem of multiple operating systems, protocols, and databases, and they are moving away from mainframe-centric environments toward flexible, cost-effective distributed environments. As this trend continues, demand will grow for robust middleware products that can help tie the pieces together.

The pieces themselves are getting more complex. Better data communications technologies and the emergence of high-speed, high-bandwidth networks are now allowing not only voice and data to be transmitted around the network, but color images and full-motion video as well. Computers are improving in speed, efficiency, and storage capacity, which is opening doors to many new types of capabilities: 3-D interfaces, integrated fax and voice mail, sound and video sources, expert systems instruction, and fast dynamic data exchange among multiple programs.

Information systems are becoming widely distributed, physical borders are giving way to virtual ones, and the links between computers will soon become as prevalent and ubiquitous as the links among today's telephones.

But no matter how advanced our computer systems become, there will always be the need to shield the user from the complexity of the machine. This middle ground between user and computer is where middleware fits in. It will continue to evolve to carry more of the load as the machines grow more powerful, the networks grow more expansive, and the user interfaces grow more human.

AUTHOR BIOGRAPHY

John Senor is the Vice President of Information Builder's EDA/SQL Division for client/server solutions which enable enterprise wide access to relational and non-relational data. During his 20 year career in the data processing industry, Senor has held sales and marketing management positions with IBM, Cullinet Software and Applied Data Research. He joined IBI in 1988. Mr. Senor's primary responsibility as general manager of IBI's middleware division is to develop leading-edge middleware products which support customer needs for open, interoperable computing. Since founding the EDA Division in 1991, he has also been involved in developing standards for client/server software through supporting SQL Access Group and X-Open's efforts to define database interoperability standards and IBM's Information Warehouse Architecture Council.

Senor holds a B.S. from the United States Military Academy at West Point.

CHAPTER 12

People Get Ready: The Coming Transformation of Information Management

by Dale W. Way

When mainframe computers, along with client/server systems and networked PCs, are discussed in the mass media ("Rethinking Your Mainframe," *Newsweek*. June 6, 1994), one can assume a certain sense of urgency and some confusion is afoot. Although it correctly noted the fallacy of the reduced-cost argument for eliminating mainframes, the article did little to illuminate the situation, instead making vague references to a "PC backlash."

The underlying motivation for clinging to the mainframe has yet to bubble up to the surface for the media or even some industry watchers to see clearly. Likewise the barriers that are blocking the process are not yet visible. The transformation process is underway, however. It is inevitable, and it is inexorable. It will have profound consequences for commerce, national competitiveness, and the daily work lives of many people. With this in mind I am attempting here to shed some light on this very complex phenomenon. The subject matter is heavy: the tone is a bit light.

• *It's not the economy, stupid. It's the paradigm.*

Mainframe computers are at the heart of a paradigm of information processing that is increasingly ineffective in the modern world. Mainframes per se are not the problem; the halo surrounding is what is most limiting and ultimately doomed. There are real "cost-per-instruction-executed-per-second" economies to be gained with the more modern chip technologies that are not found in older mainframes.

But the software technology used in these older machines, much more than the "iron," is the root of the problem. It is in the necessary support organizations for this software base that the greatest expense lies, not in the machines themselves. But considering the *expense* of doing what you are doing now begs the question "Are you doing the right things?" In the case of the mainframe paradigm, the right things are not being done, so the economic aspects are not very relevant. There will always be "big iron." The question is what is the true nature of the old paradigm, why must it be displaced by the new and what is constraining the process. More than control of IBM's $50 billion a year customer base is at stake.

- *Lethal instinct.*

At the top level, the role of information systems (IS) in large organizations is undergoing a rapid shift from *automated accountant* to *strategic competitive* weapon. Not only is the information component of almost every product and service going up rapidly, but some industries produce nothing but information: insurance, financial services, banking, etc.

Senior management now realizes that the evolving goals and requirements of the business must drive the information technology, not the other way around. Unfortunately, the glorified accounting systems of the mainframe paradigm have been driving the business.

> Example: The CEO of a very large east coast insurance company decided he wanted to see all the information his firm had on their two largest customers. Obviously if they could "see" this information in the practical way, it would be useful in knowing how to better serve those customers. It took *six months* to do! They had to search through hundreds of individual files on many different computer systems to pull together the information. It turned out their information systems were all geared around the policy and its number—an accounting instrument—not customers. Now they must change to stay competitive. Can they do it?

- *How do you keep them?*

Unfortunately for the corporate IS management, there is an established, competing "vision" of how information management should be done. The primary impact of the PC has been to make computers fairly easy for non-technical people to use effectively. The incredible power of these devices (enough to power the mainframe of only a few years ago) has been mostly dedicated to driving a point-and-click user interface that is "intuitive" once one gets the hang of it. Bright colors, instant response—it's the MTV of computing. Once an information worker has "seen Par-rie," you can't get them back on the farm of slow response, monochrome, character-based, dumb terminals of the mainframe paradigm.

This is not the critical issue, however. What really puts pressure on IS management is that if a PC user needs his or her computer to do new things, he or she can walk down to the local software store, pick software from hundreds of titles for a few bucks, and have it up and running in a day. In contrast, if a mainframe user (even the CEO) wants some new capability from IS, his or her request goes into a queue where an *army* of application programmers (look here for the real expense!) can take *two* years to get the job done. This is nontenable in the modern world. Information workers, who are aware of what is possible, are very frustrated from the non-responsiveness of IS.

> Example: It took *18 programmer-years* (at a burdened expense rate of approximately $100,000 per programmer per year) for the California Department of Motor Vehicles merely to add a field for social security numbers to the driver's license IS in preparation for support of the motor-voter law. This did not include any programming time to actually do anything with SSNs or voter registration.

- *Let slip Cerberus.*

The real problem of the mainframe paradigm is a three-headed monster that guards the gates of the software base. Head #1 is inflexibility. In contrast to the continual availability of new PC software as noted above, the home-grown application software of virtually every legacy system has been "accreting" over years and decades, implemented by different people at different times for different purposes with different standards in place at the time, sometimes working in concert with other programs, sometimes working independently.

Further, each program in and of itself has evolved by constant modification into something, in many cases, far afield from its original purpose. Yet, like the reptilian and mammalian brains latent within each human, the older roots of these programs, lost in the mists of time, can have an unwanted or unknown impact on the future capabilities of the program. Furthermore, each program most often overlaps with others in the use of the same data (however, to add to the confusion, they often call the same data element by different names).

Thus, making a change, even a simple one (see California DMV example above), takes a long time and a lot of money. Hundreds or even thousands of these poorly-documented and to some degree incomprehensible, programs must be studied just to *assess* the impact of the considered change. Additionally, the impact on the shared data must also be analyzed. Then, after the change is made (usually a fairly straight-forward thing), an almost equal time is spent testing the change; not just testing the new capability, but making sure the change has not inadvertently screwed up some other operations of these complex, spaghetti-code systems.

This last point leads to the next Head, but it is already clear that these inflexible legacy systems are not the basis of new, adaptive, "turn-on-a-dime" information systems needed for survival in the future.

Head #2 is brittleness. As a corollary to inflexibility, brittleness adds the demon of unreliability to the situation; these systems are very prone to breaking. The programs involved are not always well understood. The data structure, as it relates to all the programs that rely on it, is not ever completely understood. So the fact that more and more changes are going into something less and less certain, means errors will be inevitable. Such errors are statistically certain but unpredictable.

In fact, it is in the nature of the procedural application program technology of these systems, called third-generation languages, or 3GLs (COBOL and FORTRAN are the prime examples), to be mathematically non-deterministic (that means there is no way to *prove* they are correct). They may look like they are working right, but that only means that the right combination of inputs has not yet impinged upon them in the right sequence to send them spinning out of control.

Such legacy systems not only "pump the information life-blood" of all the large organizations in the world, they manage all the air traffic control, international finance and military defense systems as well. This is only annoying when your phone bill is screwed up, or when you are told your outstanding balance of $0.00 must be paid immediately or your name will be turned over to a collection agency, or if you temporarily lost your health insurance coverage when your boss left the company and all his or her staff was inadvertently eliminated from the group policy. Two airliners in the same space at the same time, or lost global currency transactions, or even phantom missile responses are more than annoying, they are dangerous.

Head #3 is inaccessibility. Large organizations can spend hundreds of millions of dollars a year acquiring and managing the data that is the life-blood of their operations. Yet because of the rigid way the data is organized and the difficulty in rearranging and presenting it in new ways (see the insurance company example above), this valuable asset is not available to all the people who need it, when they need it, nor is it presentable in a way that is useful.

An army of application programmers (sometimes referred to as a "priesthood") sits between the information worker/user and the data and information he or she needs to do his or her job effectively Besides being wasteful, this inaccessibility inhibits or outright blocks needed attempts at business process reengineering. Business process reengineering, which deserves its own thoughtful coverage, is key to the transition underway in enterprises of all kinds as they convert from an industrial business model to an information business model. It is more than the business "fad-of-the-month."

To deal with this shortcoming, information workers demanded and to some extent received links between their PC networks and the corporate mainframe databases. These gateways, as they are called, allow users to pull data down into their PC for use. This is both unsatisfying and alarming. It is unsatisfying because the links are one-way read-only: the user cannot put new data back in the corporate database, and no other worker can directly benefit from the results of his or her work. It is alarming because, from the point of view of IS management, data that leaves the corporate database is no longer controllable; the PC user can incorrectly change it or, after the PC user has employed it, it could be changed in the database by a mainframe program, and the PC user would not know he or she now has bad data.

This ad hoc attempt at integrating legacy mainframes with PC and client/server networks is not the answer, especially when the fact that legacy systems themselves are un-integrated, fractured into many databases addressed by hundreds of separate programs, is considered.

Cerberus, the three-headed dog who guards the gates of hell in Greek mythology is an apt metaphor for the obsolete software base that both fatally limits the effectiveness of the mainframe paradigm for the future and also, by its complexity and incomprehensiveness, blocks any attempt to transform these legacy systems into flexible, extendible, modern, client/server information systems.

- *It's how you play the game.*

I could spend 500 pages describing the nature and benefits of the new, non-mainframe paradigm (there is not yet a good name for it, and all the ones used, like client/server, are incomplete and focus only on one or two aspects of the phenomena). What is important about it, however, is how it addresses the shortcomings of the old paradigm. True, you get modern GUI (Graphical User Interface, pronounced "gooey") interfaces for worker productivity and you get modern system price/performance improvements, but the critical driving force is *liberation* from the 40-year-old accumulation of third-generation software technology.

This stuff is incomprehensible, brittle, and takes armies of high-priced priest-programmers to maintain. Also, as discussed above, it cannot keep up with modern competitive requirements, it cannot be adapted easily for competitive or business process re-engineering purposes, it cannot provide different "views" of the information to different workers, it cannot allow information workers to work directly with it, and it cannot be depended on to not break.

Understanding the technical foundation of the new paradigm will enable one to see the incredibly difficult barrier that has actually prevented the new paradigm from displacing the old (and it is not a "PC backlash"). Explaining something that is technically complex in the extreme in a non-technical way, is not easy but I will attempt it. The rest of this point and the next will cover it—more or less. Here goes.

There are new programming methods beyond the COBOL/FORTRAN third-generation languages (3GLs) of the 1960s and 1970s; a newer, more advanced software technology base has emerged. One of new methods may be familiar to you—the ubiquitous spreadsheet, like Lotus 1-2-3 or Microsoft's Excel. "But wait," you may say, "that's not a programming method, it's a program itself that people buy from the software store." Yes, but it is more than that.

What users of spreadsheets do is get computers to process information for them by putting numbers and formulas into little boxes (cells) on a matrix worksheet on a computer screen. Then they link the cells with pointers within the formulas and let the program calculate ("calc") the worksheet. Writing a 3GL COBOL program can do this, too. But look at the differences! First, normal people can use spreadsheets. They do not have to be specialist programmers. Secondly, it is fast; a few minutes versus hours or days.

Thirdly, specialist programmers have to know not only how to write programs, they also have to know all kinds of details about the particular computer operating environment the program will run in. These details have to be accounted for in the program. Spreadsheet "programmers" do not have to worry about machine details.

Fourthly, and perhaps most importantly and most overlooked, spreadsheet programming is *declarative*, not procedural like 3GLs. You simply "declare" what you want in the cells; the order or sequence that you put them in does not matter. In procedural languages (also called sequential languages), the order of the program statements is *extremely* important. In fact, most of the "knowledge" in the program is expressed in the sequencing of the statements (like meaning in human languages is in the sequencing of words). Yet the complexity lies in sequencing, since each step is dependent on the one that went before, and that on the one before that, all up the line.

Any program that is not trivial has many, sometimes hundreds or thousands of branches, forks in the road of the program execution. If the program is to work correctly, predictably, and reliably, the programmer must *completely* anticipate all the possible states that could exist at all the branch points at any time. To do that, he or she would have to analyze all the possible inputs to the program in all the possible input sequences. Because this is impossible for any non-trivial program, or for the hundreds or thousands of programs in large mainframe systems, these systems are statistically certain to break.

It is only a matter of time until an unanticipated sequence of inputs or branches is encountered. (Remember, our missile defense systems are controlled like this.)

Declarative programs do not generally have this problem, although some "procedural-ness" can sneak into them if certain kinds of procedural program segments (called macros) are used. Even then, it is not nearly to the same degree. The lack of a time dimension greatly simplifies declarative programs.

Declarative programs are sometimes called "structural" programs. You can visualize this by looking at a spreadsheet and imagining the pointer links that connect the cells being visible; the structure emerges. It is not the structure in physical space that matters; the cells can be rearranged and as long as the pointers still connect them correctly, it has not effectively changed.

The logical structure is what matters; the "meaning" of each cell and the "meaning" embedded in their links. The "knowledge" in this kind of program is expressed in the structure. This concept of structure will be used later.

Why not use spreadsheets as a replacement for large-scale information systems? There are many problems with that. A really big one is that each spreadsheet has its own data embedded in it and this data has to link properly with the formula cells to work properly. What happens when many users, maybe hundreds or thousands, want to interact with the same data? What if they all want to add new kinds of data or new links? It will not work.

This is the same problem legacy mainframe systems have. Most 3GL application programs were written at a time when each program processed and managed its own data. Most, perhaps 70% to 80% of the data stored in legacy systems is still in this form.[1]

The progress in handling shared data has been a major achievement in the evolution of information systems. It is critical to understand where we are and where we have been in this process. The current state-of-the-art for large systems is to separate the data-handling aspects of the system, which are common to many programs, away from the program aspects themselves.[2] A separate kind of program, called a Database Management System (DBMS), surrounds and manages the data control aspects of the data. These DBMSs are complex products supplied by a separate software industry segment called, not surprisingly, the database industry (although some big computer companies sell their own database products as part of their machines). DBMSs unburden the programs from critical navigational tasks and, more importantly, from the data integrity tasks; data can easily get screwed up if they are being handled in a thousand different places by

[1]Let's define "managing data" like this: specifying input data "rules", specifying how to place/find data in storage [navigation], specifying data integrity "rules" for protecting data from getting corrupted, and specifying output data "rules." "Processing data" means: specifying input mechanisms, specifying any calculation "rules," specifying any decision "rules" (If.., then..), and specifying output mechanisms. "Rules" are the what-to/how-to's and what-not-to/how-not-to's of system operation. There are two kinds of rules then, data and process. Enforcing both kinds of rules is another way of saying what information systems do.

[2]It surprises everyone (especially software people who are, after all, programmers) that most of the work done in commercial information systems is data-related. The program tasks—calculations and decisions and I/O mechanisms (see footnote 1) are a small part of the whole, both in time the system spends and in the number of "rules" enforced by the system. 80% to 90% of the rules are data rules and more than 90% of time spent is just in moving data around. (It is somewhat more balanced in scientific and graphical computing.)

a thousand different people at a thousand different times. Database management systems focus the data management function in one place (or with more than one DBMS, a small set of places).

However, the DBMS product is only an "engine." It has to drive the customer's data design. The data design involves deciding what are the data elements the customer wants to keep track of and how they are related to each other. **Important point**: Data design is the customer's responsibility because it is a unique reflection of the customer's organizational identity and needs. These concepts will be visited later, too.

We have taken a detour from the subject of modern programming methods. We now reconnect with that path. There are a number of newer, higher than third-generation, programming languages available. Called (big surprise) fourth-generation languages, or 4GLs, these languages are characterized by several very powerful features: First, they are declarative languages, not procedural. You simply "declare" what you want the system to do.

Secondly, they are often graphical, meaning you do not have to type the commands in on a keyboard, you just point and click on menus or icons with a mouse (a GUI front-end) with perhaps a few keystrokes to name things. What you want gets automatically converted into instructions the computer can understand.

Thirdly, many 4GLs are portable and machine-independent; you can move programs generated with them between different kind of computers, and the programmer does not have to concern him- or herself with details of the machine's operating environment.

Fourthly, some 4GLs are very context-specific, having almost artificial intelligence aspects to them; they have a lot of knowledge of the application area of interest embedded in the languages themselves. Examples include financial 4GLs that "know" accounting and financing principles and guide you along, or report writers that understand page layout and reporting conventions necessary for generating an invoice or a shipping document.

It should be clear that 4GLs are not exactly languages in the traditional sense, yet they perform the same functions, only faster, more simply, and with much more flexibly. They are also much more reliable. You can see 4GLs are approaching the point where normal everyday people could do all their own "programming." This is a very powerful concept, and it promises to transform the work world and alter the competitiveness of whole industries and even national economies. Some would call today's spreadsheet programs, which make us all our own programmers (for some things anyway) a 4GL. I would, except for the next point.

The last characteristic of 4GLs is the most important for understanding the barrier between the old paradigm and the new, and the factor that underlies the reason mainframes (and their attendant paradigm) are still hanging on in spite of all short comings. It turns out that almost all 4GLs used for large-scale information systems can only work when they sit on top of a newer type of database management system, called a *Relational* Database Management System (RDBMS).

Now, I do not want to get too deeply into the nature of this beast; RDBMSs are very complicated and many books on the subject are available. What is important about it for our purposes are three things: First, the access method, the way a programmer puts data in and gets data out of the database, is uniform across all relational databases no matter from which vendor the RDBMS comes. (This is theoretical. In practice all vendors "extend" the standard method with "enhancements" which activate the special features of their products. These minor violations of standards are unavoidable in a competitive industry and not disruptive in the overall scheme of things. Having even a standard "approach" to data access is a big improvement over the current legacy hodge podge.) The access method uses a simple, higher-than-third generation language that shields the programmer from needing a detailed knowledge of the "innards" of the database system.

Secondly, unlike pre-relational databases where each and every navigational path through the data had to be predetermined at system design time, relational databases allow ad hoc paths to be created dynamically at any time during system operation. This means that the data can be "sliced and diced" or "viewed" in any way appropriate to any application or user. In the insurance company example above, if all the data been stored in a relational system, shifting the "view" from that of the policy and its number to customers and their needs would have been straightforward. New programs might have to be written to explore this new view and extract and apply the knowledge gained therein, but the data would be available in an easily retrievable manner. (Also the new programs, could be written quickly and easily using 4GLs and point-and-click programming tools, not low-productivity, hard-to-maintain 3GLs.) If the need for yet newer "views" emerge, it would be able to easily adapt to those as well.

Thirdly, relational databases more completely and with more certainty that any other approach, assure the *integrity* of the data from goof-ups by the programmer or program. That is because relational databases have a formal mathematical basis underlying it. That means if "done right," it is predictably correct (or *deterministic* as scientists and technologists like to say) and reliable, and is mathematically *provable* as being so.

Imagine an information system where 80% to 90% of the work and the "rule" enforcement is done automatically in a mathematically correct way—no matter how the application programmers or their programs mess up. Would not you sleep better knowing your country's defense systems were based on this new technology? Or your bank? Or your brokerage company? Or your hospital? And wouldn't it be great if every worker in any organization, commercial, governmental or home, could easily be an instant programmer and have access (subject to security controls) to all the information in his or her enterprise (and to some degree outside of it) *and* be able to put it together in any way they need to at any moment?

No more two-year waits for information you needed yesterday. No more priest-programmer armies between information and the people who need it. No more spending

80% of data processing budgets to keep old, brittle mainframe monsters alive. You would be able to adapt information systems rapidly and easily and to reengineer your business processes to address new opportunities or threats. You would also be able to increase your organization's "learning velocity" many-fold.

In short, you would be able to make your information systems the servants of the enterprise and not the other way around. With the continued advances in 4GL programming tools and the propagation of relational database technology, this *is* the new world. PCs are important. New super fast, super cost-effective computers are enablers, networks are facilitators and distributed client/servers are the organizers; that is the infrastructure; that is the "how." What follows is the "why."

- *It's also the data, stupid.*

All well and good, you say, but what is it that has to be "done right" for this all to work? Well, the RDBMSs engines come off-the-shelf; they are what they are. No, what has to be done right is for the customer data to be properly organized inside the RDBMS. "Wait a minute," you say, "this whole marvelous new paradigm all hinges on how the customer data is organized?!! That doesn't seem right." Well, it is.

Remember above where I say that the customer's data design—the data elements the customer wants to keep track of and how they are related to each other—uniquely reflects the customer's organizational identity and needs? "Reflects identity and needs" is a loose way of expressing a particular kind of knowledge about the "rules" that the customer wants enforced by the system regarding the handling of data. Remember further above when I say that a spreadsheet was an example of a "structural" program—where the "knowledge" is not expressed in a sequence of steps but in the structure of the spreadsheet—the data elements (cells) of interest and the links between them? That is exactly how a relational database works—*the knowledge is in the data structure*—the customer-defined data elements and the relationships between them. That is why it is called *relational.* And remember, 80% to 90% of the "knowledge" of the new systems is moving into the database system, away from the application programs.

Relational database technology gives the world the advantage of a structural approach to programming (and eliminates much of the procedural) while handling better than anything else the shared data problems of the past.[3]

[3]There is a newer yet kind of database technology on the horizon called an object-oriented database (OODBs). Without going into any depth about OODBs, it can be said that those that emerge victorious will be built upon a relational foundation. That is the only way to get the mathematical certainty necessary for the mission-critical systems of the future.

But to work correctly, the data design, the structure of the customer's data, has to be put into a mathematically *precise* form. It has to be "boiled down" to its least redundant, most complete form. In technical jargon this is called *normalization.*

The data design has to be correctly and completely normalized to be "done right." Normalization is a double-edged sword; if done right the database will work correctly with mathematical certainty, if *not* done right, then the database will work *incorrectly* with mathematical certainty.

Unfortunately for the early arrival of the new paradigm, there are two big problems with doing normalization. First of all, for any practical system design normalization is *impossible* for a human to do with mathematical precision; the combinatorial complexity is too great. Even teams of humans can not do it because it can only be done *en toto*, it cannot be split into pieces; some single "intelligence" must analyze the entire design all at once.

Yes, this means that every relational database system out in the world is not done correctly. They work, sort of, because the programs, in a trial-and-error manner, have been re-worked to take back the responsibilities of operating and maintaining the integrity of the database system—exactly negating one of the principal reasons for going with a relational approach in the first place. Although the data integrity in this case is no better than in a pre-relational system, the value of having dynamically alterable "views" of the data and a higher-level access method is well worth it.

Secondly, and more crippling to the emergence of the new, the "knowledge" needed to normalize a relational design based on a mainframe legacy system is *unavailable.* Guess what! That knowledge is in the structure of the legacy data and it is locked up in a convoluted mass of interdependent programs and data files. Nobody knows how to get it out. They look at the programs, but they run into the same all-possible-combinations-of-all-possible-sequences-of all-possible-inputs barrier that we talked about above.

They look at the data-about-the-data in documentation or the data definitions parts of programs, but that does not tell them about the relationships *between* the data from different programs. (Don't forget, many of these legacy systems involve *thousands* of programs.) Finally they look at some of the data itself; they hire high-priced consultants to "analyze" databases containing thousands of files with a hundred or more data elements (columns) each and millions of records (rows) for each file manually. You can see the structure there, can't you?

So you cannot re-structure legacy data. No data perestroika. The new software technology base depends on new data structures. You cannot get there from here. Thus, you cannot abandon the mainframe. You can build new systems out of modern technology where legacy mainframe data is not used, but not using the main store of data in the enterprise, limits new uses to the fringe. That is where all the client/server hoopla has been, at the fringe, at the outer ring of organizational activity.

Forget the mainframe, forget the old software, and start over. There is conservatively *$1 trillion* invested in mainframe application software and its data structures. That stuff contains the accumulated information knowledge of the entire enterprise through all time. It is the *beating* heart of the enterprise. And there are no heart-lung machines. Walk away from it? No way.

What to do, what to do. Punt. This is the essence of why mainframes linger and networked PC client/server systems are stalled in penetrating the market. Nobody knows how to engineer the transition from the mainframe system to a modern one; the old paradigm to the new. Those that have tried have gotten burned (see California Department of Motor Vehicles debacle—$44 million down the drain or the new FAA air-traffic control system—$1 billion over budget and three years late. Those are the public ones).

With 80% of DP budgets going to maintenance and an unclear, open-ended, high-risk path to getting out from under their old systems, no wonder the flood gates to the future remain closed.

- *There's a train a comin'.*

There is good news. Most large U.S. organizations are to some degree committed to the change. Even in the government. All they need is help to do it. Finally new tools that address this gap are emerging; from Silicon Valley mostly, but from Texas and Route 128 as well. There is hope.

The further good news is that the Japanese are still in deep denial. Having a substantial base in mainframe computers (Fujitsu, Hitachi, and a big share of U.S. maker Amdahl) and absolutely zero share of client/server and network market and little in the PC, they are mightily clinging to the old centralized paradigm. Once again their authoritarian, people-as-intelligent-machines culture, which helps them so much in large-scale widget manufacturing, is killing them in seeing the importance of the democratic, information-sharing values necessary for the rapid rate of business adaptation needed to win in the next century.

All of our industries that can manage the transformation first can kick their Theory Xs.

The organizations that can master this transition will have a gigantic advantage over their slower competitors. Effectiveness returns of ten to twenty times are not unusual (See Hammer and Compy's *Re-Engineering the Corporation.*) The map of commerce will be re-made. IBM's $50 billion a year installed base is up for grabs and so is the leadership of every industry there is.

In a nutshell, we are ending the industrial age of computing and entering the information age of computing. It has already happened on the desktop with PCs and departmental networks. But these "sensor and motor" nerve pathways out in the body of the organization have yet to be fully linked with the "brain"—the major computing resources of the organization. Furthermore, they have yet to be linked to the "mind" of the organization—the race memory and knowledge contained in the data of the enterprise. Knowledge is information in context; wisdom is knowledge with experience. Both can only be gained in a learning organization and both must be available to every information worker. Only then can large organizations adapt and evolve successfully in an environment of accelerating change.

AUTHOR BIOGRAPHY

Dale W. Way is currently the Executive Vice President at DBStar, Inc., a San Francisco-based provider of automated data engineering software tools. Way has over 20 years experience in the computer, data communications, and information industries. Way was a partner in his own strategic consulting firm where he advised clients such as International Business Machines, Digital Equipment Corporation, the Open Software Foundation, National Advanced Systems, Canstar Communications, and others on future trends and opportunities in their respective businesses.

Way was a co-founder of Excelan, Inc., an early entrant in the Local Area Network industry later acquired by Novell, Inc. He has held marketing positions for medium and large domestic and international companies such as Memorex Corporation, Olivetti and Zilog, Inc., then a subsidiary of Exxon Corporation. Way has authored numerous articles and has spoken at numerous conferences and seminars on leading edge technologies and markets. Prior to his marketing roles, Way was a computer engineer. He holds a BSEE from Michigan State University.

CHAPTER 13

Total Process Management: The Case for Holistic Change

by Mark D. Youngblood

A gale of change is sweeping across the corporate landscape—new competitive pressures from every point of the globe are dramatically redefining the model for the successful business. The suddenness of these changes has left many companies poorly positioned to compete effectively.

These companies are bloated with management and administrative overhead; centralized decision making strangles responsiveness; quality is poor; customer awareness is weak, and costs are high. In a frantic effort to restore a competitive edge, companies are embracing a wide range of change methodologies. Business process reengineering (BPR), Total Quality Management (TQM), Time-Based Competition and technology-driven "rightsizing" are being used in an attempt to delayer top-heavy organizations, drastically reduce expenses and increase customer responsiveness.

With recent headlines proclaiming that as many as 70% of business process reengineering efforts fail and with similar results reported for TQM, you can see that few companies are succeeding.

The problem is not with rapid change itself, but with the fact that companies do not manage change effectively. Most companies are ill-equipped to handle organizational change.

The *Harvard Business Review* reports that "the problem for most executives is that managing change is unlike any other managerial task they have ever confronted. When it comes to change, the model he uses for organizational issues doesn't work."[1] This is supported by the results from a recent Gallup poll. The survey showed that many people believe that their organization is changing rapidly, but more than half of them think that their company cannot handle the change.[2]

Traditionally, corporate managers have responded to the need to improve bottom-line performance with one of three approaches: institute sweeping layoffs, automate, and restructure. By themselves, these approaches have proven ineffective in achieving sustained performance improvements over an extended period. The reason was understood to be that most of these techniques are implemented without any significant changes to the underlying processes and people systems.

The emergence of TQM and business process reengineering were supposed to correct this inadequacy. Unfortunately, as the results indicate, they do not.

Both the old and new techniques struggle to deliver significant, sustained results because they are grounded in an outdated view of business. In the 17th century, Sir Isaac Newton and Rene Descartes combined to create a mechanistic view of life that prevails in business even today.

They viewed the universe as a giant machine that with sufficient research and scientific analysis could be reduced to ever smaller components. This mechanistic, reductionist model is still how we view business. We believe that companies are merely the sum of their parts, and if the business is performing poorly, we need only to replace a part, and the problem will be resolved. This is why we try in futility to solve problems with a short-term, reactive, "component" approach.

Just bring in a new CEO, implement a new computer system, create self-directed teams, focus on quality, or reengineer business processes, and performance will magically improve. We now know that companies have spent *trillions* of dollars on these approaches and have only seen modest improvements in overall business productivity in America.

The new science, quantum mechanics, is giving us another model of the universe—one in which "the universe is no longer seen as a machine made up of a multitude of objects, but has to be pictured as one indivisible, dynamic whole whose parts are essentially interrelated and can be understood only as patterns..."[3]

The emerging model of business is that of an organic system where everything is related to everything else. We cannot hope to make a change to one part of a system while neglecting the rest and hope for positive results. Change in complex systems

[1]Duck, Jeanie Daniel. "Managing Change: The Art of Balancing," *Harvard Business Review*. November-December, 1993. p. 109.

[2]*Quality Digest*. November 1993, p. 13.

[3]Fritjof Capra. *The Turning Point*. Bantam Books. 1982.

requires that the whole organism be taken into account. We are learning now that effective change must be *holistic*—it must address the entire organism and not just a small part. Under the new model, successful performance improvement methodologies must address the following aspects of change:

- The requirements of *all* significant stakeholders: customers, suppliers, owners and debtors, employees and their families, our communities, and even the environment
- Corporate identity: shared vision, culture, values, tradition, and ritual
- The entire "change continuum" from incremental to dramatic improvements
- All four organizational components: process, technology, structure and people
- All three dimensions of performance: cycle time, productivity, and quality
- The entire organizational hierarchy, from top to bottom

TOTAL PROCESS MANAGEMENT

Total Process Management, an integrative approach to change that was first introduced in the book *Eating the Chocolate Elephant* (Micrografx, 1994), is representative of the new wave of holistic performance improvement methodologies. It addresses the key points described above, and is integrative in the sense that it works with existing efforts such as TQM and BPR instead of replacing them. The "constellation" shown in Figure 13.1 graphically represents the scope of the Total Process Management approach. Each element of the constellation is discussed in more detail in this section.

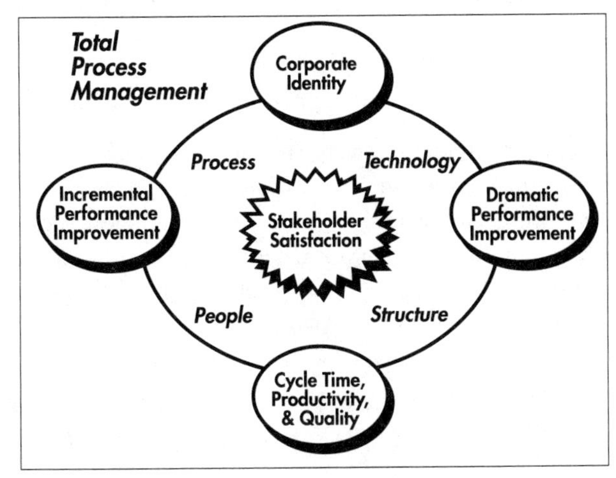

FIGURE 13.1. Stakeholder requirements

Stakeholder Requirements

The current focus of almost every performance improvement methodology is on the customer—and companies would be foolish not to seek to exceed their customer's expectations. Customers are stakeholders vital to our success, but there are other stakeholders that are equally vital.

These include our company's owners, employees and their families, suppliers, corporate lenders, our communities and the environment. Happy customers do not necessarily mean that the needs of the other stakeholders are being met. Although certain stakeholders, such as owners and debtors, will usually benefit in relation to customer satisfaction, other stakeholders who are essential to excellent corporate performance may not.

If our intention is to achieve optimal corporate performance, then it is essential that we attend to the needs of *all* of our key stakeholders. If we treat suppliers as adversaries and constantly shop for the best price, then we forsake the benefits to be gained through partnering. Many companies today are reaping the advantages of long-term supplier relationships through their participation in reducing costs, increasing quality and reliability, and in responsiveness. Employees must be viewed as stakeholders with important needs. If we view employees as assets and not humans, we may achieve short term costs savings at the expense of high employee burnout, absenteeism, and turnover.

The concerns of employee's families (another important stakeholder group) often limit the employee's effectiveness. Our communities depend upon the companies in their area to provide stable employment and to act responsibly toward the concerns of community. Environmental damage from destructive corporate practices rebounds to reduce employee health and productivity. Holistic change requires that the needs of each of these constituencies be considered.

Corporate Identity

Having a clear sense of corporate identity is vital to sustained performance improvement. Without it, we may "climb the ladder of success" only to find that it is propped on the wrong wall. Corporate identity is achieved through clear communication of a shared vision (mission and strategy), through cultivating a healthy culture, and through effective use of tradition and ritual. Employees must understand what the company's purpose is and internalize it as their own. The mission tells them where the company is going. The corporate strategy tells them how to get there.

Corporate culture (our beliefs and values) communicates what we *care* about as a corporation, and how we behave in achieving our objectives. Tradition is important in shaping understanding of a common history and in reinforcing the corporate culture.

Tradition and the related rituals give us a sense of meaning and of belonging. Ultimately, having strong corporate values lends employees the only real security they have during times of dramatic change. While the *form* of the company is changing, the company's *essence* (its values and ultimate purpose) remain the same.

THE CHANGE CONTINUUM

A partner at the Big-6 consulting firm where I once worked frequently admonished his consultants with the quote: "If the only tool you have is a hammer, everything will look like a nail." His point was that if you only know a single technique for solving a problem you can be blind to other solutions that might work better. This explains in part why TQM and business process reengineering frequently fail. Both are excellent approaches for implementing organizational change, but rarely are they effective by themselves.

Total Quality Management is meeting with limited success due to several inherent drawbacks. TQM, as companies such as Xerox can attest, can deliver remarkable results. However, executives needing quick results are finding TQM a bitter pill to swallow. TQM requires a great deal of time, often several years, before it begins showing results. Also, it requires significant investment in the early years with no matching offset in cost savings. It is no wonder that executives facing cash flow problems or angry stockholders are opting for quicker results through business process reengineering.

However, over time, BPR is not faring any better. First, it is agonizingly tough to implement. As *Fortune* magazine states: "All change is a struggle. Dramatic across-the-company change is a war."[4] Also, companies are discovering that the results of BPR efforts tend to erode over time. This seems to be because BPR is frequently viewed as a "one-time shot." The employees who are responsible for sustaining the improvements are not afforded training in the principles and techniques required to continue adapting the organization to ever-changing competitive circumstances.

In order to illustrate the weaknesses of these approaches as independent solutions, consider the game of golf. Imagine trying to play a competitive game with only one club. If you had to choose just one club, you would probably select one that would give you the most versatility, like a 5-iron. However, given just this choice, organizations are selecting putters (TQM) and drivers (BPR). Putters give you control and small incremental advances—drivers deliver dramatic advances in a short time. Both are important but can be dangerous if we rely on them exclusively.

[4]Stewart, Thomas A."Reengineering, the Hot New Managing Tool," *Fortune*. August 23, 1993, p. 42.

For instance, a driver will be of little use on a short, narrow fairway with a dogleg or two, and a putter would be useless if we had to cross a broad ravine or lake. Similarly, in a business environment, TQM cannot be used to reduce product delivery by 75% in six months, and BPR cannot be used to solve widespread quality problems and organizational waste. What we need is something that combines the strengths of both, while minimizing their weaknesses.

Total Process Management solves this problem by providing both the ability to achieve rapid, dramatic improvements in the manner of BPR and also incremental, gradual improvements as achieved through TQM. Total Process Management views dramatic and incremental improvements as the same in approach, but differing in *scope.* TPM is scalable so that it can adapt to the appropriate level of change required by environmental forces. In relation to our golf analogy, it is the equivalent of having the entire range of clubs available to contend with constantly changing competitive challenges.

Four Organizational Components

There are four organizational components that must be addressed in order to achieve a comprehensive approach to change. These are process, technology, structure, and people.

Process

Companies are shifting to a new management model, often referred to as "the horizontal corporation." In this model, it's about managing across, not up. Companies identify their key processes, assign "process owners," align employees from multiple functions into teams and assign them responsibility for satisfying customer expectations. This arrangement takes an organization's focus off its own internal structure and puts it on meeting customer needs, where it belongs. The fundamental shift for businesses in the 1990s is from managing departments to managing processes.

"Getting horizontal" is not easy. Untangling years of bureaucracy to uncover the underlying processes is a time consuming and painstaking exercise. For instance, at AT&T 's Network Services division, they identified around 130 processes before narrowing them down to 13 core ones.[5] As difficult as it seems, documenting processes must be done—companies that wait will simply find themselves that much further behind the competition. Every process should be documented and fully understood by everyone participating in the process. Only then can every employee achieve the understanding necessary to make effective business decisions—a prerequisite for any real form of empowerment.

[5]Byrne, John A. "The Horizontal Corporation," *Business Week*. December 20, 1993, p. 78.

Technology

Information technology is one of the key enablers for performance improvement. Today the pace of development and increasing capacity of information technology are expanding exponentially. The potential for leveraging technology for competitive advantage is unlike any other time in history. Although the range of technologies is virtually endless, a few of the more common technologies being employed today for performance improvement are listed below:

- Distributed client/server computing
- Electronic Data Interchange (EDI)
- Image Processing and document management
- Expert systems
- Groupware
- Workflow
- Mobile and remote computing
- Multimedia
- Wireless telecommunications
- Executive information systems (EIS)
- Bar code printing and scanning

Structure

This organizational component relates to both the physical layout of the work environment as well as the organizational hierarchy. A poorly-conceived or unplanned physical work environment layout can be a significant barrier to productivity. Substantial performance improvements in speed, accuracy, and efficiency will usually result from better workplace design. In addition, organizing around process almost always results in some changes to the physical work layout.

Organizational hierarchy reveals excessive levels of management and managerial spans of control that are too narrow. Targeting these areas for improvement can result in large returns in productivity and organizational efficiency. Total Process Management addresses both the physical layout and organizational hierarchy dimensions of structure in providing an overall solution for performance improvement.

People

Performance measures and rewards, job definition, hiring and termination, employee competencies, shared skills and style, and training systems are included here. This organizational component has the most profound impact on the success of a change effort and it is frequently the most neglected. Employee resistance is given as the single greatest cause of failed change efforts. This results in large measure from the fact that we ignore people considerations when developing and implementing performance improvements. Some key considerations are discussed below.

Every company has a core set of skills and of management styles that are shared throughout the organization. Without them, there is no common way for conducting the business. Shared skills include problem solving, teamwork, conducting meetings, communicating, project management, and performance improvement techniques. The expression, "shared styles," refers primarily to the company's management and communication styles. It is essential to overall employee effectiveness that organizations develop these shared skills and style.

Performance measurements and rewards are additional elements of the people component. The old adage is still true: you cannot improve what you do not measure. Although some companies measure non-financial performance, almost none measure *process* performance. In addition to measures, companies must pay attention to performance rewards. People will do what you reward.

People must have the skills to operate effectively in this changing world. For instance, people comfortable with a command and control management style cannot transition to an empowered, process-oriented style without significant retraining. Corporate training and education systems, requirements for employee retooling, must be established or enhanced to accommodate effective change.

THREE DIMENSIONS OF PERFORMANCE

There are three dimensions to superior performance. These are cycle time, productivity, and quality. Cycle time, also referred to as process speed, is determined by the elapsed time for all activities in the process. Productivity is measured as the combination of labor, direct cost and indirect cost consumed for an activity to produce a given amount of output.

Productivity can be improved by increasing output at the same cost, decreasing cost for the same output, or both. Quality is determined by how well the process produces goods and services that meet or exceed customer requirements. Quality can be measured internally through number of customer service calls, number of product returns, percentage of defective products and other such indicators.

Current performance improvement methodologies tend to emphasize just one of these performance dimensions and downplay the others. For instance, Time-based Competition views improvement through the "filter" of cycle time. Proponents of time-based competition claim that quality and productivity are automatically improved when you perform work in a shorter time.

Business process reengineering focuses largely on productivity (reducing cost) with a secondary emphasis on cycle time, and little on quality. Total Quality Management (as you can imagine) views performance improvement principally as a quality issue. Taking an exclusive approach such as these introduces the risk of "being a hammer looking for a nail." Cycle time, productivity and quality are each important to performance improvement. For certain processes, one may be less important than the others, but that is driven by the nature of the process and not the methodology being used.

CHANGE THROUGHOUT THE ORGANIZATIONAL HIERARCHY

So often in implementing performance improvements, changes begin *below* the executive level and extend to the front-line workers. In asking for change, many executives mean that what they really want is for the organization to change but for themselves to remain untouched. Many executives have the attitude that the problem is "out there" in the organization, and not "in here." In order to be effective, however, change must include the entire organization, from top to bottom—no exceptions.

Executive management can have a tremendous impact on the success of their organization's performance improvement efforts by first looking to change themselves. The first rule to follow is "walk the talk." Employees will do what management does, not what they say. Secondly, executives must be willing to act in the best interests of the organization, even when it is not in their own best interest. For instance, management may have to relinquish power and certain privileges, be willing to listen with an open mind to unpleasant feedback from subordinates, attend extended amounts of training in new skills, dedicate substantial personal time to the change effort, or learn to coach instead of command. Voluntarily relinquishing the perquisites that come from positions of power requires wisdom, insight and courage. However, the response from the organization will be well worth the sacrifice.

THE SEVEN PART TPM METHODOLOGY

There are seven parts in the Total Process Management methodology, as shown in Figure 13.2. The first two parts create the foundation for performance improvement. The remaining five parts comprise the process improvement cycle. A brief overview of each part follows.

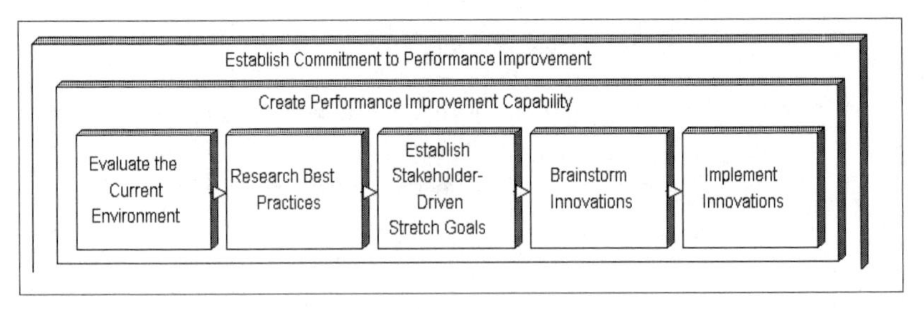

FIGURE 13.2. The 7-part TPM methodology

- *How do we cultivate the motivation to change?*

Part 1, "Establish Commitment to Change"—addresses the will to change. It involves defining the market forces that are driving the need to change, refining mission, strategy, and culture, communicating with the organization, removing barriers to change, and rewarding performance. This portion of TPM is known as "change management" when implemented as a standalone methodology.

- *How do we obtain the skills needed to implement change?*

Part 2, "Create Performance Improvement Capability"—addresses this question. Once the organization has the will to change, it needs the skill to change. This part of TPM involves training on the skills required for performance improvement and for organizing around processes. It includes the performance improvement principles, problem analysis tools and techniques, and creative brainstorming, but can include any skill required for a given organizational change.

- *What are we doing today?*

Part 3 is "Evaluate the Current Environment"—before you can hope to improve your organization's performance, you must first understand your business. In this part of TPM we document and evaluate each of the four organizational levers for change: process, technology, structure, and people. This is done through a series of interviews and/or team documentation sessions that include people involved in performing the work along with customers and suppliers.

- *What are other companies doing?*

Part 4, "Research Best Practices"—provides us with a view into business techniques being employed by our competitors and by world-class companies. These are a vital source of ideas for how our own business can be improved. The best practices information helps us to set stretch goals and feeds innovation in our creative brainstorming.

- *How much improvement do stakeholders want?*

Part 5, "Establish Customer-Driven Stretch Goals"—we meet with key corporate stakeholders, both internal and external, and determine their requirements for process performance in the dimensions of cycle time, productivity, and quality. This information, coupled with the results of best practices research, is used to develop stakeholder-driven stretch performance goals for use in creative brainstorming.

- *What ideas do we have about how to improve performance?*

Part 6, "Brainstorm Innovations"—we conduct "paradise visioning" where we seek to reinvent our business. This involves a series of creative brainstorming sessions followed by increasingly analytical "focus group" sessions. The final products from this step are "change actions" which document what will be done, by whom, when it will be done, and why we're doing it. These change actions drive the implementation of improvements in the final part of the methodology.

- *How will we turn ideas into results?*

As Cervantes said, "The proof of the pudding is in the eating." Part 7, "Implement Innovations"—is where results begin to be realized. This part of TPM addresses the activities required to implement the change actions generated during creative brainstorming. It incorporates practical methods for testing and rolling out innovations, and for tracking them to completion to ensure that expected results are achieved.

CONCLUSION

Organizations will be forever more in a perpetual state of change. For the first time in history, change is the norm with stability the exception. Unfortunately, we are only just now awakening to the reality that we must do more that react to change—we must weave it into the fabric of our businesses. As Ronald E. Compton, CEO of Aetna Life and Casualty, said: "Change is not something that happens. It's a way of life. It's not a process, it's a value. It's not something you do, it engulfs you."[6]

Leaders have the responsibility to find healthy ways in which to respond to change. Our current approaches achieve performance improvements that usually are not sustainable and which come at a high price in human suffering. The role of executive leadership is now emphasizing, more than ever, the ability to effectively shepherd an organization through stormy transitions.

Leaders must look to new approaches to change that recognize the indivisible, organic nature of their organizations. We must recognize that companies are like societies and not machines, and embrace change methodologies that reflect this reality. Successful performance improvement will quite literally require a *quantum* step toward holistic change.

AUTHOR BIOGRAPHY

Mark Youngblood is a principal in The Renova Corporation, a professional services and training organization headquartered in Dallas, TX that specializes in leading companies through all aspects of the change process. He is the author of the book *Eating the Chocolate Elephant: Take Charge of Change Through Total Process Management* and is a contributing author to several other anthologies. Youngblood conducts seminars and speaking engagements on the subject of Total Process Management and effective corporate change to audiences throughout the world. His clients include Chubb Insurance, NationsBank, Mary Kay Cosmetics, Frito Lay Corporation, Lexmark, Army & Air Force Exchange Service (AAFES), Union Pacific Resources Corporation, and Boy Scouts of America.

Youngblood joined Renova from Micrografx, Inc. where he served as vice-president of management systems responsible for information services and for implementing Total Process Management worldwide. Prior to that he was a senior manager in the Price Waterhouse Management Consulting Services Change Integration group. Youngblood is a Certified Public Accountant, Certified Managerial Accountant, and a Certified Data Processor.

[6]Rifkin, Glenn. "Reengineering Aetna," *Forbes*. ASAP, p. 81.

CHAPTER 14

The Wired World

by Andrew Forbes and Fred Seelig

Bandwidth is a measure of the ability of telecommunications facilities to transport information from an information source to an information sink. A high bandwidth system, like a cable TV or a fiber optic cable, can transmit information at millions or even billions of bits per second. A low bandwidth system, like the Plain Old Telephone System, or POTS, can transmit at a few thousand bits per second, thousands of times slower than a high bandwidth system.

Over the next ten years, bandwidth to consumers and businesses will increase by two to three orders of magnitude while continuing to cost about the same as current telecommunications service. The reality of this basic economic fact is going to reshape our society as surely, and as completely, as the Industrial Revolution reshaped the western world in the 1850s. Today we stand at the beginning of a new era. Future historians will call this the dawn of the Information Revolution.

But information is not the only important thing. The speed at which it is delivered is as important as the quality of the information itself. Imagine turning on your faucet and watching your sink fill in four seconds. Now imagine turning on the same faucet and watching a whole swimming pool fill up in the same time. And finally, imagine that all that extra water does not cost you anything extra! That is what high bandwidth service will give you: lots of information—fast—at the same rate you are being charged for your phone service or cable TV service.

CUTTHROAT COMPETITION

Until recently the telephone industry was a single utility, with all the lack of market competition and dynamism that the word "utility" suggests. The landmark 1987 U.S. District Court decision to break up the American Telephone and Telegraph (AT&T) company resulted in the creation of 22 Bell-operated companies (BOCs) organized into seven regional Bell holding companies (RBHCs), commonly called Regional Bell Operating Companies (RBOCs). In today's (1994) world, consumers have seen fierce long distance telephone competition. By the turn of the century there will be the same war fought for local telephone access service. The divestiture was the first shot fired in the Information Revolution.

Local access will require Congressional involvement. The 1934 Communications Act established by Congress in the Roosevelt administration defined the rules of the telephone game. This included oversight by the Federal Communications Commission (FCC) in such categories as rate structures, telephone interface, and service and provisioning. The current economic philosophy is driving Congress to allow for more free-market competition into local access communications.

The reasoning is as follows: Telephone companies are hampered in their ability to bring new technologies to market because their switching systems are amortized over 40 years. Yet the "half-life" of a new technology is somewhere between three and ten years. They do not want to commit to a new communications technology because of the rapidity of change of the underlying electronic components. Also, Silicon-Valley business leaders and Nobel-prize winning economists have said that federal government's tampering with emergent technologies will stifle, rather than stimulate, growth.

Congress and the Senate have gotten the picture: loosen the Fed's grip on the telecommunications industry, let new technologies emerge, let survival-of-the-fittest free-market competition decide which technology survives, and let rates float to the level which the market can bear. (Congress, of course, is not entirely altruistic in this hands-off approach, since it stands to gain a tax windfall with a newly revitalized, and very large, industry.)

THE PLAYERS

There are five major groups of players in the telecommunications game. Each of the groups has its own identity, strengths and weaknesses. Each brings to the table an existing infrastructure of equipment and corporate identity. Some groups are prepared to make the transition to the new telecommunications world, while others are not. Some will have to reengineer their companies, while others are unprepared to do so. We discuss each of these market players in turn, to give you an idea of where they stand.

The players are the RBOCs, the long distance carriers, the wireless communications companies, the cable television companies, and, perhaps somewhat surprisingly, the power companies.

RBOCs, as previously mentioned, are responsible only for local telephone service. They are currently precluded from involvement in long distance service. In the decision to break up Bell Telephone, eight new companies were created, the RBOCs and AT&T. The seven RBOCs were each limited to providing telephony services within a specific area, or region, hence the eponymous name.

AT&T was created from the old Long Lines division as discussed below. RBOC management, company philosophy, switching equipment, and home access circuits were largely inherited from the old Bell Telephone Company. It has only been recently that each RBOC's management team has seen itself in true competition with other RBOCs. The telephone company industry's pace is much slower than parallel electrical engineering industries, but this too, is changing as the reality of market competition is sinking in. A positive asset which RBOCs have is its top-quality switching equipment, and its massive investment in both telephony hardware and software. The RBOCs know communications technology down into their very bones.

Interspersed with the RBOCs are the so-called "Ma and Pa" Bells, small local exchange carriers that fill in the gaps where RBOCs do not provide service. Between the RBOCs and the Ma and Pa Bells, almost every home in the U.S. has the option of phone service. Ma and Pa Bells traditionally have had poorer service and poorer grades of equipment, which came as a result of being undercapitalized. These companies may be acquired by RBOCs. But they are generally not located in high-revenue communications-intensive areas.

Long distance carriers are also an amalgam of old Bell Telephone and independent companies such as MCI, Sprint, GTE, and WilTel. AT&T Long Lines became simply AT&T. It is responsible for overseas connections and major cross-country telephone trunks. The overseas connections are a hybrid of undersea oceanic cable (both metallic and fiber optic) and geostationary satellite.

AT&T is barred from competing in provision of local telephony services. No other company has as much market share as does AT&T in long distance service. MCI is a close second. The other companies have a combined 10% market share of the remainder. MCI and AT&T have a mixture of telecommunications equipment, switching equipment, and access, although none of their infrastructure is positioned to be used as a local service provider.

The feather in their cap is that they are accustomed to working with high bandwidth signals. As computer-to-computer communications and video on demand services increase, high-bandwidth technology will dominate. So, long distance companies can exploit their knowledge of high bandwidth technologies in the coming Information Revolution. It remains to be seen if they have the corporate resolve to do so.

Wireless communications providers include cellular telephone and paging companies. Although some are independent business entities, most are not. Some are owned by RBOCs, some by long distance companies. Their owners are gambling on the wireless market being a future source of revenue. Wireless access reaches two kinds of users: mobile and non-traditional remote users.

The mobile community is dominated by business users. Non-traditional users include farmers, who use cellular telephones from their tractors to radio back to the farmhouse; and schoolchildren in major metropolitan areas, whose pagers are used by working parents to coordinate after-school schedules.

How can wireless companies be expected to adapt to compete in the future Global Information Infrastructure (GII)? The two kinds of wireless providers are divided on this issue. The wireless cellular telephone companies have, and continue to increase, their presence in major metropolitan areas. They have switching and access infrastructure, and they connect to local and long distance companies. They are well-positioned to grow into GII-type companies. Paging companies are not.

They have the wrong kind of physical link structure geostationary satellite to earth surface, one-way only), the wrong kind of switching equipment (typically, not high-bandwidth telephone company equipment), and the wrong kind of information content (one or two lines of text, maximum). In short, paging companies are dead ends in the evolutionary tree of the global information network of the future.

Cable television companies' historical business roots are in one-way entertainment dissemination, not in two-way information transmission. Their genesis as humble earth-bound competitors of satellite Direct Broadcast Service (DBS) companies such as Home Box Office did not hint at what they are today: a potent force challenging local telephone companies for the future of local access to the GII.

What cable companies have is bandwidth, raw bandwidth. The ability to transmit 10 Mb/sec of data over a line into the house is a capability telephone companies wish they had. They also have access to a large percentage of homes. At the time of this writing, it is estimated that 65% of the homes in the U.S. have cable service, and 95% have the option of cable service. What cable companies do not have is, well, lots of things.

Cable companies do not have a reputation for reliable service. They do not have, in any sense of the phrase, a capable switching system, or anything remotely resembling a telephone company's central office (the last switching office before a call is routed to a user's home phone). They do not have two way information flow, nor does their equipment allow them to have home-out transmission. They do not have much penetration into the business market.

THE WIRED HOME

Almost every home in the U.S. is wired for A/C power. This means that the power companies own a right of way into almost every home in the U.S. They do not currently use these rights of way for telecommunications infrastructure, although they are investigating the possibility of doing so. There are two reasons that electrical power companies are regarded as players in the telecommunications world of the future. First is the obvious advantage of already-existing access to homes and businesses, as well as access through major metropolitan areas. This access is simply not available any more. Second is that optical fibers can be run in the same cable, or strung next to a high voltage power line, with no interference between the two signals.

Light signals in optical fiber are relatively immune to high voltage electrical lines. This means that optical fiber can be strung over the entire electrical grid, using the same access enjoyed by electrical companies and can be switched and run right into a home or business with no additional access considerations. Although power companies may seem like dark horse candidates to be future telecommunications leaders, the single huge factor called access makes them a very real and important player.

Each of the players: RBOCs, long distance carriers, wireless companies, cable companies, and power companies evolved to meet a specific set of restricted challenges. In all five cases, the challenges, and the solutions, were to one degree or another regulated by the U.S., state, and local governments.

But the U.S. government is rewriting the regulations. The net result will be to throw open the local telephony, long distance telephony, wireless, and cable television markets to all comers. Almost all the companies in all five groups would like to be active in all the markets that Congress is about to open. But there are hurdles that each of the players will have to clear to succeed.

The RBOCs must replace old wiring with new wiring. Phone service wiring that runs into most homes and businesses is six wire twisted pair (i.e., three sets of twisted pairs of wires per telephone line). Even with the most advanced compression and modulation techniques, twisted pair is capable of the transmission of a very limited amount of bandwidth, on the order of tens to hundreds of kilobits per second. Plain old telephone service (POTS) wiring is simply incapable of delivering the two to three orders of magnitude worth of increased bandwidth that will be required to transmit at video rates. To compete, RBOCs are faced with a major rebuild of their networks.

The cable companies wired up the U.S. with coaxial cable. The potential bandwidth of coax cable is several orders of magnitude greater than that of twisted pair. Unfortunately, the cable systems are broadcast systems. The cable television signal runs from the cable system hub out to all the homes the system serves. This is a one-way system. To summarize, current cable video systems have no switching and offer only one-way service. To compete, cable companies, like the RBOCs, are faced with a major rebuilding of their networks.

Since the commercial introduction of cellular technology, the wireless companies have built systems that provide coverage to a majority of people in the U.S. The problem that these companies have is somewhat analogous to the problems facing the RBOCs. But while the RBOCs can solve their bandwidth problems by installing more and better cable the wireless companies face a different problem. Spectrum is a finite resource.

There are new modulation and compression techniques that increase the effective digital bandwidth of a fixed amount of spectrum, but these new technologies are not currently in commercial use. Current cellular service can handle voice and low bit rate computer to computer data, but it is not meant for high bit rate traffic. Future cellular systems, which are capable of higher bit rates, would require replacing all base stations and handsets with new equipment. Thus, once again, to compete, the wireless companies are faced with a major rebuild of their networks.

The power companies are, where telecommunication services are involved, operating in a vacuum. They have no investment in existing telecommunications infrastructure, and they have no experience operating telecommunications services. If they want to be telecommunications competitors, they are faced with building telecommunications networks from scratch. They, more than any other group, are faced with not just a rebuilding, but first-level construction to create a network.

In some respects, the long distance companies are in an enviable position. They do not have a major investment in home and business telephony services infrastructure that requires a major rebuilding. On the other hand, they do not have a telecommunications path to every home and business either. They do, however, have money to spend and experience with telephony networks. This would make them inclined to purchase, or to partner with, one or more of the companies in the other groups. Their ideal partners would provide right of way, switching infrastructure, and a local business presence, while the long distance companies would provide capital and telephone systems experience.

The telecommunication industry is currently in the throes of merger mania. Since every player is faced with major capital expenditures to be competitive, each is trying to find partners to share expenses. After the mergers a time for consolidation and competition will ensue. Following that, we will see an industry-wide shakeout to three to five major telecommunications firms. Regardless of which firm "wins," the competition and deregulation will result in a high bit rate telecommunications system at low rates to telephones, televisions, computers, cellular phones, and to wireless devices of all kinds. The combination of high bandwidth and low rates will hasten the coming of the Information Revolution.

This idea has been addressed by technological visionaries over the last decade. Having articulated the vision of GII, is not the same as creating it. What force will propel the GII into existence?

The answer to that can be stated in a two-word phrase: marketing communications. Advertising pays a majority of the cost of producing newspapers and magazines. Viewers pay nothing to watch broadcast television situation comedies. Marketing communications subsidize almost every mass market method of content delivery. It is important to understand that marketing communications will pay most of the costs associated with building the GII.

BUILDING THE GII

Because all the major players understand that the GII will be financed by marketing communications, the GII will be designed to facilitate the delivery of marketing communications to consumers. The impact of that statement cannot be underestimated. It costs money—lots of it—to wire houses and businesses, to build fast, wide bandwidth switches, and to build in network intelligence for the GII.

The GII will allow information consumers to have access to an unimaginably rich variety of information. Consumers will want to get the information, they will even be willing to pay for it, but not if the rate has to be so high as to completely fund the building of the GII infrastructure. This cost can be shared by GII advertisers. Advertisers will barter their ability to advertise in a new medium, which allows them to choose a target audience very precisely and use pointcast advertising techniques, for the consumers' ability to get information at an affordable cost.

There will two kinds of GII businesses: access providers and content providers. Access providers will be the firms that are a consumer's gateway into the GII, and content providers will be the firms that provide the reason consumers use the GII. All the telecommunications firms mentioned previously will be access providers. They may dabble in content, but the main thrust of their business will be access provision.

The main source of marketing communication income for access providers will be in targeted advertising.

The consumer is targeted based on information the advertiser and access provider have about that consumer. Such information is apt to be indirect in nature. A consumer profile can be built from such indirect information. Building such a customer profile, and then building a database of customer profiles by region, income, etc., allows an access provider to pointcast his or her marketing communications more effectively.

How will the access provider learn about each consumer? The access provider will track every action that a consumer takes while on the GII. What channels are viewed? What did this consumer order in the past? With which companies did this consumer have transactions in the previous year? Because information about every activity of each consumer will be valuable, access providers will offer consumers bundled services.

Access providers will try to persuade consumers to buy their telephone, television, computer, and wireless (voice and data) services as a bundle. By bringing together all access points into the GII into one billing center, the access provider can track all the ways a consumer uses the GII. This allows the access provider to build a more complete picture of a consumer than if, for example, it were only to be involved in providing telephone, but not computer, service.

In return for this information the access provider will offer consumers savings on bundled services. The early networks will look like those shown in Figure 14.1.

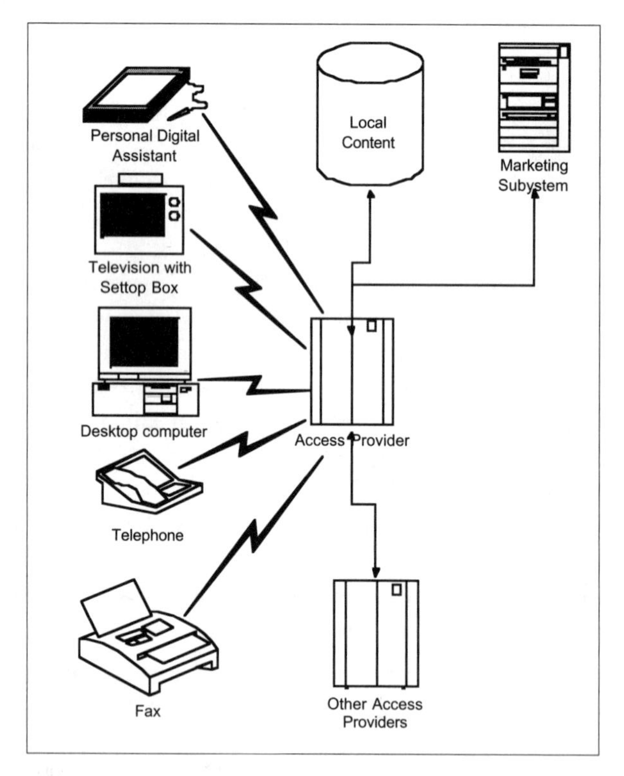

FIGURE 14.1. The early access provider network

Financed primarily by the income from presenting marketing communications to consumers, access providers will build and maintain the GII. A parallel effort will be occurring as content providers build infrastructure to offer up specialized libraries of information. Recall that content providers are information "sources," and the consumers are information "sinks," and that the GII is in between, conveying information between sources and sinks.

What is a content provider? To answer that question requires a brief review of the endpoints of the GII. Consumers will be able to interact with the GII in a myriad of ways. They will be able to use their television, their computer, their telephone, their cellular phone, their wireless personal digital assistant, and, if current research pans out, even their home thermostats and major appliances. In other words, almost any object that uses or generates data will potentially be on the GII. This suggests that the phrase "content provider" is a generic term for accessible data of almost any type.

Some content providers will make textual data and scanned photographs available, a virtual online library service. Some will sell audio/visual material (e.g., movies, old TV shows, music videos, news clips, archival speeches). Some will operate virtual reality multi-user multimedia games. Some will administer mailboxes that store and play back messages, regardless of the original form of the message and the current playback device. Some will sell houses, automobiles, and catalog merchandise. Universities will offer online, remote graduate school programs. Building management firms will adjust heating and air conditioning levels remotely, even into separate areas of the same building. Banking will be able to be done from home computers. As will stock transactions, airline reservations, hotel and business meeting reservations, teleconferencing, voting, and working.

Because marketing communications will build and maintain the GII, the cost of equipment to operate as a content provider will be minimal. In fact, it will be in the best interest of the access providers to encourage content providers to be on the GII. The larger the number of content providers, the more reason consumers will have to use the access provider's services.

Thus, the end state of the GII will be a telecommunications world of access providers and content providers. How is today's mélange of RBOCs, long distance companies, cable companies, wireless companies, power companies, and potential content providers going to transform themselves into the GII? The answer is with partnerships, mergers, and outright purchases of each other.

The telecommunications industry today has no integrated approach to telecommunications. The networks that provide these services are not integrated because they are usually not owned and operated by the same company. That is going to change as the industry moves to the concept of access providers.

Access providers will take a horizontal business model approach. When a set of consumers in the same region buys their telephone, television, and wireless services from a single access provider, the only path to those consumers for interactive marketing communications is through that access provider. Content providers will know that one access provider has information regarding all means of communications used by a particular consumer. Therefore they can be assured of a complete picture of the behavior and purchasing profile of that customer. This enhances the value of the access provider in the flow of information about the consumer. The owner of access will be the holder of the "keys of the kingdom." The partnerships, mergers, and purchases that roll across the telecommunications industry will be aimed at building firms capable of offering bundled services. Long distance companies will merge with RBOCs and/or wireless communications firms and/or cable companies and/or power companies. The result will be a handful of huge horizontal firms.

The U.S. telecommunications industry will eventually coalesce into three to five access providers. A logical question would be "how many access providers will there be worldwide?" The "Global" portion of the GII is not a meaningless word—the GII will be a telecommunications network that spans the globe. Other governments will want to control their citizens' access to the GII, much like citizens' passport control physical movement across national boundaries. Therefore, there will be at least one access provider for each country on the GII. So the total number of access providers worldwide will be somewhere between two hundred and fifty and one thousand.

The initial architecture of the GII will be many content providers connected to a single access provider, in a hub-spoke arrangement. This means that your access provider will be connected to local or regional content providers, such as local banking services, businesses, and video entertainment services. As time goes on, there will be continuing innovations in the manufacturing and storage of content. At first the access providers will build and maintain content storage systems, but as the cost and difficulty of content storage drops, independent content providers will start to come online, as shown in Figure 14.2. Each access provider will set up a network of content providers.

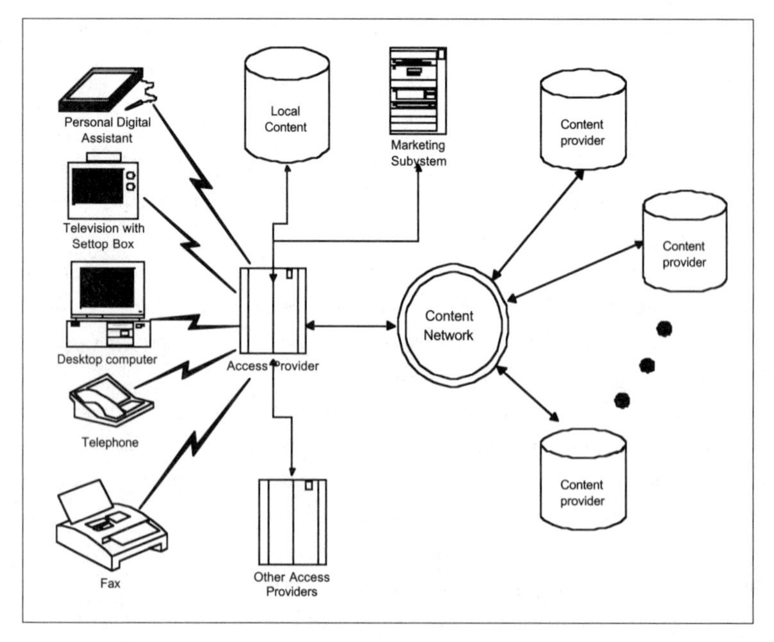

FIGURE 14.2. The content provider network

It will be in best interests of the access providers to have exclusive rights to a content providers content, but it will be in the best interest of the content providers to be on as many access provider networks as possible. Figure 14.3 is the first step towards the true GII.

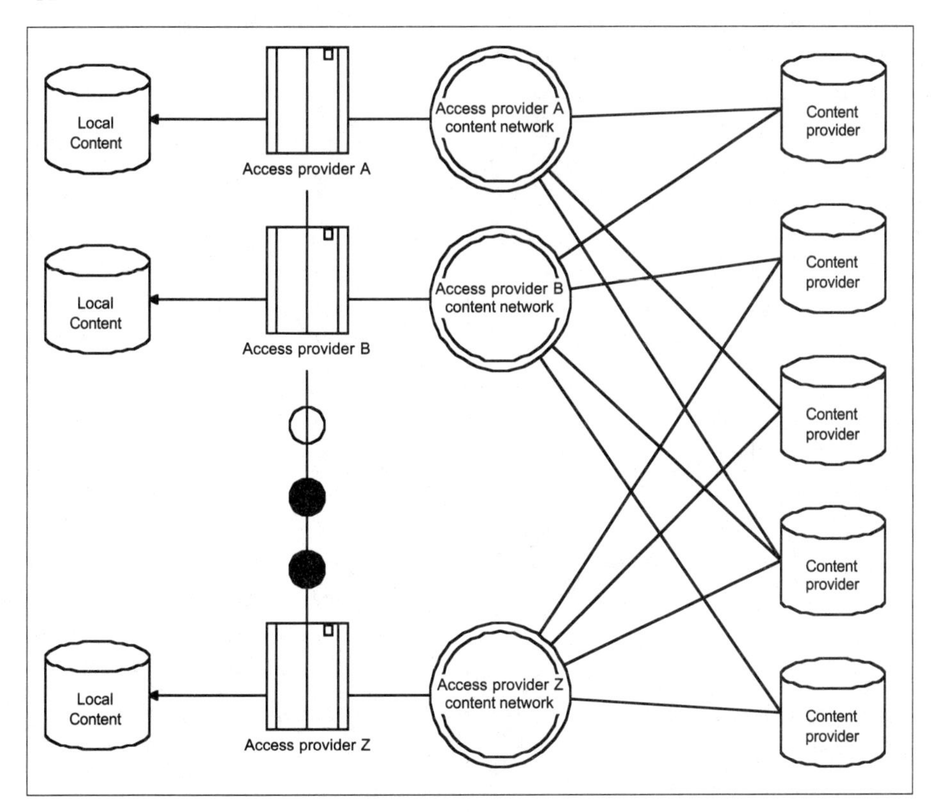

FIGURE 14.3. The initial convergence

Mergers will affect content providers as well as access providers. As the number of content provider networks shrinks to double digits and then single digits, most major content providers will be on all of the remaining networks. In fact, as the number of networks begins to stabilize, gateways between each of the remaining networks are likely to emerge. This will, for all practical purposes, give consumers access to almost all content on all networks. The access provision will become very transparent. The consumer will only be conscious of the wealth of information available, not the location of the source of that information. The Information Infrastructure will indeed become Global in nature, as in Figure 14.4. Everybody will have access to all information.

How many content providers will there be on the GII? Hundreds of millions of content providers will exist worldwide. It is estimated that there are currently more than fifty thousand computer bulletin boards in the U.S. So, there are already fifty thousand content providers in the U.S. alone. More importantly, there are fifty thousand content providers for a presentation channel (computers) that most people find confusing and difficult to use. As additional presentation channels (televisions, personal digital assistants, thermostats, etc.) come into popular use the number of content providers will go through a period of exponential growth.

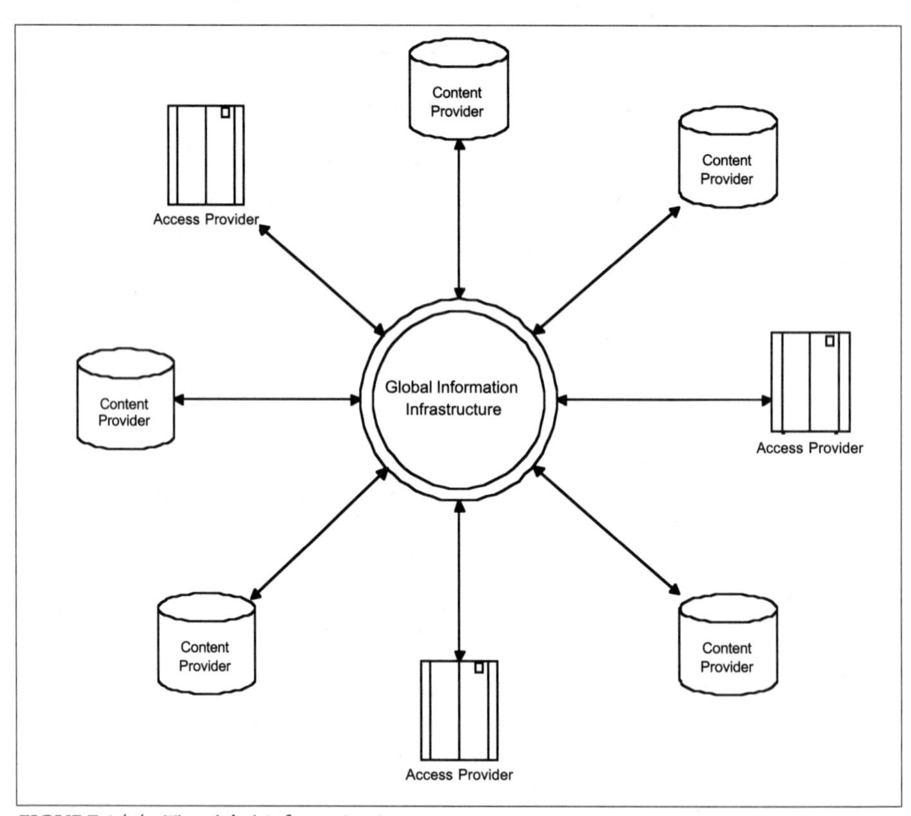

FIGURE 14.4. The global information instructure

THE UNIVERSAL BOX

To discuss content provision in detail requires taking a small detour to the idea of the universal "in box." A universal in box is an electronic device, physical or virtual, which is connected the GII and which can store incoming content regardless of the type or format of that content. Text, fax, still photo, audio, audio/visual, multimedia, should all be able to be stored and the content should all be available for playback, regardless of the playback presentation channel. If the presentation channel is a TV set, the contents of the universal in box are transmogrified to be displayed on a TV. If the presentation channel is a computer, then the contents of the universal in box are transmogrified to be displayed on a computer. A universal in box should store any kind of content for playback on any presentation channel.

Obviously, it will not be possible for a universal in box to convert all types of content for playback on all types of presentation channels. But it will be possible for most types of content to play back on most types of presentation channels. This is why the concept of a universal in box is important to any discussion of content provision. Content format, as a rule, will be presentation channel independent. Content providers will create content that can be used by as many presentation channels as possible.

Initially there will not be a content provider network. The access providers will set up content storage equipment and solicit content producers for content to sell to the access provider's customers. This means that the content available through an access provider will be presentation channel specific. It also means that early access providers will be involved in content provision. Movies and television shows will still be targeted for televisions, text and still photos for computers, and voice for telephones or portable handsets. As the cost of content storage equipment drops, content providers will start to set up their own content provision systems. This will be agreeable to access providers because it gets them out of the content business and lets them concentrate on access provision.

As the access providers begin to establish content networks, the early content providers will still create presentation channel specific content. Film studios and television sitcom producers will provide audio/visual content servers. Online computer services and bulletin boards will provide text and still photos. Phone companies will provide voice mail services. But outside forces at work are going to change the way content providers do business.

Presentation channels are going to start evolving at a furious rate. The functionality currently defined by the words "telephone," "fax," "VCR," "computer," "television," and "personal digital assistant" are going to merge, mix, and match in new and interesting ways. Some computers, such as the Apple Macintosh Aptiva, already have television reception capability built in. Television image capture is possible while you watch. Interactive set-top boxes attached to televisions make them more computer-like, as well. What is less obvious is that telephones are going to become more computer-like. Personal digital assistants are going to get personal communications services and become like phones. As interactive television comes online one of the first new services consumers will be offered is a fax in-box. Just as the phone company now operates "voice mail" boxes, in a year or two they will be operating "fax" boxes. Consumers will be able to view their in-bound faxes on their computer or television.

The presentation channel will become much less important than the content being presented. This means that over the next few years content providers will begin to concentrate less on the presentation channel and more on content, and on creating content that is reasonably channel independent.

As the access providers and the content networks continue to merge, the number of content providers available to consumers will grow dramatically. How are consumers going to find information on the GII when there are hundreds of millions of content providers?

Consumers will use software based intelligent agents. Intelligent agents will search the GII for information the consumer desires. Intelligent agents will filter the incoming data, limit the data to what the consumer wants to see, and format the data for the consumer's current presentation channel. The consumer's intelligent agent will interact with the access provider's marketing communication system to make sure that there is a match between the consumer and specific marketing communications.

THE INTERNET

Where does the Internet fall in all of this? The Internet and the emerging commercial content networks are currently at opposite ends of the spectrum. The Internet is currently an open decentralized content provision network with many access providers. The emerging commercial content provider networks will be, in their early stages, proprietary closed networks with a single access provider per network. Both have features attractive to the other that will eventually result in their merger.

The Internet already has, for all practical purposes, more content than any one person could use in a lifetime. The Internet is experiencing explosive growth at the current time, both in users, content providers, and sheer quantity of content. What the Internet does not have is a clear way for content providers to profit from the provision of their content. The Internet is currently a wide open freebooting environment where most users feel they should not have to pay for content, or for that matter, respect trademark, copyright, and libel laws.

The emerging commercial content networks will, at first, have a limited amount of content (compared to the Internet.) But from day one they will have the mechanisms in place to track content usage and reimburse content providers. Their systems will have centralized control of users, content providers, and the interactions between users and content providers. Most importantly, the commercial networks will enable the presentation of marketing communications, the main source of financing for the GII. The convergence of the two types of systems will be the true GII.

What are some of the first order effects that the GII is going to have on our society?

The GII is going to enable telecommuting. The same infrastructure that enables interactive shopping, video malls, and online ordering will enable video dial tone, and audio/visual teleconferencing, and wide area networks that reach into employees' homes and computers. Any job that does not require physical contact with co-workers or customers, or only requires an occasional face to face meeting, will be a candidate for telecommuting. Telecommuting employees no longer need to live near their employers. An obvious class of candidates for telecommuting are software developers. They already use computers comfortably, and they are accustomed to the concept of information networking and e-mail, so telecommuting will not be a major adjustment in work lifestyle.

The GII will impact the educational system. GII services will enable home schooling. Tele-education will allow children's education to be tailored to the needs of each and every child. Math and reading skills can be sped up or slowed down to meet each child's individual requirements. Children with Attention Deficit Disorder can be taught from their homes. Handicapped or disabled students will be able to receive a full day's schoolwork at home. Students can receive supplemental tutoring at home, as they need it. Tele-education will mean that students in remote locations will no longer face long daily bus rides. In fact, no child will face long daily bus rides. When a quality education exists on every television or computer screen, where the children are located when receiving that education becomes much less important.

The GII is going to normalize the standard of living around the world. High technology enterprises will have as their available labor pool the entire networked world, not just local talent. They will hire skilled workers regardless of physical location. This will be a dual benefit: companies will make use of a floating pool of globally available talent, and employees will be able to live where they want to, rather than having to live near their employer. U.S. companies are already making use of programmers in Russia, India, and southeast Asia.

Author Biographies

Andrew Forbes has been working with computers and telecommunications systems for 10 years. He received a BS in Computer Science from the University of Maryland, Baltimore County. He has worked on NASA, DOD, FAA, and commercial projects. He worked for ODSI where he developed spacecraft attitude display software for NASA. While working for PSI he worked on DOD projects, and then he went to work for Stanford Telecommunications, Inc. where he modeled Federal Aviation Administrations telecommunications networks, designed and built a GUI for a Differential Global Positioning System testbed, and designed a GUI for a system used to monitor and test

Intelsat's satellite network. Forbes is currently working as a consultant for firms that want to be "players" in the emerging Global Information Infrastructure.

Fred Seelig has been involved with communications systems engineering for 15 years. He received a B.S. in Electrical Engineering at Purdue University and an M.S.E.E. from Georgia Tech. He has worked on military, NASA, and FAA projects. He began his engineering career at Harris Corps Government Satellite Communications Systems Division in Palm Bay, FL, where he worked on a U.S. Army anti-jam spread spectrum modem. He has also worked at M/A-COM Linkabit in Vienna, VA, where he performed military satellite communications systems engineering analysis for various government agencies. He is presently a Senior Engineer with Stanford Telecommunications, Inc., in Reston, VA working on FAA air-ground radio engineering projects. His work has resulted in letters of commendation from the National Security Agency and the Federal Aviation Administration. Seelig is pursuing a doctoral degree in electrical engineering at the George Washington University, Washington, D.C.

CHAPTER 15

Teleworking

by David Helfrich

Teleworking is a growing business practice that is redefining the boundaries of the traditional workplace. At home, on the road, or from other remote sites, teleworking moves work to the worker, not the worker to work. Using telephone company services and remote access technologies, teleworkers can produce as effectively offsite as in the office even more so, according to results from a large number of pilot projects conducted across the United States.

Though environmental and governmental influences provided an early impetus for exploring teleworking, experience has proven that teleworking is simply good business. Teleworking not only helps attract and retain good employees, but it can also lower expenses to increase bottom-line profits.

As with any new business process, a successful teleworking program does not happen accidentally. Thoughtful consideration of the issues, good planning, and a solid implementation scheme are critical factors in creating an atmosphere in which teleworking can thrive.

When it comes to understanding teleworking issues, 3Com draws on three different types of experience: its internal teleworking program, its participation in the Smart Valley Telecommuting Project, and its technical understanding of connectivity issues as a leading networking company.

Internally, 3Com is building a global remote access network so that any 3Com employee can connect to corporate resources from anywhere at anytime. Teleworking is a major part of this effort. To that end, 3Com has recently completed a pilot teleworking project and is now implementing a broader teleworking program. The 3Com pilot was part of larger Smart Valley project involving many Bay Area businesses. Through the Smart Valley project, 3Com gained valuable insights into teleworking issues facing other companies.

As a networking vendor, 3Com views teleworking as business practice enabled by the personal office connectivity system, a key component of its High Performance Scalable Networking (HPSN) strategy. HPSN takes a holistic approach to networking. It views all functional areas of the network the building and campus backbone, the workgroup, wide area connections, remote offices, even personal offices as an integrated system. HPSN allows corporate customers to extend the full power of the network to teleworkers, and to other individual remote users through a connection strategy called personal office internetworking.

The aims of HPSN are to extend the reach and improve the performance of corporate networks by offering manageable future-proof technology serving the entire spectrum of network users around the globe. Studies show that the most successful teleworking projects take a similar, big-picture view widening the corporate umbrella to include the teleworker and evolving corporate policies and technologies accordingly.

This chapter looks more closely at the managerial, technical, and user-related issues of teleworking from such a standpoint. The intended audience includes corporate management, human resources personnel, and information services professionals. Throughout the chapter, the term teleworking implies performing normal business functions from a remote location on a regular basis, from one to three days per week.

WHAT IS TELEWORKING?

Teleworking is a business practice that allows employees to perform normal job responsibilities outside their traditional workplace. Teleworkers keep in contact with their supervisors and co-workers and gain access to corporate resources on an as-need basis using traditional telephone company services and remote access technology.

Software engineers, technical writers, and graphic artists telework to focus on challenging projects. They can download or upload files from servers, use electronic mail to ask questions of their colleagues, and even send their results to their supervisor's printer. Yet removed from the hubbub of the office environment, they can concentrate on the problem at hand.

Telemarketers, customer service employees, and others whose duties involve computer-centric telephone interactions can work as easily from home as in the office. To locate information or to file reports, they connect to the same corporate databases that they normally use over dial-up links.

For managers, teleworking provides an opportunity to complete time-consuming reports and proposals at home or after hours, thereby freeing up more time for supervisory functions while at the office. If a problem arises that requires their immediate attention, they can send and receive communications by phone, fax, voice mail, or electronic mail.

As defined in this chapter, teleworking does not include everyone who works in remote locations—only those who dial into corporate networks to accomplish their work, as the "tele" in teleworking implies. Teleworkers are those employees who fulfill their conventional job commitments from a remote location on a regular basis, from one to three days per week. Low-cost telephone services, high-speed interface devices, and sophisticated remote networking hardware and software allow teleworkers to maintain transparent electronic links to the corporation for two-way communication and complete access to critical corporate resources.

MOTIVATING FACTORS

While the concept of teleworking is not new, it is only in the last decade that large corporations and government agencies began studying its benefits systematically. Several key factors are now motivating companies around the world to view teleworking as a viable business practice.

Perhaps the most important reason for the increased interest in teleworking is that the technology is now available to support it cost effectively. The widespread availability of inexpensive telephone services and innovative connection technologies are making teleworking economically feasible.

Improved quality of life for its employees is a commonly cited reason why companies are exploring teleworking. Employees thrive on the flexibility that teleworking provides. Pilot projects completed over the last two years have shown an increase in job satisfaction, company loyalty, and productivity among teleworkers, with reduced stress. Businesses can use teleworking as an incentive to attract or retain valuable employees, even those who relocate to other geographic areas. Companies can use teleworking to expand after-hours customer service.

Teleworking can also cut overhead and real-estate costs by allowing teleworkers to "time-share" a common workspace. Teleworking offers an alternative for U.S. companies to comply with the 1990 Clean Air Act's Employer-Directed Initiatives. By this law, employers with more than 100 people within certain high-pollution, heavily populated metropolitan areas must increase average vehicle occupancy to their worksite during peak travel hours by 25%. Other companies are encouraged to comply on a voluntary basis.

International countries who participated in the Rio Earth Summit view teleworking in a similar light as a strategy for reducing greenhouse gas emissions.

Teleworking also provides an avenue for companies to provide meaningful work to employees taking advantage of the Americans with Disabilities and Family and Medical Leave acts.

TYPES OF TELEWORKERS AND CONNECTIVITY ALTERNATIVES

Teleworking is characterized by flexibility. This flexibility is most easily demonstrated by considering the different types of users that telework, the variety of remote locations that can accommodate them, and the alternative connection strategies available to them, as shown in Figure 15.1.

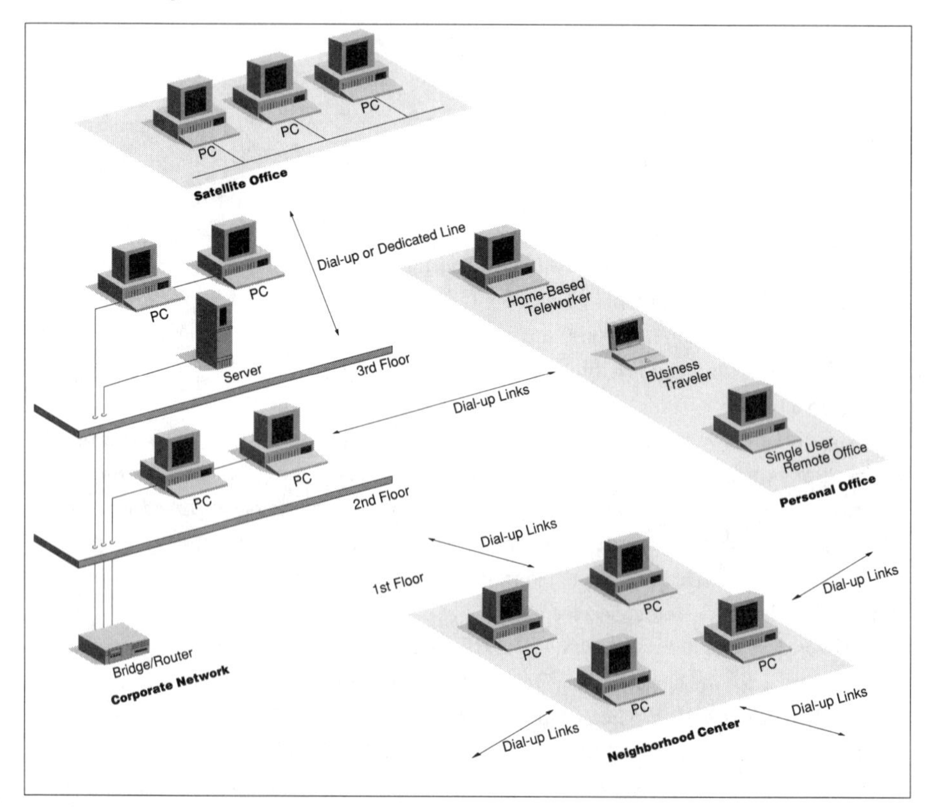

FIGURE 15.1. Types of Teleworkers

The most common teleworkers are full- or part-time employees who work from their personal offices at home a few days a week. These workers can be engineers, administrators, accountants, writers, artists, marketing personnel, customer support personnel, salespeople, and even managers. Just about any corporate function that involves information processing can be performed at home with the right equipment. Home-based teleworkers usually work from a desktop or laptop computer and gain access to corporate resources over standard or digital telephone services, depending on the amount of information they have to exchange with their colleagues on any given day and the phone services that are available to them.

When working at home is not a suitable alternative, some teleworkers perform their work from single-user remote offices, such as leased executive suites. Regional sales offices are often setup this way to allow salespeople to work outside of headquarters while maintaining pleasant accommodations for client visits and meetings. As with home offices, these teleworkers use portable or laptop computers and connect to headquarters via standard or digital dial-up connections.

Teleworkers can also be business travelers. From the airport or hotel, they use analog modems to dial into the corporate LAN from their notebook or laptop computers.

Teleworking can involve groups of workers as well as individuals. Two types of facilities can accommodate multiple teleworkers at regional sites.

Satellite offices are remote offices established by a corporation in a regional location close to employee homes. These offices offer many of the same amenities and resources as the corporate office, to which they are connected over dial-up or dedicated links.

Neighborhood centers are clustered worksites established by the community or by multiple organizations in outlying areas. They provide common work facilities on a time-share basis and offer business connectivity using available dial-up technology.

TELEWORKING ROI

The costs of a teleworking program must be weighed against the projected benefits to determine the overall return on investment.

Some benefits, such as less traffic pollution, have environmental benefits, but they do not directly affect a particular company's bottom line. The value of other benefits, such as increased productivity and decreased office space, can be quantified by looking at the results of actual programs.

Productivity Increases

The California Telecommuting Pilot Program reported 10-30% productivity gains from its participants. Hewlett-Packard's Response Center, which offers technical phone support to customers, reports that employees who work at home field 20% more calls on average.

US WEST has more than 600 employees who participated in its teleworking program. Their records show productivity increases as high as 40% for some employees. Customers say teleworkers provide better service than in-house employees. Managers indicate that working with telecommuters has improved their management skills.

Space Savings

In December 1993, *Business Week* reported that 5,000 IBM employees share space while in the office, and work at home, on the road, or at customer sites the rest of the time. IBM needs only one in-house desk for every 6-8 of these employees. Teleworking can help IBM eliminate almost 20% of its office space over the next few years, for savings in the tens of millions of dollars.

An AT&T spokesperson in the same article estimates that for every $1 invested in technology, they save $2 in real estate. JALA Associates, which has conducted several pilot programs with California governmental agencies, estimates that large-scale implementations of telecommuting policies could reduce office space requirements by 30%.

Estimating ROI

Taking such benefits as these into account, employers might be surprised at how quickly startup costs are recovered. Costs include:

- Identifying existing teleworking
- Projected cost/benefit analysis
- Selections
- Training teleworkers
- Training managers
- Counseling households
- Designing offices
- Furniture and equipment
- Legal costs
- Measuring and monitoring

Initiating a Teleworking Program—A Collaborative Responsibility

Until recently, one of the largest obstacles to the growth of teleworking has been corporate resistance, but the concerns are gradually being overcome.

Management fears have included loss of employee control, lower productivity, and increased cost, yet documented results contradict these perceptions. IS administrators have questioned the reliability of telecommunications technology and have feared an increase in security risks and network management headaches. Remote networking technology has matured to address all of these concerns. Human Resources managers have worried about the legal and administrative issues surrounding teleworking. The successful completion of numerous teleworking studies and pilot projects has paved the way for expanded guidelines in these areas. Employees have feared being cut off from their coworkers and managers and eventually being forgotten. Good teleworking guidelines can remove these concerns.

Effective planning can address all of these corporate issues directly and smooth the way to a successful teleworking program. Because teleworking affects many levels of the organization management, information services, human resources, and the individual worker it is important for everyone associated with the project to be involved in the planning. In this way, most issues come to light before the program begins, thereby reducing unwelcome surprises.

The following sections introduce some important teleworking planning issues to consider, grouped by corporate function. The bibliography at the end of this chapter lists other sources of information about developing and administering a teleworking program.

Management Issues

The success of any teleworking program depends heavily on the support and cooperation of management, from the highest levels to the departmental supervisors. Management must not only define the goals and limits of the project, it must also participate directly in monitoring the progress of the program, establishing strong communication, and adjusting those parts of the program that do not work as effectively as planned.

Once the program is under way, managers should communicate regularly with teleworkers to keep them in the departmental loop. The book Teleworking Explained suggests some ways teleworkers and managers can keep in touch, without overdoing it:

- Voice calls, for real-time interaction
- Electronic mail, voice mail, or fax for things that do not require an immediate response

- Scheduled calls to formal meetings
- Face-to-face in the office, when issues can wait until the teleworker returns

By minimizing the number of real-time interruptions, the teleworker can enjoy longer periods of concentration, and the manager might find his or her workday is less hectic.

It is incumbent on the manager to set clear goals and objectives for the teleworker and to measure the resultant productivity accordingly. In fact, management by objectives, rather than by visual monitoring, can benefit all employees, not just teleworkers.

Management Guidelines

Guidelines include the following:

- Establish a teleworking implementation committee
- Work with Human Resources, Information Services, and employee unions to develop written teleworking policies and procedures
- Define program parameters
- Analyze program costs and expected return on investment
- Work with Human Resources to select appropriate teleworkers
- Set specific performance objectives and develop evaluation criteria
- Define role of co-workers' interaction with teleworkers
- Communicate with and motivate teleworkers
- Establish regular meetings with teleworkers under supervision to monitor the progress of the program
- Measure performance against mutually defined criteria
- Make adjustments as necessary

Information Services Issues

From the IS perspective, teleworking can be viewed as another networking service. Rather than connecting local users to corporate resources over a LAN, teleworking involves remote users of varying levels of computer sophistication working from home on their PC of choice, from LAN-based facilities at a corporate satellite center, or from another remote location using a personal, portable or laptop computer. While teleworkers rely on dial-up or dedicated phone services to use enterprise network facilities, they expect the same level of networking service as when locally attached to the corporate LAN.

The first IS challenge is finding the right teleworking technology. For full productivity for both the teleworker and the IS manager, the right technology must accommodate all types of remote users, integrate easily into the existing infrastructure, support a variety of connectivity options, grow with the program, and be effectively managed and secured.

The second challenge is to gear the support staff to help teleworkers troubleshoot problems from their personal offices. While it might not be necessary to established a dedicated resource for the program, depending upon the size of the teleworking population, it is important that support staff members be trained to deal with teleworking issues.

Guidelines for IS are as follows:

- Assess equipment and technology needs
- Establish a written support and repair policy
- Publish troubleshooting contact list
- Educate technical support staff on teleworking issues
- Educate teleworkers on the best ways to report problems
- Develop and distribute technical reference materials
- Train teleworkers on the proper use of equipment and applications, including making file backups and checking for viruses

Human Resource Issues

Often the responsibility for teleworking program policies, training, and evaluation belongs to a corporation's Human Resources department. In addition, Human Resources must deal with teleworking issues regarding benefits, liability, and employee well being.

As defined in the section "What is Teleworking?", on page 206, teleworkers are considered regular employees and conform to standard corporate guidelines with respect to health benefits and worker's compensation. However, other areas can be somewhat fuzzier.

The best strategy is to be explicit about legal responsibilities, insurance responsibilities, and other issues affecting both the teleworker and the parent company. Setup clear guidelines for reimbursables, such as phone line installation, remote access and other telephone charges, electrical usage, and the like. Both the employee and the employer should understand at the outset all aspects of accountability.

Ergonomics play as important a role in work productivity at home as they do at corporate headquarters. Lighting, seating, work surface, and storage all have an effect. Human Resources can offer guidelines for setting up a home office and assist the teleworker in implementing the recommendations.

The following are guidelines that the Human Resources department should follow:

- Develop and disseminate teleworking policies and procedures
- Develop criteria to select participants and recruit participants
- Present teleworking orientation sessions
- Develop a written agreement for managers and employees participating in the program
- Train teleworkers and their supervisors
- Administer the pre-program evaluation
- Analyze and report results of the evaluation
- Research the legal and tax-related implications of teleworking and provide easy-to-understand explanations and disclaimers; review these with all participants
- Consult the corporate insurance carrier and provide complete explanations of policy restrictions for both the employer and employee
- Provide guidelines for establishing ergonomically correct remote offices
- Monitor the progress of the participants and modify policies and procedures as necessary

End-User Issues

Not everyone is suited to teleworking. Some employees rely on the structure of the workplace to help them to be productive; teleworking is not a good solution for these employees. The teleworker has to define his or her own work structure and have the discipline to stick to it. Self-motivated, well-organized individuals can adapt well to teleworking, although it sometimes can take them up to 90 days to find their optimum teleworking regimen.

Communication is key to a positive teleworking experience. It is common for teleworkers to initially feel isolated from their associates and co-workers, and to worry about losing their manager's mindshare. Good communication can alleviate both of these concerns. Working from home should not prevent a teleworker from sharing information with or seeking assistance from colleagues via electronic mail, fax, or telephone.

Teleworkers should also communicate the boundaries of their work environment to family members and friends. It is important to differentiate work time from family or social time.

One drawback of the convenience of teleworking is the opportunity for overwork. While it might be expedient to work longer hours some of the time, knowing when to stop is crucial for good performance in the long run.

Guidelines for the teleworker:

- Make certain that to understand management expectations and objectives
- Establish well-defined boundaries with family members
- Designate an "official" work area and secure the appropriate equipment
- Organize work that can be effectively performed at home
- Become familiar with the equipment and software
- Work independently
- Setup limits to avoid overwork
- Communicate regularly with supervisors and co-workers
- Produce results that are measurable
- Address concerns promptly
- Ask for help when you need it
- Give yourself time to adjust

THE 3COM TELEWORK PILOT PROGRAM

3Com began a pilot teleworking program involving 37 employees from its Santa Clara offices. The pilot was part of the Smart Valley Telecommuting Project, of which 3Com CEO Eric Benhamou was a co-director. Some of the Bay Area's most prominent companies and organizations—Silicon Graphics, Pacific Bell, Tandem Computer, Hewlett-Packard, Cisco Systems, Deloitte & Touche, Regis McKenna, Gray Cary Ware & Freidenrich, and Stanford University—conducted pilot teleworking programs as part of the Smart Valley project.

3Com's program was a collaborative effort between its Site Services, MIS, and HR departments, in partnership with line management and other staff organizations. 3Com's Telework Task Force included representatives from MIS, HR, Site Services, Finance, and Legal departments. To launch its program, 3Com began with a half-day orientation and training program for teleworkers and their managers. Each participant received a comprehensive *Teleworking Policy Handbook*, including the following information:

- Goals and benefits of the program
- Program overview
- MIS technical support standards
- Recommendations for ergonomics, safety, and security
- Legal, tax, and insurance ramifications of the program
- Tips for successful teleworking
- Tips for managing teleworkers

Workers and managers were encouraged to sign a Telework Agreement, documenting their understanding of the communication issues, expense issues, employee responsibilities, and human resources issues surrounding the program.

When the pilot project ended, the Telework Task Force issued a full report on the results of the program, including program successes and areas for improvement. Each of the 37 participants teleworked one day per week and saved, on average, six hours of commute time (152 miles) per month. Participants used that extra time to relax, be with their families, run errands, or perform extra work. The pilot teleworkers noted a learning curve for reaching peak effectiveness and reported that anticipating slow periods, organizing their time, and planning for the workday helped boost their productivity. Tasks requiring concentration and focus were the most adaptable for 3Com teleworkers. While maintaining good relationships with co-workers was an issue for many of the participants at the outset, those concerns faded as everyone became accustomed to the practice. In June, 3Com began recruiting participants for the regular program, whose target population is expected to reach 110-120 teleworkers by the end of fiscal year 1995.

MATCHING TECHNOLOGY TO FUNCTION

The role of technology in the teleworking application is to allow remote employees to work as productively as local workers. The trick is to find a solution that adapts to a broad population of teleworkers, performing a wide variety of different functions.

For example, software engineers might need concurrent access to UNIX-based file and database servers, development applications, and electronic mail. To work effectively, they need a solution that supports reliable, real-time, high-bandwidth data transfer.

Graphic artists working with multimedia applications or complex bitmapped graphics files might need similar computing and data-transfer power. Technical writers or managers, on the other hand, might perform most of their work from applications installed locally on their home PC. They might only need access to corporate resources once or twice a day, to check in or retrieve text files and send and receive electronic mail. Thus, their bandwidth and response-time requirements might be significantly less demanding than the engineer's. Accounting and finance teleworkers might need to update database records periodically on Novell or IBM servers to perform their work most effectively from home, while marketing and sales professionals might want occasional access to forecasting tools, reporting tools, and sales databases on Macintosh servers.

IS managers have two alternatives for resolving these need differences. They can purchase a different solution for each user—a management nightmare. Or, they can purchase a single solution that accommodates the differences automatically.

The goal is a transparent, high-performance network utility that extends backbone resources to all employees working outside the normal bounds of the enterprise.

In choosing a solution that works for all types of teleworkers, IS managers must address all of the core components of the teleworking application, shown in Figure 15.2.

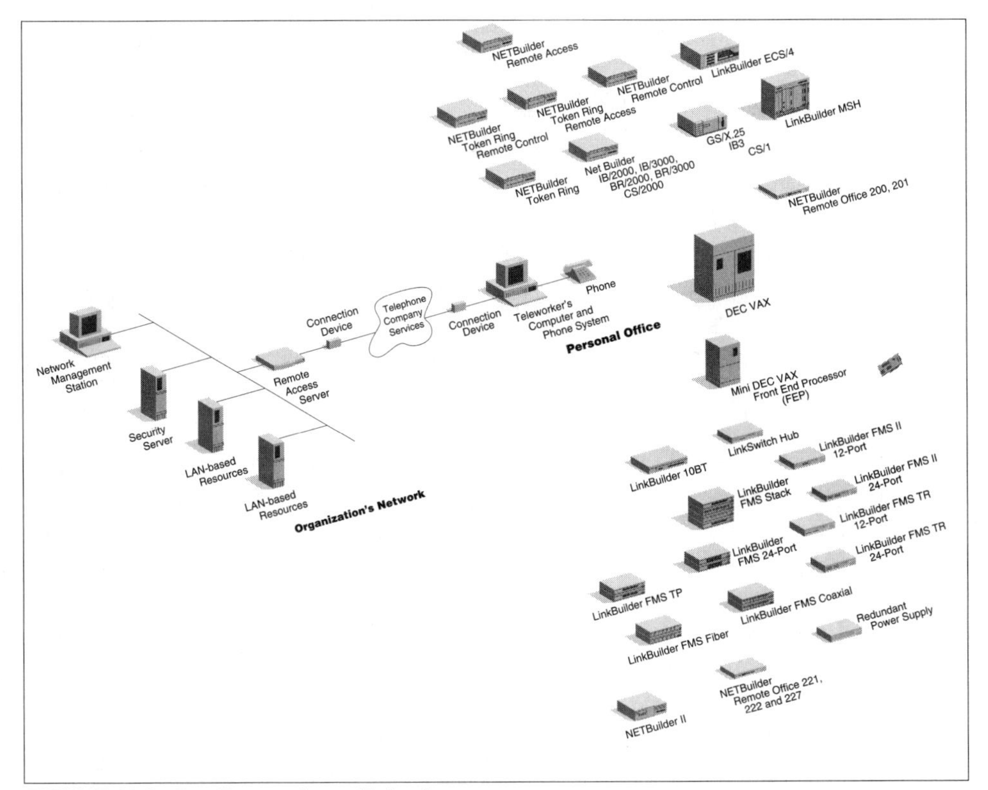

FIGURE 15.2. Core Remote Access Technology

Components of the teleworking application are:

- Teleworkers' computer systems and applications
- Telephone company services (type and availability)
- The remote access server that teleworkers dial into at corporate headquarters
- The existing corporate network infrastructure

TELEWORKER'S COMPUTER SYSTEM AND APPLICATIONS

Equipment needs of personal offices vary with the tasks each teleworker intends to perform. There are a few basic requirements:

- Telephone
- Communications link and modem (or other connection device)
- Personal computer and software
- Voice mail, electronic mail, and/or answering machine

The home office might also include a facsimile machine or fax modem and a printer. Community teleworking centers and commercial photocopying shops provide this type of equipment along with photocopying machines, scanners, and other devices on a usage-fee basis.

A teleworker might have a different computer system at home from the one he or she uses in the office. In choosing a home system, teleworkers should consider screen size, access speed, disc space, RAM capacity, and cost effectiveness. They must also address compatibility issues that might arise from sharing work across multiple computer systems.

For example, a development engineer might normally work from a UNIX-based workstation at the office but might select a high-speed PC with high-capacity RAM, high-speed drives, and CD-ROM storage as a workable alternative for a home system. Most personal computers can run software that emulates a wide variety of other devices. Software can provide terminal emulation, even X-Terminal emulation, from a Macintosh or PC. Conversely, UNIX-based workstations can emulate PC or Macintosh environments. To be most productive, teleworkers should be able to run the same applications seamlessly across both work environments.

A cost-effective solution for many teleworkers, especially those who travel for business, might be a docking station. This alternative allows the convenience of portability without sacrificing screen quality, disk space, and expansion slots.

In practice, the different types of computer systems that populate personal offices might be as varied as the teleworkers who use them. However, all teleworkers have one technical challenge in common gaining access to resources on the corporate LAN. If corporate resources are not fully available to these remote users, teleworking becomes a barrier, rather than an enhancement, to productivity.

To accommodate a heterogeneous and growing teleworking population, the IS manager should look for a remote access product that meets these key requirements:

- Works on a wide range of client platforms (workstations, PCs and compatibles, Macintosh)
- Supports a wide range of operating systems (DOS, Windows, Windows NT, UNIX, Macintosh)
- Supports a wide range of network operating systems (Novell NetWare), Microsoft LAN Manager, Windows for Workgoups, Banyan VINES), DEC PATHWORKS, IBM LAN Server, PCNFS)
- Supports a wide range of protocols (TCP/IP, IPX, AppleTalk, DECnet, XNS, NetBIOS, NetBEUI)
- Supports all available dial-up alternatives, as shown in Figure 15.3

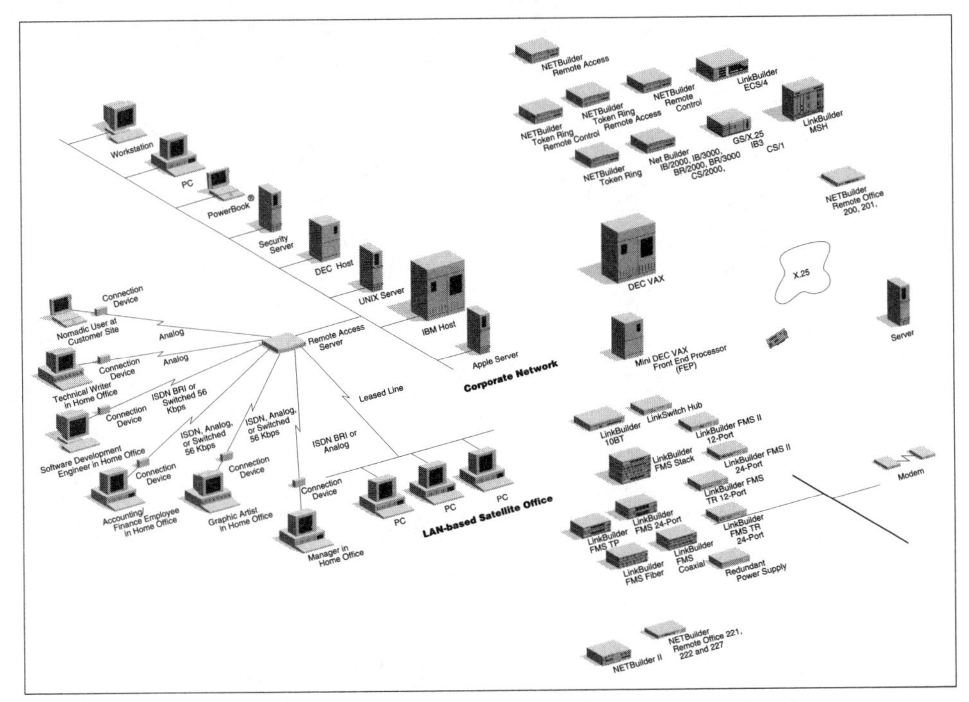

FIGURE 15.3. The remote access server should resolve the differences among teleworker environments

For maximum uptime, the client software should be easy to install, configure, and manage; the following section provides more information on ease-of-use issues. To support performance levels comparable to those in the office, the product must offer compressed data transfer over a wide variety of connectivity alternatives.

REMOTE ACCESS CLIENT SOFTWARE—KEEP IT SIMPLE

One of the most notable differences between working in an office environment and working at home is that teleworkers have much more direct responsibility for their computing systems and applications. Many teleworkers are unaccustomed to loading applications, connecting to networks, and logging onto servers it just seems to happen automatically at the office! Yet these are realities for the teleworker.

IS managers can greatly reduce the number of support calls from frustrated teleworkers if they choose a single solution that accommodates all remote users and has an easy-to-understand, graphical user interface. The software should be easy for even a novice computer user to install and should step remote users through dial-up and troubleshooting activities.

Teleworkers want their daily logons to happen as transparently as possible. Some remote access products support utilities that allow users to preload network logons to servers and printer connections. Thus, teleworkers can gain authorized access to the requisite LAN resources simply by dialing into the remote access server at the corporate site. For accounting purposes, the utility keeps track of the resources each teleworker uses and the duration of the session. Not only can such a feature help track the ongoing costs of teleworking, but it also helps network administrators evaluate the types of resources teleworkers need the most so that they can plan for growth.

Traveling teleworkers need a facility to store scripts for frequently dialed numbers that automatically configures the client-side software for the environment dialed into. In this way, the teleworker's system is always ready to connect to mission-critical resources, whether dialing from a personal office, hotel room, or the airport.

TELEPHONE COMPANY SERVICES

Four factors influence which type of phone service is most appropriate for a given teleworker:

- The amount and types of data to be transferred
- The frequency of data transfers

- The distance the data must travel from the personal office to the corporate LAN
- The types of services that are available from the telephone service carriers

While a decade ago the only affordable dial-up alternative for the teleworker was standard analog transmission, today switched digital services are widely available from public telephone network providers at attractive rates. Since teleworkers can choose the service that best suits their work requirements, the remote access server at corporate headquarters must be able to support different dial-up alternatives simultaneously.

Teleworkers who run most of their applications locally from personal-office computers and occasionally connect to backbone resources to read electronic mail or transfer text files can expect reasonable performance from a standard analog phone services. High-speed modems, such as the new V.FAST modems, are capable of transmitting raw data rates of up to 28.8 Kbps and compressed data speeds of up to 115.2 Kbps, depending upon the line quality.

Switched digital services offer better quality transmission, higher speeds, and greater reliability than standard phone services. While not a requirement for the average teleworker, high-speed digital dial-up links can provide the power user with the bandwidth necessary to perform large or bulk file transfers, repeated database transactions, and X-Windows client access to servers and hosts at the office. Two options are well suited to the personal office environment:

- Integrated Services Digital Network (ISDN) Basic Rate Interfaces (BRI) is a new digital phone service consisting of two 64 Kbps bearer channels (B-channels) for voice and data transmission and one 16 Kbps data channel (D-channel) for signaling. With the right remote access equipment, the B-channel bandwidth can be combined to offer 128 Kbps of uncompressed bandwidth.
- Switched 56 provides 56 Kbps digital data transmission.

The increased speed and reliability that digital services offer as compared to standard phone service can make a big difference to the teleworker who works with very large files. There are definite differences in transfer times for a 10 Mbps file transmitted by modem over standard phone lines, using switched 56 transmission, and using two aggregated B-channels from an ISDN BRI line as shown below.

14.4 Kbps modem	11.6 minutes
28.8 Kbps modem	5.8
Switched 56 Kbps	3.0
ISDN at 128 Kbps	1.3

Because it consists of three separate logical channels, ISDN BRI can support both voice and data transmission simultaneously over the B-channels, while data devices, such as PCs or FAX machines, use the packet-switched D-channel on a contention basis.

Switched digital transmission is charged on a usage basis. Thus, for many teleworkers, it is a cost-effective solution for getting high bandwidth at low cost. The following section compares the cost of standard, ISDN, and switched 56 dial-up connectivity.

Satellite and other remote offices that transmit steady streams of data to the corporate backbone might consider leased 56 Kbps and 64 Kbps lines. Leased lines are charged at a fixed monthly rate, regardless of the amount of traffic between locations. Once this link exists, it can be cost effective for personal office teleworkers to dial into branch or satellite offices at local access rates and allow a branch office bridge/router or a remote access server to pass the data the traffic along to the corporate enterprise over the dedicated link.

Some remote access servers can take this application one step further by offering dial-on-demand routing. In this case, if the traffic is destined for a different LAN than the one a teleworker dials into, the remote access server can dial the other LAN automatically, route the data, and then close the connection when transmission is complete.

REMOTE ACCESS SERVER AT CORPORATE HEADQUARTERS

One of the most characteristic aspects of today's corporate networks is their heterogeneity. A mixture of computing platforms, operating systems, network operating systems, and protocols provide a wide range of network resources to users of all levels of sophistication—software engineers, accountants, writers, graphic artists, administrators, management. The teleworking population usually consists of a cross-section of these users, who need access to the same types of resources when working from their personal offices as they do when directly attached to the corporate network.

The remote access server is the teleworker's conduit to the corporate resources. To be an effective interface between the telephone and corporate networks, the remote access server must support a wide range of both telephone services and computing environments. Teleworkers can expect reliable, high-performance access to the full range of corporate resources if the remote access server meets the following requirements:

- Provides reliable, high-performance transmission
- Supports all popular PC and workstation platforms (PC, Macintosh, and UNIX)
- Supports all major operating systems, network operating systems, serial line protocols, network protocols, and NIC drivers

- Bridges and routes information to and from users concurrently, so that teleworkers can run a variety of applications simultaneously (electronic mail, database, spreadsheet, word processing)
- Secures the backbone against unauthorized user access
- Integrates with centralized, standards-based management systems
- Can scale to larger populations as the program grows

The following sections look at some of these issues in more detail.

Performance Considerations

Teleworkers can connect to the backbone network at a variety of speeds, depending upon the type of telephone service they choose. But even the fastest of these dial-up technologies cannot compare with the bandwidth capacity of 10 Mbps Ethernet or 4 or 16 Mbps Token Ring networks available at corporate headquarters. Remote access servers must provide ways to compensate for these bandwidth differences to deliver performance levels to the personal office that are comparable to those local backbone users enjoy.

IS managers should look for two key features in selecting a remote access server that optimizes performance, as shown in Figure 15.4.

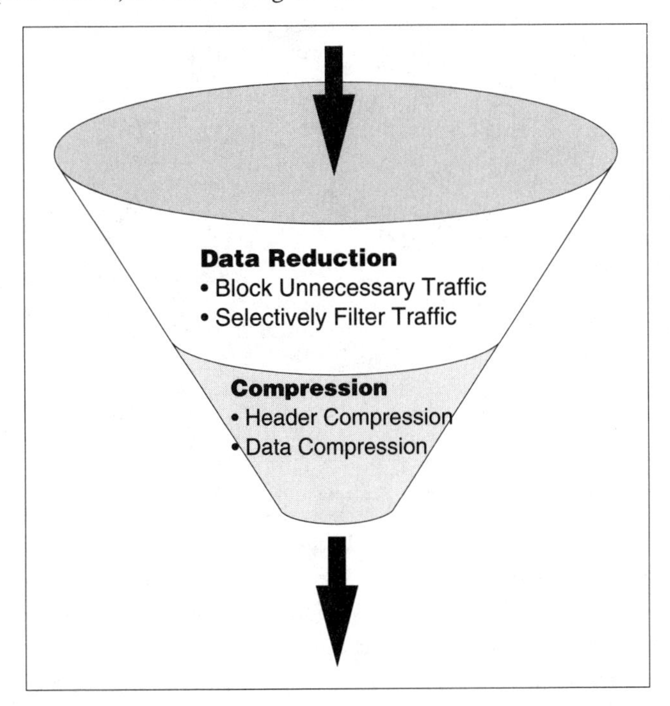

FIGURE 15.4. Optimizing bandwidth over dial-up links

These two features are:

- Filters that block unnecessary outbound traffic, so that no dial-up bandwidth is wasted
- Compression that squeezes data traffic into the most efficient bundle, to maximize available bandwidth

The combined action of these alternates can increase the effective throughput more than fourfold.

Security Considerations

One of the largest challenges facing IS managers adapting their network to teleworking is the potential security risk that dial-up access poses. While all of the most popular remote access servers today support security features such as password protection, user IDs, and automatic callback, these security solutions address only the narrow group of users dialing into each server. But the security requirements of the enterprise network extend way beyond the population of teleworkers (see Figure 15.5). A centralized, standards-based solution that addresses every user on the network—including the teleworker—is the more logical alternative for most IS managers.

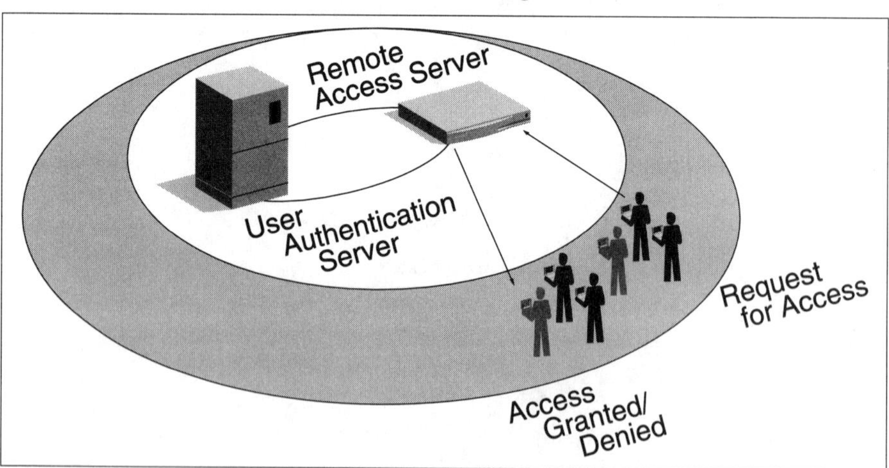

FIGURE 15.5. Securing corporate resources against unauthorized access

A centralized security solution allows IS managers to administer accounts and passwords, add and remove users, and manage the security of the entire system from a single location. In the unlikely event that a hacker intrudes, managers can move in quickly to disable the unauthorized access. Without centralized management, locating the source of security problems can take precious time that increases the risk of breach with each passing second.

Industry-standard security protocols, such as Kerberos and OSF/DCE, have been developed to allow centralized servers to control access to all enterprise network resources. Other security and naming databases are available from independent vendors to run on a wide range of platforms. In addition, companies such as Security Dynamics, Digital Pathways, LeeMah, and Millidyne offer external security devices that can authenticate large populations of remote users. Remote access servers that support all of these security options are now available.

Network Management Considerations

From a network management standpoint, the teleworker is just another network client and should be manageable using the same procedures as other users on the network. Support for standards-based network management is a key component for integrating the teleworker into the corporate computing environment.

To deliver on the promise of SNMP-based management, remote access servers must provide a fully implemented SNMP agent. A full SNMP implementation allows an IS manager to completely configure the server from an SNMP console or network management station, without setting DIP switches or jumpers.

Most network management platforms see each device on the network as a discrete entity. However, as networks increase in complexity, this approach faces serious limits. An integrated, enterprise approach to network management, such as 3Com's Transcend™, can simplify network complexity. By managing the network as logical groups of related devices, Transcend conforms to the way the network actually works and thus provides a truer assessment of network assets.

Scalability Considerations

Most teleworking programs begin with a small, tightly-monitored pilot and grow as interest and resources permit. Thus, scalability is an important component of a successful remote access server. IS managers want a product that can grow from four concurrent sessions supporting 128 teleworkers to 16 concurrent sessions supporting 512 teleworkers without having to remove equipment from racks, change chassis, or upgrade processing power.

As the program grows, flexibility and modularity become more imperative. A stackable solution lets IS managers interconnect a wide range of media types, LAN/WAN technologies, and user sites into an integrated, fault-tolerant system that can be centrally managed. As networks expand to take on more local users, remote sites, and teleworkers, stackable networking systems can grow the infrastructure easily, cost effectively, and seamlessly.

IS should keep the following issues in mind:

- Integrates with existing corporate network infrastructure Multiplatform, multiprotocol
- Concurrent bridging and routing
- Complete telecommunications support (standard analog, ISDN BRI, switched-56 leased line)
- Standards-based SNMP management
- Comprehensive security
- High-speed, high-performance access
- Easy-to-use client interface that accommodates a wide range of operating systems and network operating systems
- Scalable solution that can grow with the program

Risk Management

As with any new program, the safest and most successful changes occur in small steps. A phased approach to teleworking incurs less risk and amortizes cost over time.

A controlled teleworking pilot project can help companies discover what works, what does not work, and what issues need further study. By regularly monitoring the progress of the project so that issues can be addressed in a timely fashion, companies can tailor the program to best suit the needs of their corporate culture.

Once the pilot program has smoothed out the kinks in the system, the program can expand to incorporate more teleworkers. Controlled growth of the program can minimize the impact of changes and maximize the benefits.

Studies have shown that the best teleworking programs are setup on a voluntary basis. That is, no one is required to telework to fulfill job responsibilities. In this way, all participants are automatically predisposed to the success of the program.

SUMMARY

Teleworking is an application that is coming of age. The technology is available to support a teleworking strategy, and the success of a large number of pilot programs has provided insights into the application's managerial, technical, and human resource issues. Through careful planning, ongoing communication, and careful evaluation, companies can phase in successful teleworking programs to improve productivity, lower costs, and increase the quality of life of their employees.

GLOSSARY

Basic Rate Interface—An international standard switched digital interface offering two 64 Kbps bearer channels and a 16 Kbps signaling channel to carry voice, data, or video signals.

B-channels—The bearer channels in ISDN communication, which carry voice, data, or video signals.

D-channel—The data channel in ISDN communication, which carries control and signaling information to setup and breakdown connections.

High Performance Scalable Networking—3Com's strategy for providing enterprise-wide, high-performance, scalable, and global data networking solutions.

Integrated Services Digital Network—An international standard technology that delivers high-quality transmission and supports a variety of digital services over switched telephone networks.

Neighborhood centers—Clustered worksites established by the community or by multiple organizations in outlying areas to provide common work facilities on a time-share basis and business connectivity using available dial-up technology.

Personal office—A single-user remote office located in a home, a hotel room, a customer site, or any other remote location needing access to corporate enterprise LAN resources.

Personal office internetworking—Providing individual remote users access to corporate LAN resources.

Remote networking—Extending the logical boundaries of a corporate LAN over wide-area links to give remote offices, telecommuters, and mobile users access to critical information and resources.

Satellite offices—LAN-based remote offices established by a corporation in a regional location close to employee homes.

Switched 56—A digital service for transmitting data at speeds of 56 Kbps over switched telephone networks.

Teleworkers—Employees who fulfill their conventional job commitments from a remote location on a regular basis, from one to three days per week, using telecommunications and remote networking technologies.

Teleworking—A business practice that allows employees to perform normal job responsibilities outside their traditional workplace.

Transcend—An integrated management solution from 3Com based on groups of logically related devices and including a powerful set of management tools.

ABBREVIATIONS AND ACRONYMS

BRI—Basic Rate Interface

HPSN—High Performance Scalable Networking

ISDN—Integrated Services Digital Network

NIC—Network Interface Card

Kbps—Kilobits per second

Mbps—Megabits per second

SNMP—Simple Network Management Protocol

V.FAST—Draft standard for 28.8 Kbps modem technology

WAN—Wide Area Network

AUTHOR BIOGRAPHY

David Helfrich is vice president for 3Com Corporation's Personal Office Division (POD). He is a thirteen-year veteran of the data communications industry. He joined 3Com in 1994 as a result of 3Com's acquisition of Centrum Communications.

Helfrich joined Centrum in 1993 from Ascend Communications, where he was vice president, marketing. There he developed the strategic plan and executed the marketing programs leading Ascend to become the top inverse multiplexer manufacturer. Prior to Ascend, he held various marketing and sales position with Newbridge Networks, the leader in T-1 multiplexers, as well as Codex/Motorola and Infotron Systems.

Helfrich holds a Master's of Business Administration from the University of California at Los Angeles and a Bachelor of Science degree from the California State University of Chico. He can be reached at David_Helfrich@3Mail.3Com.Com.

CHAPTER 16

The Information Superhighway—Separating Hype from Reality

by Joseph M. Segel

It's been called many different things—

- The information superhighway;
- The electronic superhighway;
- The communications superhighway;
- The interactive superhighway;
- The multimedia superhighway.

Some call it a highway rather than a superhighway. But no one calls it a country road.

In fact, we are not talking about different concepts or different width roads. Call it what you will; it is all essentially the same thing. The experts just have not agreed on the terminology. So you will find the animal referred to by many different names in the press.

What we are talking about is a high-capacity two-way electronic pipeline to the home—capable of simultaneously transmitting a huge number of different television programs and other electronic services—with phone-like connectability between users and services, and between users and other users.

The Information Superhighway will be used to provide many different kinds of services—including wireless services. But its principal value will be to greatly expand what you can do with your television sets. Eventually that will extend to using the TV set as a videophone as well as a way to view your daily newspaper.

Digitization and digital compression are the key elements that make it all possible. The basic technology is here—today. But the practical implementation of this technology is still largely in the experimental stage, with lots of different concepts and variations being tested and refined. You're going to hear many conflicting opinions about how soon the information superhighway will be available on a significant scale, how it will look, what it will do, what it will cost and how well it will be accepted, I add my opinions to this din, and I warn you in advance that I have adopted a devil's advocate view regarding the practicality of much of what is being currently promoted. Of course, I cannot guarantee that my predictions will come true, no one can be sure at this early stage how it will all play out. So, with that caveat, I offer to you an update about what is going on, some food for thought, and some ideas about how you, as direct marketers, might prepare to deal with these evolutionary, and perhaps even revolutionary, developments.

Okay, let us now try to separate the hype from the reality.

Reality 1: Television Is a Very Powerful Medium

Over the past decade, television has dramatically changed America's living habits. 193 million U.S. homes—98% of U.S. households—now have television.

So television is virtually universal in the United States. About two-thirds of the homes have more than one set, and about the same number are connected to cable. The figures for Canada are roughly comparable.

Americans spend nearly half their free time watching television. According to Nielsen, the average TV set is on approximately seven hours a day, and the average American watches TV between three to four hours a day.

The only activities that take up more time than watching TV are work and sleep. At almost any moment every evening, over 11% of the U.S. population is in front of a television set. That's a massive market. Nearly all of these people are at home while watching TV, not at the mall. Of course, they are not all watching the same channel. This huge TV audience is made up of a multitude of moving targets.

Sight, sound, and motion combine to make television the most powerful means of mass communication. That unique combination makes TV the most effective—and the most cost-effective—medium for direct marketing. As has been proven by the shopping networks, TV can move mountains of merchandise, very rapidly.

REALITY 2: COMMUNICATIONS TECHNOLOGY IS MUTATING

Wired and wireless are switching roles. Conversations used to be transmitted mostly through wires. Now voice transmissions are increasingly being sent over the airwaves. Television used to be transmitted mostly over the airwaves. Now video transmissions are being sent mostly through wires and cables.

New services are constantly being developed for wireless, but the big concentration of research effort is focused on maximizing the capacity and versatility of transmissions by wire and cable.

Everything is moving toward digitization. The most dramatic change now just getting under way—is converting phone and television transmission systems from analog to digital.

Digital signals offer many advantages over analog signals—including greater precision, less distortion, elimination of ghosting, more special effects, and more opportunities for two-way communication.

Digitized voice, data, and video can all be transmitted the same way, over the same lines. That is one of the key reasons why the television, telephone, and computer industries are all doing mating dances with each other.

Digital compression will revolutionize communications. Right on the heels of digitization is digital compression. Digital compression is analogous to what orange juice processors do when they remove the water, ship only the pulp, and then the water is added back in at the consumer's home.

Recently-developed computer chips automatically remove the digits that represent the repeated parts of successive video frames. To restore the complete picture, related chips in the receiving set put the redundant digits back in. A single analog TV channel takes up a fixed slice, or bandwidth, of the government-rationed frequency spectrum. With digital compression, up to 10 channels can be transmitted within the same bandwidth.

Eventually, all new TV sets will have built-in digital reception and processing capabilities. Before that happens, there will be a new generation of converter boxes that will process digital data and convert it to analog signals that are viewable on present-day TV sets.

Fiber-optic cables exponentially increase capacity. In the meantime, cable companies and phone companies are racing to rewire the country with fiber-optic cables.

Fiber-optic cables can carry thousands of times as much digitized information as standard telephone wire or coaxial cable. It is claimed that a single fiber-optic cable can transmit the entire contents of the Encyclopedia Britanica in one second.

There are, however, two bottlenecks—(1) the lower-capacity wires that run from the pole to the home, and (2) the wires within the typical home. This is much more of a problem with phone wires than with coaxial cable. The full capacity potential of fiber-optics will not be realized until the far future, when those last few feet are rewired. Meanwhile, combining fiber-optic trunk lines with phone company switching equipment and cable company home connections will enable consumers to tap a substantial portion of the increased capacity and also to enjoy improved transmission quality and reliability.

HDTV is coming and will increase TV viewership. The next step will be HDTV—High Definition Television. HDTV is now being transmitted on a limited basis in Japan. Standards for the U.S. are due to be approved by the FCC next year. Within a few years, HDTV should get rolling in this country, paralleling and merging with the digital revolution.

HDTV will make large TV pictures almost as sharp as a photograph. You will see much more detail, with better colors, so it will be more pleasurable to watch television. The result will surely be a further increase in the amount of time that the average person watches TV.

Reality 3: The Television Landscape is Changing

For the past decade, there have been more cable networks than channel capacity to carry them. Channel scarcity has been a big problem. Within the next 3 to 5 years, that situation is likely to be turned upside down, with far more channels than programming available.

Channel scarcity will thus gradually change into channel abundance, as capacity opens up, the major occupiers will be:

Video on Demand

"Video on demand" services are likely to be the first—and eventually the most prolific—users of the Information Superhighway.

You will be able to select a movie from a large library of old and new movies and have the movie start on your TV set at a time of your choosing. That should certainly be an appealing option for many consumers, especially those who now go out to a store to rent videotapes. Several competing video-on-demand systems are in an advanced stage of development and testing. There are still some rough edges to be worked out, but video-on-demand is definitely coming.

Video Malls

There will surely be an expansion of televised shopping channels. Macy's, Nordstrom, and Spiegel have already announced their intention to enter this business. The more big names on television, the more people will begin shopping this way.

Special-interest shopping channels will also be tested, and some may be successful. Right now, for example, QVC is like a department store where only one department is open at a time. In the future, with multiple channels, it is possible that QVC will be more like a mall, where you can click on a QVC sporting goods channel or a QVC photo equipment channel, and so forth.

I emphasize the word "possible," not "definite," because it has yet to be proven that such specialized channels can be produced profitably. The cost or running a shopping channel is far greater than most people realize.

Games and Voting

Several companies are developing interactive games for the new television environment. In the future, kids will be able to play Nintendo-like games against a central computer or with other viewers across the network—that is, until their parents get the communications bill. However, downloaded games may partially solve that problem.

Voting will not take up much channel capacity, but it is another service that will be offered in the future. It is possible that voting from your home by clicking on a remote control will expand from voting on issues to actually voting in elections.

Reruns Rather than Original Programming

What you are not likely to see is much expansion in the way of elaborately-produced entertainment programming. It costs too much—especially in a new environment where the ratings will probably be lower as a result of audience fragmentation.

Educational Programming a Sleeper

Very little has been written about the possibility of using expanded channel capacity for educational programs—probably because there is not much commercial potential in it.

Consider the possibilities. If only it could be financed, we could put the most talented teachers and professors on TV, with channels categorized by subject matter. We could have educational programs for every grade level, and even a television university. What a marvelous way to uplift the educational level of great numbers of Americans.

Sadly, the teacher unions would probably consider it a threat and oppose it. Also, the viewership might be dismal. But it would be nice to see it happen, and it may happen.

It was reported last week that Michael Milken is interested in personally financing such a public service project.

Special Interest Niche Channels

A number of special-interest niche channels that have already started—such as The Travel Channel, a food and cooking channel, a comedy channel, a couple of game-show channels, a couple of live courtroom channels, a computer channel. Others are in the wings, just waiting for channel capacity to open up.

REALITY 4: CHANNEL ABUNDANCE IS A MULTI-EDGED SWORD

Reaching more narrowly-targeted audiences will be possible. After channel capacity expands, it should be easier and cost less to buy time on television. So, for many products and services, it will be possible to reach a narrowly-targeted audience.

For example, consider the proposed Golf Channel. Obviously, if you're selling golf clubs, the viewers of that channel are the people you want to reach. TV should be less costly than direct mail, because you do not have the cost of paper, ink, and postage and you may reach many previously unknown prospects. On the other hand, the number of people tuned into a particular channel when your commercial runs is likely to be a small fraction of the number you could reach by direct mail: it is a trade-off.

Audience fragmentation will be a problem for all channels, especially the general-interest channels.

As the Information Superhighway grows in breadth, the increased number of choices will seriously fragment the total audience, creating a viewership problem for all channels and a ratings problem for those who sell advertising.

The major broadcast networks deny the danger. But as viewing choices increase, ratings for traditional broadcast networks and local TV stations will probably continue to slide downhill. If that happens, reductions in advertising revenue are likely to follow. Some broadcasters will undoubtedly be resourceful enough to find ways to survive and prosper. Those who do not adapt will surely become casualties of progress.

500 channels are likely to evolve into one channel. People have enough trouble choosing among 50 to 100 channels. It would take far too long to scan several hundred channels.

The way the press has focused on the number "500" is unfortunate. It is not a magic number, and it has no solid foundation. With digital compression, there may be room for 400, 600, or 1000 channels or more. The point is that many different signals can be run through the electronic pipeline at one time, not that there will actually be 500 channels to scan.

A future television remote control is likely to look like a Nintendo controller, with a joystick and two or three buttons. The joystick will move a cursor on the TV screen. The main button on the remote control will select what the cursor is pointed to.

The so-called "500-channel" system will essentially evolve into a one-channel system—your personal channel. When you turn on your TV set, stored intelligence will display an abbreviated menu offering a selection of the specific channels you normally watch, and the specific electronic services you normally use.

To access the other networks, programs or services that are available, you will probably have to go through one or more sub-menus.

The Information Superhighway will offer a cornucopia of services. The key questions are: Will consumers want these new services? And will they pay for them?

REALITY 5: THE NEW SERVICES MAY BE DIFFICULT TO SELL

Many of the concepts being promoted are solutions in search of a problem. Where is the problem?

Everyone knows how to turn on a TV set and—with one or two clicks of a remote control—how to select the channel they want to watch.

Everyone knows how to dial a phone to call an 800 number and place an order. So, present-day systems are about as simple as they can be. People are very comfortable with the way TV and phones work. If a new system requires more steps to do essentially the same things, consumers may resist it.

Creative geniuses must not forget that people always gravitate towards doing things in the easiest possible way. Consumers are generally satisfied with the range of choices now on cable TV. Fairly or unfairly, the main complaint with cable TV is not lack of choices, but the cost of service.

If new services are very easy to use, very inexpensive, and very appealing, consumers may buy them. Focus groups and limited-market tests seem to indicate that is the case. But there is no crying demand on the part of consumers for additional services or for interactive TV.

It is not consumer demand, but, to a large extent the specter of having a huge number of channels available and not enough programming to put on them, that has spurred the communications and entertainment industries to develop new services. It should be noted, however, that a corps of very bright businessmen—such as Barry Diller, John Malone and Bill Gates—who are totally convinced that interactive services will be gobbled up by the public, in big numbers.

Text and non-moving pictures will not satisfy the public. Moving images are more appealing than still images, and listening is more appealing than reading. Those are two powerful reasons why every text-based and still-picture based interactive television experiment in the past has failed. That is also why no computer terminal-based service has ever generated anywhere near the volume of retail merchandise orders that the televised shopping channels have achieved.

Few people will be interested in a television interface that reminds them of a computer. To gain maximum acceptance, the visual appearance of interactive services needs to be more television-like than computer-like.

Ideally, an interactive shopping program should produce full-motion demonstrations of the selected products, but doing that in a practical and economical manner is yet to be developed. Television viewers are naturally passive. Some system developers and their software programmers seem to assume that consumers are itching to be converted from passive to interactive TV watchers. Certainly, there are people who fit that mold. But I venture to say that the great American public prefers to just watch television and let someone else do the work.

Being computer-literate does not change human nature. Our younger generations are more comfortable using computers than older people. But that does not mean that they would prefer to do something with a computer-like interface that they can do in a more natural way. Ordering merchandise with a computer-like interface is not as much fun as playing games.

Although I say that the new services may be difficult to sell, I believe that because so much is being invested in the information superhighway, the telecommunications industry will also invest whatever is required to eventually sell the new services to a substantial number of homes.

Up to this point, I have focused mostly on what I perceive to be the realities of the situation.

WHERE'S THE HYPE?

Two things are being promoted that I believe are inherently impractical and not likely to get off the ground:

Unreality 1: One-Button Merchandise Ordering

Almost all the simulated demonstrations of futuristic interactive TV systems assume that you will be able to actually order merchandise by simply pushing a button on a remote control. In my view—and I put it very bluntly—ordering merchandise that simply is a pipe dream. It's easy to produce a simulation on a demo tape, but that ignores many of the steps that are required to reliably process and fill an order in the real world.

Inventory issues. First, to serve the customer properly, if the item is sold out, you should inform the customer right away—not with a back-order notice several days later. On the other hand, if the item IS in stock. You must be able to instantly assign that piece to the customer and remove it from available inventory. Otherwise, you are going to have a disappointed customer who knows he or she does not have to put up with such problems.

Database access and compatibility issues. Unless you are able to instantly access all the computers of all the direct response vendors who are likely to advertise on the Information Superhighway—on a real-time basis, with compatible software—you are not likely to be able to match the kind of service that customers normally get by calling an 800 number. When you call an 800 number, you are usually connected directly to an operator who has instant access to the merchant's inventory and database.

Credit card approval issues. Next is the problem of credit card approval. Most of these proposed new systems assume that the customer is willing to place his or her credit card number in a master database and is willing to pay for any merchandise that is ordered by someone pushing a button in his or her home. I think consumers will be instinctively reluctant to do that.

Credit card fraud issues. The possibilities for disputes and fraud are mind-boggling. A PIN number is not an ideal solution, because that starts complicating the procedure, requires verification, and is not likely to be as secure in a home as it is at an ATM machine. Think about a child watching Mom order something and then deciding to do the same thing. A skilled order entry operator can tell when a child is on the phone. A machine cannot.

Other potential problems are card expirations, credit turndowns, and other credit card problems. From a customer's viewpoint, such problems are best resolved by an order entry operator who is able to get a real-time response from the credit card clearance service while the customer is still on the phone and the order is being processed.

A partial, but not complete, solution would be to include a built-in credit card reader with each converter box or future TV set. That may come. But that makes the equipment more complex and expensive.

Customer service issues. Additionally, many situations arise where the customer has a question about the product, about the color or size, or wants the product shipped to another address, or wants expedited delivery, or has one of a myriad of other things in mind that can only be resolved by talking to an order entry operator.

My view, therefore, is that ordering merchandise by simply pushing a button on a remote control device is inherently impractical and is thus not likely to ever get off the ground—despite the tens of millions of dollars being invested in developing such systems.

Unreality 2: Conversing with a TV Set

Talk to a television set, and it will understand you. A wonderful dream, but nearly an impossible reality. If—and it's a huge if—it were truly practical to talk to the TV set, and the TV set would understand EXACTLY what you want—as simulated on one of the recent interactive TV simulations—that would be a fantastic service. It could be a great success, unless of course it cost too much.

Reliable voice recognition by computers is, in my opinion, nearly an impossible dream. There is a limit to what can be done with software. Now, that is a pretty strong statement. Considering the enormous progress we have seen over the past decade, one should never underestimate what can be done with computers and software. Nevertheless, there are limitations. In fact, computers cannot do many things—and computers will never do certain things as well as human beings. Understanding speech is one of those things.

Speech variations and syntax rules are deadly. I maintain that cognitive, broad-based, unstructured speech recognition by computers is a fantasy because of the infinite variety of accents, dialects, inflections, vocabulary, syntax, idioms, and other elements that make up human speech.

Anyone who has experienced any kind of voice-recognition system with a phone, a remote control or even a powerful mainframe computer knows what I mean. You have to train the machine to recognize certain words spoken in your normal voice, and even then you have to follow certain rules of syntax.

Once again, I'll be blunt. I say that there is no software now existing—or likely to exist in the foreseeable future—that can understand the infinite varieties of human expression as well as a halfway-intelligent human being. Some companies, such as AT&T, have made remarkable progress with voice recognition. But unless it works easily, and relatively flawlessly, it will frustrate people. Just a little frustration is enough to turn people off.

Voice recognition is not the same as voice response and voice recognition should not be confused with computerized voice response—or "VRU" as it is called. There are many direct marketers, and QVC is one of them, who have successfully introduced VRU ordering systems—not voice-recognition, but voice-response systems where you can place an order with the keypad on your phone.

That system is precise. Computers can recognize the twelve tones generated by the standard phone push buttons flawlessly. Admittedly, many people groan when they run into a VRU or a voice-mail system. Often they just hang up. But a growing minority are getting used to it. Some repeat customers actually prefer to order that way because it can be done pretty fast, and you do not need to wait on the line if all operators are busy.

Computerized voice-response is not a one-button process, however. It is not even a five-button process. To order things by simply pushing one or two buttons on a remote control or by talking to their TV set is, in my judgment, oversimplified and woefully impractical. I am not against it. I would really like to be proven wrong and use those systems myself. But I just do not think it will happen the way it is currently being demonstrated.

There is a way that I believe it could work—after phone and cable services are merged. Use one button on the remote control to automatically dial-up the 800 number of the merchant whose product is shown on the screen at that moment and then utilize the future telephonic function of the TV set to order by voice or with a keypad on the remote control.

CONCLUSION

Nothing is going to happen overnight. The Information Superhighway is NOT just around the corner. It is going to be a while—probably several years—before the Information Superhighway is finally paved for general use. It will take longer still for it to attract a significant amount of traffic.

You will continue to see many different pilot programs and tests with small numbers of homes. There will be several different and incompatible systems that are likely to confuse more than inspire the consumer. With so many different players and conflicting interests, I think it will be several years before the smoke clears.

The recently-announced merger of TCI and Bell Atlantic could accelerate the process by enabling that highly-focused company to establish de-facto standards that others may be willing to follow, however. Then again, Time-Warner is well along the way in establishing a substantially different system, which will be tested next year in Orlando. There is no need to bet on any one interactive system. It is too early to waste much time or energy trying to determine which interactive television system will end up being dominant. Perhaps none will dominate. It is all guesswork at this point in time. I would suggest that you wait until the smoke clears.

As with most exciting new technology, there is an early ramp up of expectations, which is where we are now, with reality lagging far behind. After a while—in this case I would say in a year or two from now—people will get tired of waiting for miracles, and the expectations are likely to have a sharp fall-off. Then you will not hear so much about the Information Superhighway. It will be yesterday's news. The technology will continue to move forward, and eventually the reality is likely to cross over and surpass the expectations. Perhaps in an entirely different way than now envisioned. It is still very early in the game.

AUTHOR BIOGRAPHY

Joseph M. Segel is the quintessential entrepreneur. During the past 30-plus years, he has founded more than a dozen companies in businesses as diverse as publishing, aviation, photography, minting, software testing, and televised home shopping.

Retired in January, 1994 as Chairman of QVC Network, Inc., Segel is active as a Board member and retains the title of Chairman Emeritus. In the past, Segel was best known as the founder and Chairman of the Franklin Mint Corporation, one of the world's largest direct marketing organizations.

Established in 1986, QVC is now the largest cable-exclusive televised-shopping network in America, and its popularity is increasing every day. The Network's original channel, The QVC Shopping Channel, is broadcast live, 24 hours a day, and reaches more than 42 million cable homes nationwide.

CHAPTER 17

Building a Network Infrastructure: Managing the Corporate Network

by Lionel de Maine

Currently network management solutions focus on individual devices rather than groups of interconnected devices. This single-device approach forces administrators to fragment their activities and work with a number of interfaces. Management tasks become increasingly difficult and time-consuming as the network grows, resulting in inefficiency and higher administrative costs.

Several leading vendors offer an integrated solution designed to view different parts of the network—groups, buildings, campuses, remote offices, and wide area backbones as collections of logically-related devices that may be managed together using SmartAgent management agents and comprehensive management tools. Intelligent agents are being implemented across internetworking products, allowing management at the adapter, hub, and internetworking levels.

With comprehensive network management, administrators can improve performance at every level of the network while also keeping costs under control and protecting investments in existing equipment.

Barriers to Effective Enterprise Management

Network management continues to evolve toward greater integration, control, and automation (see Figure 17.1). But even with industry acceptance of standard management protocols and consolidated management platforms, managers still face a number of obstacles:

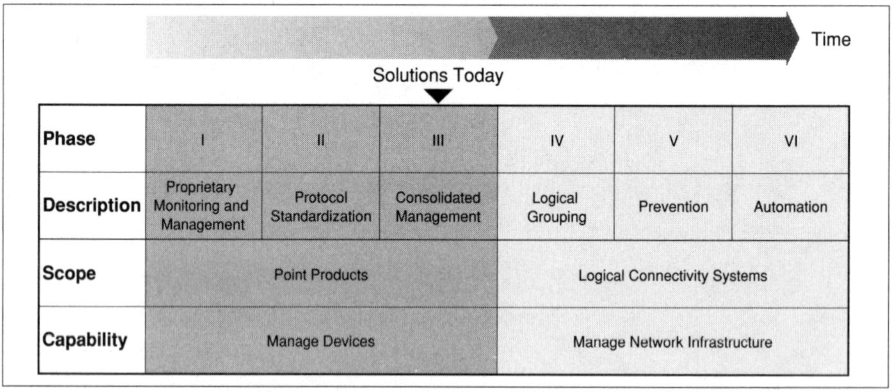

Time

Solutions Today

Phase	I	II	III	IV	V	VI
Description	Proprietary Monitoring and Management	Protocol Standardization	Consolidated Management	Logical Grouping	Prevention	Automation
Scope	Point Products			Logical Connectivity Systems		
Capability	Manage Devices			Manage Network Infrastructure		

FIGURE 17.1. Evolution of network management

- Management applications for various network products usually have different user interfaces, increasing administrative workload, and training time
- There is a lack of database integration and a lack of databases that smoothly integrate information from various applications
- Individual device management requires extensive product and protocol knowledge
- Sophisticated management tools can be difficult to learn and use
- LANs based on diverse technologies often must be managed separately

Because management tools and network products are not yet part of a single, cohesive structure, managing a network still requires the administrator to manipulate and correlate information from disparate and far-flung sources manually.

Contributing to the problems caused by lack of integration is network complexity. Even smaller enterprise networks grow in complexity as they add users and remote sites. Figure 17.2 shows how complexity in networks as measured by the number of users, number of protocols, and geographic coverage increases exponentially until it reaches a complexity barrier.

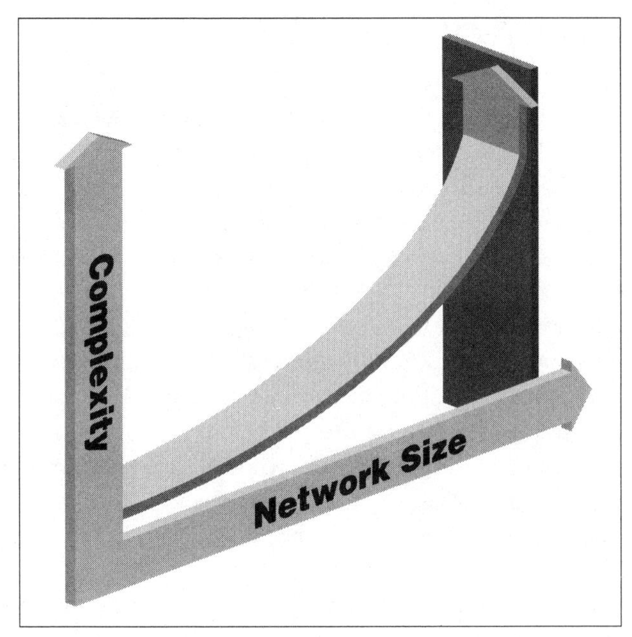

FIGURE 17.2. The complexity barrier

Despite all these obstacles, network administrators are still expected to control costs. To do this, they need to look at more than just equipment. According to a recent Gartner Group study (see Figure 17.3), network components account for only about 16% of the cost of LAN ownership, while management and operation comprise 56%. Thus one of the best ways to curb expenses is to increase staff productivity with a system that allows them to spend most of their time on proactive network management instead of routine administrative tasks.

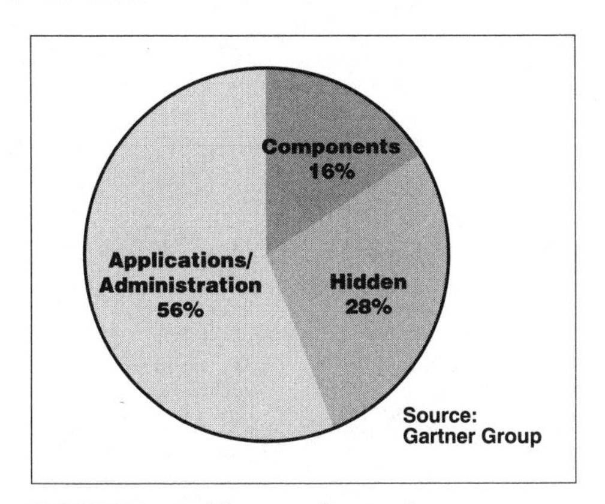

FIGURE 17.3. The cost of network management

THE IMPACT OF TECHNOLOGY TRENDS

In implementing a successful network management strategy, managers need to take a number of technological trends into account.

First, there is the fact that Simple Network Management Protocol (SNMP) has emerged as the management protocol of choice worldwide. Any multivendor management system needs to be able to make full use of SNMPs Management Information Base (MIB) data.

A second trend in network management technology is the popularity of RMON, the SNMP Remote Monitoring MIB for Ethernet and Token Ring LANs, as a method for collecting network statistics. RMON improves traffic management, fault analysis, and administration at both local and remote sites.

Though it is possible to manage end nodes with SNMP, a third trend is represented by an industry consortium called the Desktop Management Task Force (DMTF). This organization is in the process of defining a standard for the in-depth management of end nodes.

A fourth technology trend is the variety of management platforms now available for integrating management applications from multiple vendors. Whatever platform is chosen, it defines the environment in which management applications run. Managers can choose platforms from vendors such as Sun Microsystems, Hewlett-Packard, IBM, Microsoft, and Novell.

Finally, one trend with especially broad implications for today's management systems is object-oriented design. Object orientation offers several advantages as a design model:

- Distributed objects make it easier to scale networks according to organizational needs
- A complex network may be divided into parts that are more understandable and manageable
- Hard-to-manage heterogeneous network elements can be consolidated into easy-to-manage objects

For these reasons, the importance to network management of object-oriented design is likely to grow as networks increase in complexity and size.

MANAGERS REQUIREMENTS—A FRAMEWORK

When asked to state the requirements that are crucial for enterprise network management, network managers generally agree on the ones described below. Figure 17.4 presents these requirements in the form of a three-dimensional framework of needs.

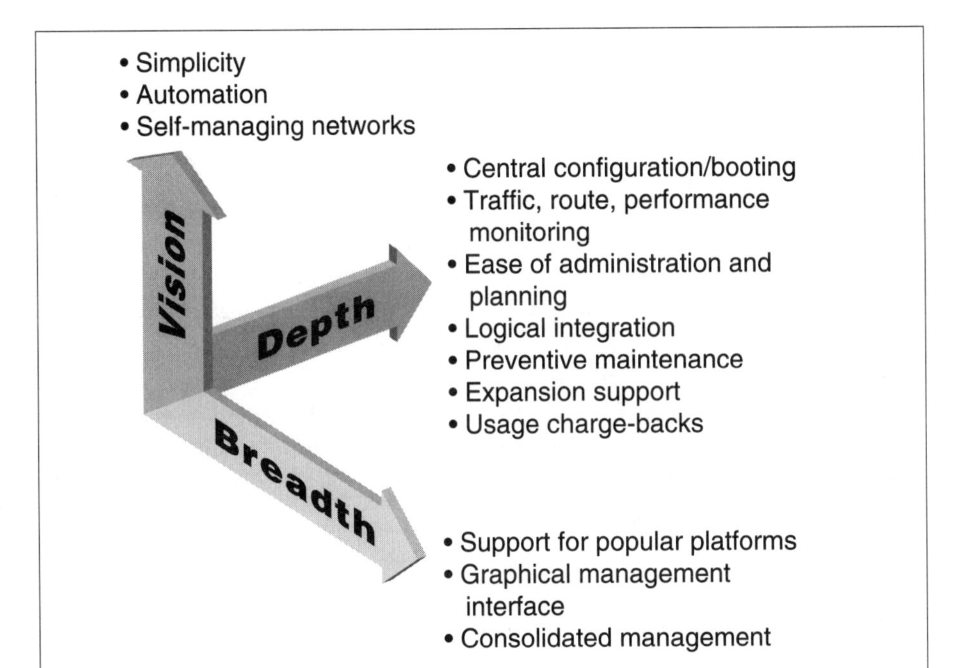

FIGURE 17.4. Framework of management needs

The first dimension, breadth, is related to comprehensive, consolidated support for management platforms and devices. Most companies have already decided on one of the industry-standard platforms and do not want to change platforms in order to run new management applications. This requirement may be summed up in the word "support."

Popular open platforms include Sun Microsystems SunNet Manager, Hewlett-Packard's OpenView/UNIX and OpenView/DOS, IBM's NetView/6000 (Microsoft's management for Windows), and Windows/NT, and Novell's NetWare Management System (NMS).

Managers also require that their management system support a graphical user interface (GUI) that allows information about the various devices to be processed and displayed in an easy-to-read manner using pictorial elements such as meters, gauges, and graphs. Furthermore, a comprehensive management system must consolidate management for the entire range of devices on the network.

The second dimension in the framework, depth, focuses on the capabilities of the tools required to manage the entire network or portions of the topology using one or more centralized management stations. The tools should accomplish all of these functions:

- Central configuration and booting
- Traffic, routing, and performance management
- Easy-to-use administration and planning tools

- Logical integration of network devices
- Preventive maintenance support
- Network expansion and optimization support
- Resource usage charge-back capability

The vision dimension encompasses directions that network managers would like to see management systems take in the future. The overriding need is for simplification of the complexity inherent in evolving networks. This involves integrating management systems and providing common tools for administering the network, eventually leading to automated, self-diagnosing networks.

Simplification also means providing a structure that is consistent with the way managers actually work and offering easy access to management information. In addition, managers want a system that is flexible enough to incorporate existing technologies and protect customers investments.

A TOP-LEVEL MANAGEMENT SOLUTION

A unique approach to network management technology builds on a hierarchy of intelligent agents, integrating network devices into logically related groups (or objects), managed by a comprehensive set of powerful application tools. Ideally, these products are appropriate for any size network and support the major management platforms.

The Complexity Challenge

The primary challenge that network managers face today is simplifying complexity. Customers are looking for a three-fold strategy for helping to meet this challenge:

- Design architectures that minimize administrative intervention
- Offer products, such as fully functional hubs, that work automatically to shield the administrator from complexity
- Build an infrastructure that integrates and structures management logically

As networks add connections and sites, keeping things simple at the management system level is particularly critical. The system must do a good job of reducing the administrative burden, integrating network elements logically, and scaling functionality to keep pace with network growth and diversity. It must facilitate effective design and expansion while also providing inventory and asset tracking, and charge-back information for equipment usage. It must offer a high degree of maintenance and configuration support. And it must do all this without overloading the network with management-related traffic.

The right management system can be the most important factor in realizing the networks full potential as a key strategic business tool.

MANAGEABLE ELEMENTS OF AN ENTERPRISE NETWORK

Though each network is unique, most of them share certain common elements yet is designed to accommodate the differing skill levels, management priorities, and ownership issues associated with each level of the enterprise network. A simplified illustration of a typical enterprise network topology is shown in Figure 17.5.

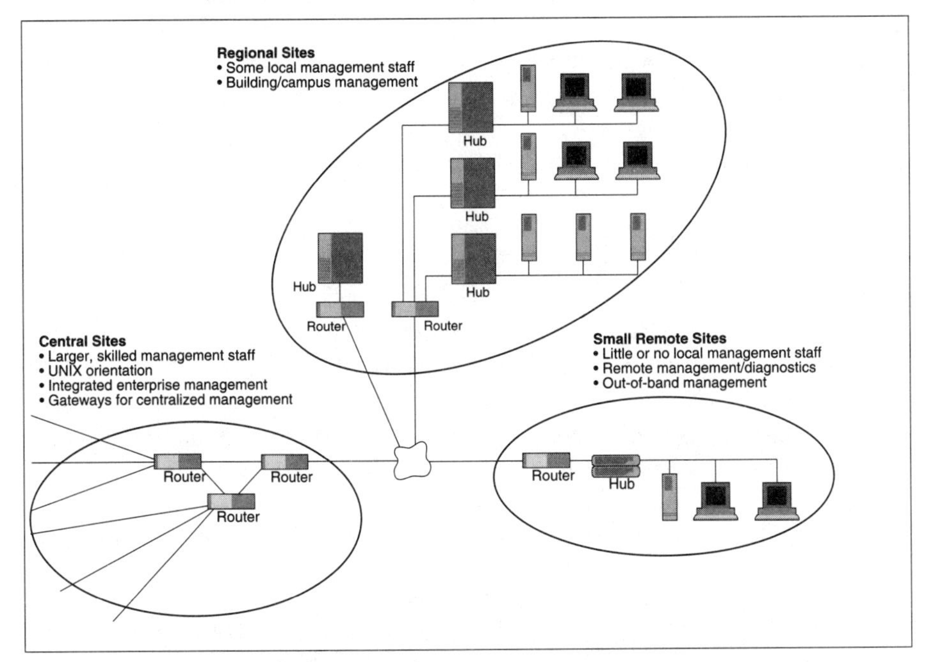

FIGURE 17.5. Topology of today's networks

Remote sites are branch offices or facilities linked to regional sites or large central sites. Each of the regional or central sites may contain a campus or building network connecting multiple LAN segments or workgroups on different floors within buildings or between buildings.

Traffic routing decisions for the small remote sites may be made within routers at the regional or central sites to minimize the cost of the remote office WAN connections. All the sites together make up the enterprises wide area network (WAN).

The staffing and skill levels required to manage the various parts of the network differ depending on how much complexity is involved. For example, a branch office is unlikely to have administrative staff resident at the location. Complex management tasks must therefore be relegated to a regional or central site, where a management station and experienced staff people are available.

Unlike a remote site, a regional site will probably have some administrative support on hand. Staff responsibilities might well include management of the backbone, the wiring closet, and one or more workgroups.

These workgroups, regardless of whether they are made up of physical or logical devices, might be managed separately from the network backbone. This is especially likely if the LANs came into existence independently and were interconnected later on.

Since the central site links the various regional sites into an enterprise network, it is generally the place where the integrated management station and gateways are located. The WAN backbone can be accessed and controlled from this central location.

A relatively large, highly skilled staff is required to administer the central site, both because of the many network connections residing there and because a failure at this critical junction would tend to have far-reaching effects on the rest of the network.

LOGICAL CONNECTIVITY SYSTEMS

A key element in a complete network management architecture is the logical connectivity system. This is simply a group of devices that are logically linked. Network nodes already operate together in a logically integrated fashion to provide the capabilities needed within different elements of the network. These elements combine to help users take advantage of these relationships to form an information management structure, giving administrators the flexibility to manage the network exactly the way they want.

Figure 17.6 illustrates how the management views of devices and logical connectivity systems combine to form an overall enterprise view.

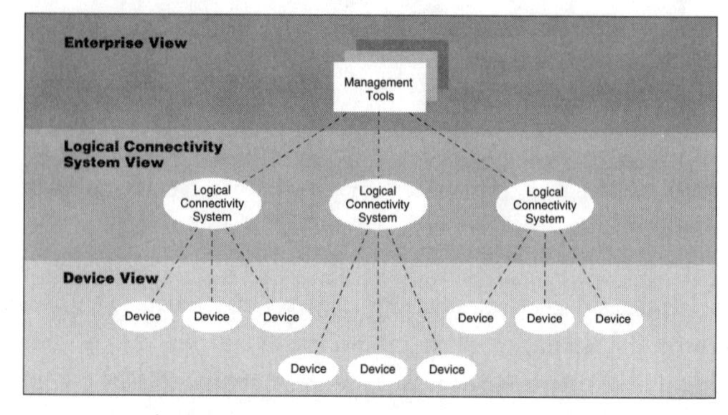

FIGURE 17.6. Hierarchical view of the network

Conforming to the way networks are actually built and managed, network management organizes the logical connectivity systems into a structure like the one shown in Figure 17.7. Managers are free to define their own structures. This particular example is based on the network elements described in the previous section.

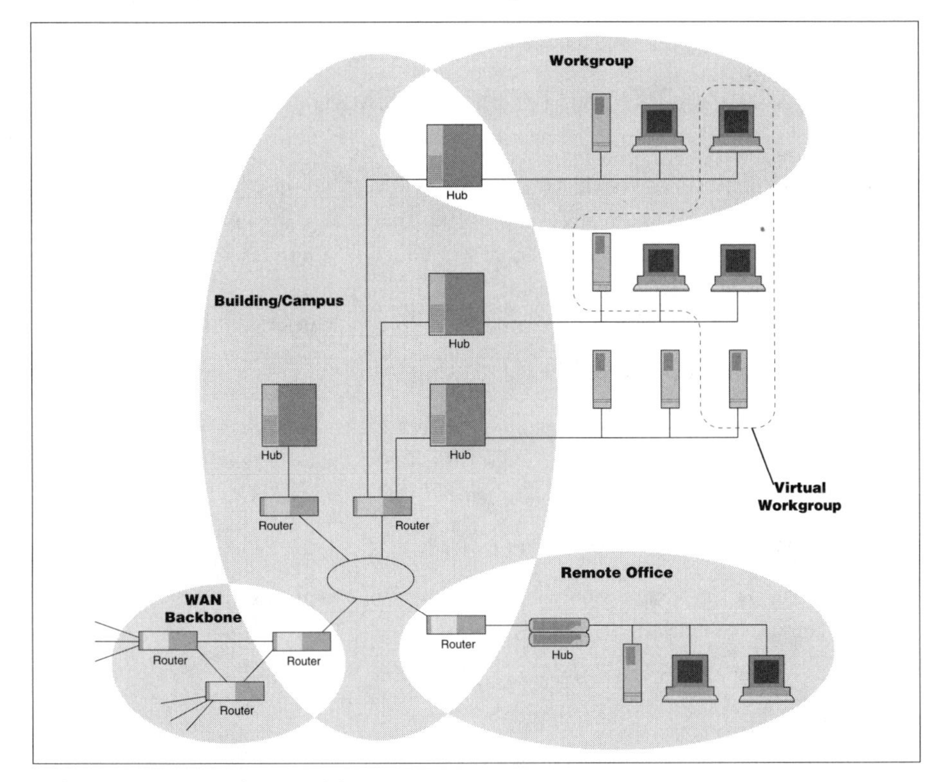

FIGURE 17.7. Logical connectivity system structure

- Workgroups
- Buildings/campuses
- Remote offices
- WAN backbone

The systems in use today manage every device within each of these network elements individually, with little consideration to related devices. By integrating management of the many devices in a logical connectivity system and treating the entire system as a single entity, and vastly simplify networking complexity, management tools may then act on entire connected systems instead of individual routers, hubs, and adapters. This shift in emphasis from managing point products to managing logical groups of products results in a more structured method of network control.

Following are descriptions of each type of logical connectivity system and associated management solutions. The solutions described are based on actual customer scenarios. Note how priorities differ according to the level of the logical connectivity system within the network hierarchy.

WORKGROUP CONNECTIVITY SYSTEMS

A workgroup connectivity system connects PC adapters, servers, hubs, and sometimes bridges, managing them as a single entity. For example, a workgroup administrator might manage a number of Ethernet segments linked to a hub, as well as one or more servers and printers in each LAN segment.

Management solutions for the workgroup administrator include:

- Viewing the network in a graphical, structured way
- Correlating information for user groups that are physically distributed among network segments or floors, but are managed as one virtual workgroup—a collection of nodes grouped according to business needs rather than their physical connections
- Thresholding devices for problem prediction
- Assisting in the isolation of noisy devices without requiring the administrator to go through a laborious process of elimination
- Manipulating device groups graphically to aid in planning

Building/Campus Connectivity Systems

An example of a building/campus connectivity system might include hubs on each floor to provide connections among network segments with links between hubs provided by a collapsed backbone in a single router. The router could be connected to routers in other buildings by a Fiber Distributed Data Interface (FDDI) ring.

Remote Office Connectivity Systems

A remote office connectivity system might consist of remote routers connected to a central router in a regional office over a WAN link. One or more hubs and associated workgroups could be connected to each remote router.

WAN Backbone Connectivity Systems

A typical WAN backbone includes routers attached to the WAN links between geographically dispersed locations.

MANAGEMENT TOOLS

Figure 17.8 shows how the tools required to manage a network and the logical connectivity systems are blended seamlessly together in a complete network management product line. This combination of integrated tools, logical connectivity systems, and intelligent device agents provides a powerful solution for managing an enterprise network.

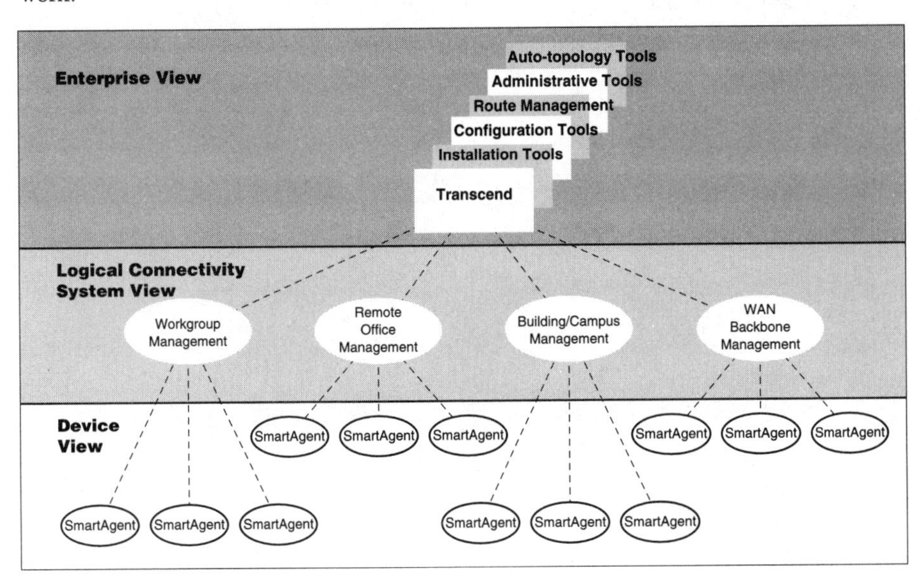

FIGURE 17.8. Management tools

Instead of being oriented to particular devices, management tools operate on one or more connectivity systems. Thus, if the scope of the tool is the campus backbone, it will encompass routers, hubs, and all the links comprising the backbone connectivity system.

First of all, an installation tool provides a central store and distribution facility for system images and configuration files relating to network devices. Configuration files can be edited centrally and downloaded to devices across the network.

In a building/campus logical connectivity system, this tool makes it possible for the administrator to load and boot all the routers from a central workstation and to change configurations when needed. In a remote office logical connectivity system, the installation tool allows central booting and configuration of remote routers without the need for network administration at the remote site.

A series of online configuration tools lets the administrator change configuration of a logical connectivity system by accessing devices directly.

In addition, a unique and powerful display tool permits the administrator to access information from the logical connectivity system in a spreadsheet-like format, then manipulate it using spreadsheet applications. As a result, the information can be easily utilized and customized.

An auto-topology tool lets the administrator draw the topology of a logical connectivity system. For example, the building/campus connectivity map would show routers, hubs, and their interconnections together with the status of each. This tool also makes it easy for the administrator to move from one logical connectivity system topology to another.

Traffic and performance tools provide the administrator with the ability to analyze the traffic flow and performance characteristics of devices, logical connectivity systems, and the entire network.

Finally, a route tool lets the administrator trace the route of a packet through the network. This is useful for debugging and locating network bottlenecks. The route tool will also allow point-to-point tracing of a packet, either within a logical connectivity system or between systems.

AN INTEGRATED APPROACH

Taking full advantage of a complete range of networking products and expertise delivers users a wealth of management benefits:

Integration

- The logical connectivity system approach structures network control and simplifies complexity, making it easier to visualize, analyze, and manage network operation
- The system manages the network rather than individual devices, offering an integrated view
- The logical connectivity systems provide high-quality information, resulting in better-informed management decisions
- A collection of heterogeneous devices can be treated as a black box (only information about the entire group is reported), greatly facilitating administration

Simplified Management

- A common tool set reduces training time.
- Centralized configuration and software distribution for local and remote devices eases administrative duties
- Logical grouping of similar devices mitigates complexity
- SmartAgent management agents distribute decision making throughout the network

Economy

- The integrated, global approach reduces administrative labor and cost by minimizing repetitive tasks
- Powerful network management tools focus on logical connectivity systems, reducing the time staff spends managing individual devices
- Intelligent local monitoring and automatic installation reduce management expenses at remote sites
- Proactive monitoring and control functions maximize up time and user productivity

Flexibility

- A modular architecture makes it easier to expand and upgrade existing systems
- The architecture is scalable to larger networks by adding logical connectivity systems with associated SmartAgent software
- Agents cooperate and can be distributed to optimize management resources

Investment Protection

- The architecture takes advantage of a breadth of current products and SmartAgent device agents, while working with SNMP-managed products from other vendors
- By building on existing products, integrated network management keeps upgrade costs low

Future-Proofing

- RMON remote monitoring can easily be incorporated into the structure of logical connectivity systems and SmartAgent software
- The system also incorporates agents for adapter management
- More intelligence can be added in the future to automate management tasks
- The flexible management structure can readily conform to future industry innovations and customer requirements

GLOSSARY

Adapter driver agents—Management protocols integrated into the network adapter.

Agent—A software routine that performs an action when a specified event occurs.

Enterprise network—A geographically-dispersed network under the auspices of one organization.

Graphical user interface (GUI)—A computer interface that incorporates easy-to-comprehend pictorial elements such as graphs and icons.

Logical connectivity system—A collection of network devices that are related logically and can be managed as a single entity.

Logical connectivity system agent—An intelligent, object-oriented agent that contains all the information needed to manage a logical connectivity system.

Management Information Base—A logical naming of all information resources residing in a network and pertinent to the networks management.

Object-oriented design—A management architecture based on logical connectivity groups and intelligent agents instead of single-device management.

Simple Network Management Protocol—The dominant management protocol in today's networks, SNMP is a set of TCP/IP network monitoring and control functions.

SmartAgent—Intelligent management agents in devices and logical connectivity systems that reduce the computational load on the network management station and alleviate management-oriented traffic on the network.

SNMP Remote Monitoring—SNMP RMON is a remote monitoring and analysis MIB that facilitates network troubleshooting.

SoftHub—Software in 3Com's LinkWatch that translates between SNMP and driver agents in Ethernet and Token Ring adapters.

Topology—The pattern of physical and logical links between nodes on a network.

Transcend—An integrated management solution from 3Com based on groups of logically related devices and including a powerful set of management tools.

Trap—An event message transmitted from an SNMP agent to an SNMP management station.

Virtual Workgroup—Nodes grouped together logically, rather than by physical connections, to better match organizational needs.

ABBREVIATIONS AND ACRONYMS

FDDI—Fiber Distributed Data Interface

Mbps—Megabits per second

MIB—Management Information Base

RMON—Remote Monitoring

SNMP—Simple Network Management Protocol

AUTHOR BIOGRAPHY

Lionel de Maine is product line manager for Transcend, 3Com's network management solution. He has also held the position of business development manager for 3Com, focusing on network management. Prior to joining 3Com, de Maine worked for ROLM where he held management positions on the international telecommunications division, PBX development group, and advanced call solution group. Prior to ROLM, he was a software engineer for Adaptive Networks, Inc., a manufacturer of power-line based communications equipment.

de Maine holds a Master in Business Administration from London Business School, a Master in Science in Computer science from the University of Toronto and a Bachelor of Science in computer science from Pennsylvania State University.

CHAPTER 18

Remote Office Internetworking: Assessing the Costs

by Duncan Potter

Remote office internetworking involves the use of a bridge or router to establish a wide area connection (either virtual or permanent) between at least two LANs. In general, unless the amount of data to be transmitted is very low, routers (together with the appropriate CSU/DSUs or modems) are the preferred device to connect remote LANs together. Routers help to restrict the flow of locally generated traffic that is not destined for remote sites from actually going across those WAN links. This is useful because WAN links are frequently many times slower than LAN links. Additionally, routers provide better security than bridges, as well as protection against broadcast storms, and they will support much higher rates of speed than simple dial-up modems where the application traffic requires it.

International Data Corporation (IDC) estimates approximately two million LANs worldwide are not yet internetworked by any form of WAN connection, and expects that 40% of these LANs will be interconnected in the next three to four years. There are many compelling reasons to extend the corporate WAN out to LANs in remote offices, such as better customer service through faster access to centrally located data, support of a flatter organization structure, or faster retrieval of data generated at remote offices (e.g., local sales data at a retail outlet).

While there is no shortage of remote office internetworking applications capable of generating positive financial results, there is a corresponding dearth of information about what a remote internetworking implementation will cost. Recently, 3Com commissioned a study of leading organizations networking budgets to provide a research-based picture of remote networking costs. This study is unique in that tens of hours were spent with each participating organization to produce useful guidelines.

By taking advantage of the information contained in this study, and by matching real-world needs with cost-effective network designs and other cost-cutting solutions, it is possible to realize savings of up to 90% in important areas such as network administration.

THE COSTS OF REMOTE NETWORKING

Remote networking costs fall into three basic areas: administration, WAN services, and equipment which affect three types of networking activities: acquisition, operations, and changes. Costs include the following:

- Administration (salary and benefits)
- Planning, design, and installation
- Network Administration
- Design reconfiguration
- WAN services (line fees)
- Installation fees
- Monthly access charges and usage fees
- Equipment (hardware, software, modems, bridges, routers)
- Equipment relocation and repairs

Cost of Remote Network Ownership Study

3Com commissioned Strategic Networks Consulting, Inc. (SNCI) a leading industry consultancy to conduct an in-depth study of a representative sampling of mission-critical corporate WANs. The study evaluated the people, WAN service, and equipment costs incurred in providing router-based connectivity to both domestic and international locations.

The study provides important insights into the costs of connecting remote sites and reveals the key areas of focus for MIS in managing the transition to remote office networking without overtaxing limited staff and budgets.

Figure 18.1 shows the key cost area percentages found.

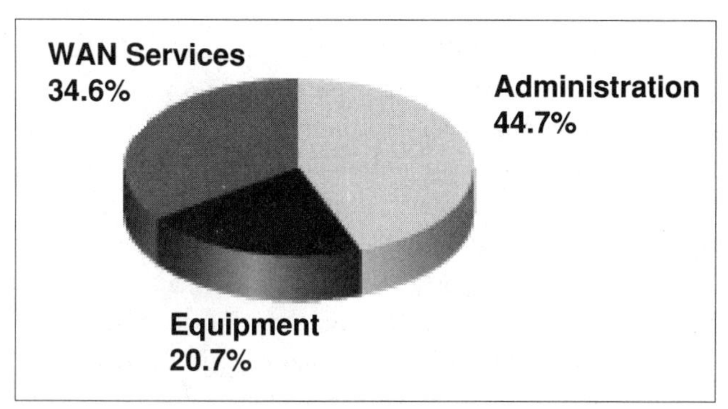

FIGURE 18.1. Cost of distribution by area

Administration Costs

The Cost of Remote Network Ownership study supports general industry findings in reporting administrative costs (people costs, including salaries, fringe benefits, and overhead) as the biggest cost component of networking and the one with the greatest potential to increase over time. In fact, the average network support staff expense for the study participants was $1.2 million per year out of a yearly networking budget of close to $2.6 million, or 44.7% of the total.

Administrative tasks are dominated by day-to-day operations, such as ongoing network administration, configuration, and performance management; changes such as reconfigurations and adding remote site routers are also significant.

WAN Service Costs

Typically, the second highest expense of remote internetworking operation comes from the cost of WAN service—often in the form of leased lines required for remote LAN interconnection. The participants in the SNCI study reported an average budget allocation of 34.6% for access charges and usage fees.

Rates and services vary widely by geographic area. In the past, most interoffice connections have been leased lines ranging from 9.6 Kbps to 64 Kbps and occasionally as fast as T1/E1 speeds (1.54 Mbps/2.0 Mbps). Monthly access charges for these lines can be as high as several thousand dollars per site. Today, the use of switched services such as ISDN, frame relay, or X.25 can contribute significantly to cost savings.

Equipment Cost

Equipment—or capital—costs usually make up a relatively small part of the total cost of network ownership, but in many cases these expenditures are a focal point because they are more easily identifiable, more negotiable with a supplier, and may be the only item requiring capital appropriation approval. In fact, the SNCI study reported equipment costs as accounting for only about 20% of total WAN expenses.

Full-function access routers with one LAN and one WAN port can cost between $3,000 and $8,000, depending upon features and performance. Total costs may seem minor when only a few remote sites are to be linked. However, this cost can become a significant factor if hundreds of branch offices are involved. Where remote LANs require the same full routing capabilities as centrally located users, the cost can escalate rapidly.

Overall, survey participants expected large increases in the number of desktops and remote office sites to be supported over the next few years. In many cases, greater than 100% growth is expected. Meanwhile, there is a limited increase in administrative headcount expected to support this additional burden.

A Boundary Routing System Architecture

Searching for additional solutions to the cost and complexity crisis restricting WAN growth, 3Com conducted a detailed analysis of typical data-flow patterns within remote office networks. After interviewing customers, 3Com found that approximately 80% of remote sites, especially the smaller sites with no more than 25 to 50 users, used only one WAN connection to the central site and had only one LAN segment within the remote office.

This configuration is commonly referred to as a hub and spoke network, with headquarters functioning as the hub and each remote site representing a spoke on the wheel. A hub and spoke solution helps lower remote office internetworking costs in three ways. First, fewer circuit miles are required so monthly WAN service charges (generally based on circuit mileage) are reduced. Second, there is no need to pay a premium for equipment with extra ports since a single active port will suffice at each remote site. And third, with fewer ports and WAN circuits to manage, administration costs are substantially reduced.

3Com looked at the challenge of simplifying the operations of the remote router by concentrating most routing decisions at the central site. By dramatically simplifying routing at the periphery, Boundary Routing allows customers to have fewer administrators because there are far fewer complex devices to administer. In fact, with Boundary Routing it is possible for organizations to add five to ten times more remote site connections with no additional administrative resources.

What Boundary Routing Does

Boundary Routing software takes standard routing software for n-way local routing and extends the LAN interface portion of that software over the WAN (see Figure 18.2). This results in a set of new, greatly simplified routing software functions for the remote router.

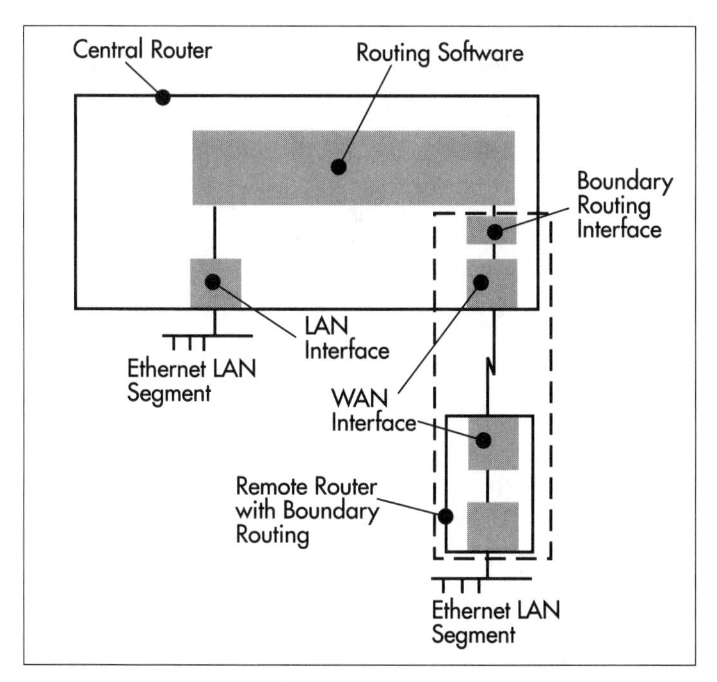

FIGURE 18.2. Central route with Boundary Routing interface

In other words, Boundary Routing software addresses the issue of routing by viewing the central and remote site devices as a single system. This innovation provides all the benefits of configuring a single interface, much like the central device in a collapsed backbone LAN environment, but in a large WAN network. There is little need for configuration of the remote router, thus reducing management complexity.

The Boundary Routing solution requires a bridge/router at a central or regional site, and a device capable of running the boundary routing software, such as a NETBuilder Remote Office router or Novell's MultiProtocol Router (MPR), at each remote location. The software innovations of Boundary Routing add a new level of simplicity and economy to the hardware advantages associated with access routers, but without limiting the remote site's choice of routing protocols.

Complete Routing Functionality

Routing functions are essential in a large network. In a purely bridged setup, the remote office receives all of the broadcast traffic from the whole network which may flood the lower speed WAN links that typically connect the remote offices to the central site. The remote offices themselves issue relatively few broadcasts and multicasts when compared to the rest of the network.

On the other hand, in a purely routed network, routing table updates and other routing protocol traffic can significantly increase the level of utilization of the WAN link. With Boundary Routing architecture, this traffic is not needed by the remote router; routers at the central site hold the tables that define the rest of the network topology.

The router at the central site in the Boundary Routing architecture therefore provides an effective interface between the remote office and the rest of the network. Equally important, Boundary Routing architecture preserves the remote offices ability to take advantage of facilities such as data prioritization, custom filters, and data compression.

It is important to realize a key fact about this approach to reducing remote office networking administration costs. Boundary Routing devices can be intermixed with routers operating in conventional routing mode within a large network. This means that Boundary Routing is not an all or nothing choice. It is an easy matter to deploy Boundary Routing software on routers at some remote sites, but not others, and to convert conventional routers to Boundary Routing mode or vice versa as needs change in each office.

SOLVING THE PROBLEMS OF COST AND COMPLEXITY FOR REMOTE ENVIRONMENTS

Boundary Routing architecture lowers management costs by simplifying router installation and configuration, reducing the complexity of network changes, and streamlining address management.

Using Boundary Routing software, remote router installation is truly plug-and-play, with no requirement for attaching consoles, entering parameters, or configuring switches. The process is almost as easy as plug it in and switch it on. Remote users do not need special training, and expert administrators do not need to travel to the site.

Boundary Routing software reduces the time it takes to connect many remote sites because configuration needs to be performed only on the central router. Central router installation requires only one additional command for each port operating in a Boundary Routing solution.

In fact, the more remote nodes there are, the larger the time savings in installation. The time- and expense-saving advantages of Boundary Routing over conventional routing would increase greatly if the comparison covered a network professional traveling to each remote site to install and configure conventional routers.

The time savings in making complex network changes, such as the addition of a new protocol, also rise when there are more remote offices. The LANQuest Labs, a leading independent testing facility, performed an analysis showing that the time required to add another protocol to a Boundary Routing network with 100 remote routers is reduced by 90% when compared to the time it takes to perform the same upgrade using a conventional access routing approach (see Figure 18.3).

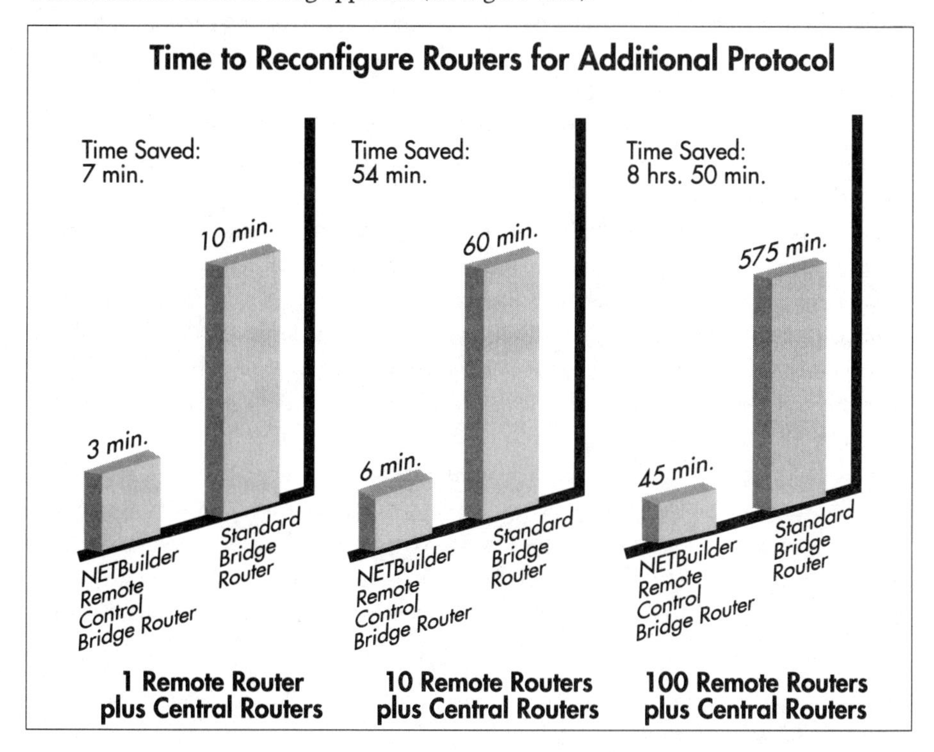

FIGURE 18.3. Reconfiguration time is reduced 90% with Boundary Routing Software

Reducing WAN Service Costs

Router vendors can provide many features that facilitate a reduction of WAN costs. These features work in both Boundary Routing and conventional access routing architectures.

WAN savings can be realized in three areas. The first is using services that are tariffed lower than conventional leased lines. Here the support for switched services and scheduled dial and dial-on-demand capabilities is important.

Virtual circuits provide users with the ability to mix and match transmission technologies such as X.25, frame relay, and ISDN—as tariffs change. By taking advantage of these services, users have the flexibility to choose the best, most cost-effective links for their requirements.

In a dial environment, a Scheduling facility can limit WAN line costs to a predictable ceiling by limiting line use to authorized times. Dial-on-demand lets a device dial the WAN link only when a packet needs to be sent. Dial-on-demand lets customers pay for a connection only when necessary. For a remote office that may require a connection for only 1 to 2 hours per day, this can result in savings of up to 50% or more, when combined with the scheduling feature.

The second area is the ability to squeeze the right kind of performance—and more of it—out of the line. Data compression, bandwidth-on-demand techniques, and data prioritization are important here.

Data compression lowers WAN costs by removing repetitive patterns of data from packets before transmission over the WAN, and restoring them on the receiving end. Data prioritization provides the capability to decide which types of traffic have a higher priority for transmission across the WAN link, allowing the lower WAN costs of single and/or lower speed link, while ensuring that mission-critical, high priority traffic, such as terminal access, takes precedence over less important traffic, such as large data file sharing on the LAN.

Finally, there is the need to provide resilience without provisioning a redundant mesh of leased lines. Dial backup capability automatically dials a backup connection when a router detects that the primary line has gone down, allowing quick recovery of a potentially disastrous break in communication. This means communication can be maintained even in the event of line failure, without the huge cost of redundant permanent links.

Reducing Equipment Costs

Today many vendors are stripping out memory and taking away features, pulling off backup ports, or selling products that cannot be easily upgraded in the field, all in order to market attractive price points. Over the life of the equipment, the hidden cost of these savings is likely to exceed the original small savings. The architectural strength of 3Com's Boundary Routing solution is that it does not force compromise in an access product.

Boundary Routing architecture reduces equipment costs by using the full routing capabilities of the central site router to allow a simple, low-cost device to operate at the remote site. For example, the fact that the central router handles most routing tasks eliminates the need for routing protocols at the remote office, reducing memory requirements for the remote site router. Services such as frame relay that permit 1 WAN port to support 20 to 70 remote connections help reduce hardware costs at the central site even further.

CONCLUSION: UNDERSTANDING AND ADDRESSING REMOTE NETWORKING COSTS

With the growth of remote networks rapidly increasing, even as budgets decrease or remain flat, it is critical for network managers to fully understand and address all the costs involved.

Remote office networking costs fall into three areas: administration, WAN services, and equipment. Of these, administrative costs are typically the highest, WAN services are also significant, and equipment is a relatively smaller factor.

To be effective, a router vendor must offer a range of solutions that lower networking costs in every area. By reducing the number of complex routers needing management, Boundary Routing architecture reduces administration and equipment costs. By centralizing and simplifying network management tasks, Boundary Routing architecture and network management software such as that in 3Com's Transcend architecture, help network support staff to work more productively. Finally, hardware products must be engineered both to take advantage of low-cost transmission services, reducing WAN service costs today and in the future, and also to minimize the need to send expert support personnel to the remote site.

Used in combination, solutions to these cost areas offer a powerful answer to the high costs of remote office networking.

AUTHOR BIOGRAPHY

Duncan Potter is a European product manager within 3Com's Network Systems Operations. He is involved with implementing 3Com's SNA connectivity and remote office internetworking strategies worldwide.

Prior to joining 3Com Corporation in the U.S., Potter held the positions of product marketing manager in 3Com GmbH with responsibility for Germany, Austria, Switzerland, and Hungary and technical support manager in 3Com U.K., Ltd.

Potter has also held various positions at International Computers Limited (ICL) in a wide range of roles and has gained wide experience in the field of European internetworking. He can be reached at Duncan_Potter@3Mail.3Com.Com.

CHAPTER 19

ISDN and Dial-up Connectivity: Switched Data Solutions for Remote and Personal Office Internetworking

by Najib Khouri-Haddad

The Integrated Services Digital Network (ISDN) is a growing worldwide public telecommunications network that will serve a wide variety of users needs. Experts call ISDN the telecommunications network of the 21st century, the foundation upon which to further build the information age. With its narrowband and broadband aspects, the ISDN data highway will evolve from today's switched telephone and dedicated leased-line networks to become a unified global network carrying voice, data, video, interactive pictures, and other services to homes and businesses.

This network technology is becoming more available on a global scale and is providing a highly reliable and flexible infrastructure that can support high-bandwidth applications and a variety of services. Asynchronous Transfer Mode (ATM) technology, which is the basis for broadband ISDN, is an ideal mechanism for carrying various types of traffic in the next few years. Today, switched services and narrowband ISDN represent the first steps in the move toward the digitization of information.

Narrowband ISDN as a means of switched WAN access in data networks is an attractive solution for branch sites or home offices because of the reliability that is inherent in its digital nature, its fast call setup (needed to support routing protocols), its flexibility in supporting different services, and its ability to support existing WAN protocols such as frame relay and X.25. ISDN provides a migration path for remote offices and individual remote users to meet the bandwidth demands of critical applications today and to evolve to higher speeds as needs change in the future.

This chapter presents an overview of narrowband ISDN technology and reports on its current deployment status around the globe. It then puts ISDN into perspective with other services, both leased and switched, for linking remote locations and remote users to each other and to corporate backbones in cost-effective ways. Next it looks at how ISDN addresses the WAN challenges for remote office and personal office internetworking. Finally, the chapter describes strategy, platforms and features that support dial-up WAN connectivity for remote offices and individual remote users.

INTRODUCTION: ISDN TECHNOLOGY

Operating under the International Telecommunications Union (ITU), the Comité Consultatif International Telegraphique et Telephonique (CCITT), Study Group XVIII has defined ISDN as a network evolved from the telephone Integrated Digital Network (IDN) that provides end-to-end digital connectivity to support a wide variety of services. (The CCITT is now known formally as the ITU Telecommunications Standardization Sector, or ITU-TSS.) ISDN combines the coverage of a geographically extensive telephone network with the data-carrying capacity of digital data networks in a well-defined structure. Figure 19.1 shows the major elements of ISDN.

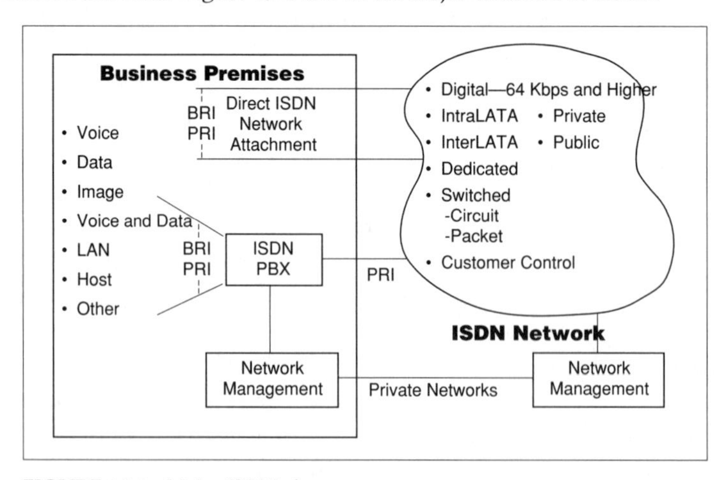

FIGURE 19.1. Major ISDN elements

In addition to true digital connectivity, ISDN supports integrated voice, data, imaging, and other services whose features are under customer control. Equally important, ISDN defines an internationally standardized user-network interface capable of providing connectivity to a wide range of network features and services.

The user connects to ISDN by means of a local interface to a digital pipe of a certain bit rate. Figure 19.2 shows the types of equipment that can connect a customer site to ISDN services through a telecommunications carrier point of presence.

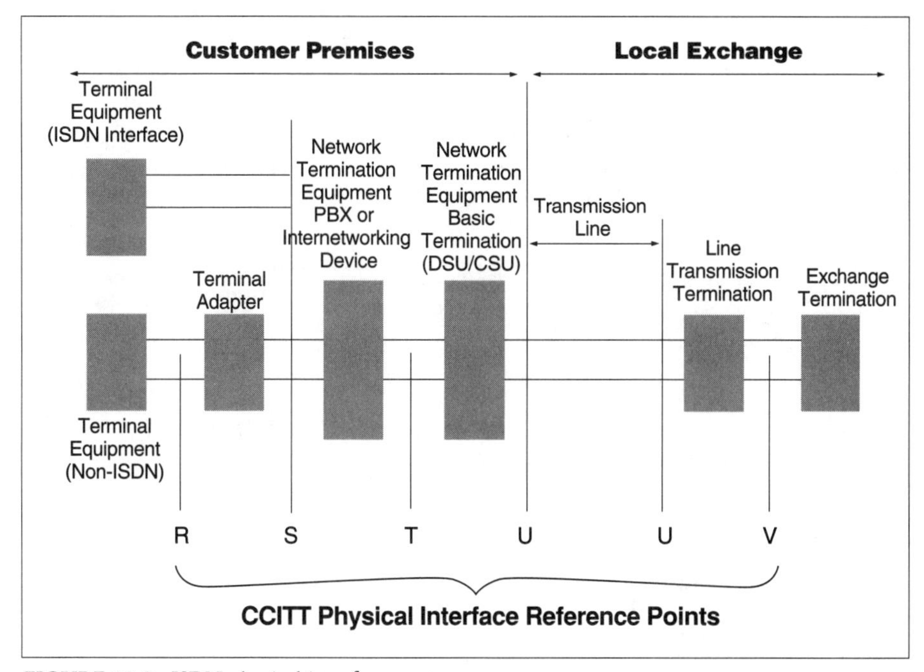

FIGURE 19.2. ISDN physical interfaces

ISDN supports digital pipes of various sizes to satisfy different user or application needs. For example, a residential user might require sufficient capacity to handle a telephone and a PC. An office most often will connect to ISDN via an on-premise digital Private Branch Exchange (PBX) or a ridge/router, and will require a much higher capacity pipe. At different times, the pipe might use varying numbers of channels, up to the capacity limit. The interface is channelized into the following types:

- B-channels are 64 Kbps channels that carry circuit-mode or packet-mode user information, such as voice, data, fax, and user-multiplexed streams.
- The D-channel transmits at 16 Kbps for the Basic Rate Interface (BRI) and 64 Kbps for the Primary Rate Interface (PRI). It carries packetized signaling information to control call setup and teardown. The D-channel also performs packet-switching for data services, allowing a more efficient use of bandwidth.
- An H0-channel is a six B-channel connection that handles 384 Kbps (6 x 64 Kbps) high bandwidth applications, such as videoconferencing or large file transfers.
- An H11-channel uses the full Digital Signal, level 1 (DS1) span and handles 1,536 Kbps in the U.S. and 1,920 Kbps in Europe. It carries circuit-mode or packet-mode user information that requires high bandwidth, such as image-processing applications or wide area backbone applications.

The CCITT I.400 recommendations define how channels are packaged within the interface. The Basic Rate Interface, also known as S0 in Europe and S/T in the U.S., consists of two B-channels and one D-channel for signaling (2B+D), as in Figure 19.3. It can support up to eight terminal equipment (TE) devices in a passive bus configuration bridged on the same wiring. In this arrangement, each B channel is allocated to a specific TE for the duration of the call.

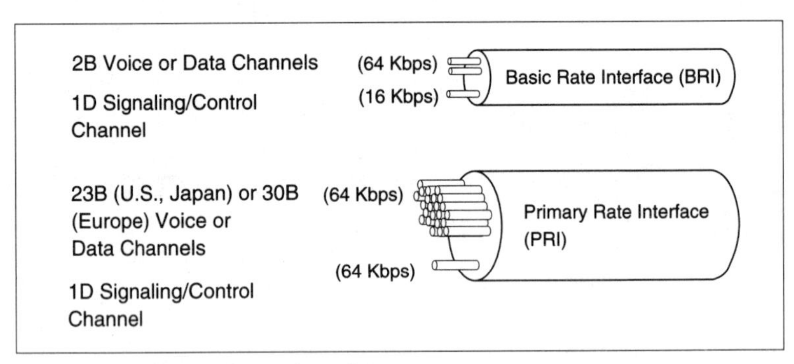

FIGURE 19.3. BRI and PRI channels

The Primary Rate Interface, also shown in Figure 19.3, is a 23B+D interface in the U.S. and Japan and a 30B+D interface in Europe, known there as S2. It is the ISDN equivalent of the existing T1 interface, the physical layer being identical for both. The D-channel is the 24th (or 31st) channel of the interface, and it controls the signaling procedures for some or all of the B-channels.

BENEFITS OF ISDN

ISDN offers many potential benefits for end-users throughout the extended enterprise network:

- ISDN provides higher speed bandwidth than traditional analog, modem-based dial-up solutions, even higher than comparably-priced switched digital services
- ISDN's end-to-end digital transmission is more accurate and reliable than analog technology for lower error rates and fewer dropped connections
- ISDN technology provides quicker connect times to better support LAN protocols, which require lower latency across the WAN connection
- ISDN can deliver high-speed bandwidth at lower costs than private leased lines

Quality, reliability, throughput, and fast call setup make ISDN an excellent medium for data applications using public switched telephone network facilities. Attractive tariffs and expanded availability make ISDN a cost-effective alternative to private leased lines for low- and high-speed data networking.

ISDN APPLICATIONS

Business customers are beginning to realize that integration of voice and data networks within the framework of international open standards will reduce costs and improve functionality. This integration will support other major trends emerging today, including the exploding use of personal computers in communication networks, the reliance on LANs for data distribution in office and factory environments, and the growing importance of distributed processing, remote office internetworking, personal office internetworking, and telecommuting. ISDN is the only technology that binds together communications and processing technologies as a coherent whole—today with narrowband ISDN and tomorrow with the addition of broadband ISDN and ATM.

ISDN enables a new set of telephony-based applications:

- *Integrated voice and data*—In its simplest form, this application provides the physical convenience of a common distribution media and a single plug in the wall. Physical integration offers the economies of sharing, flexibility for growth, and ease of management. The more complex integration of voice and data enables the interplay of capabilities between the two information types, such as high-function telephones and electronic directories, or voice-annotated electronic mail.
- *Resource sharing*—ISDN provides convenient, high-speed connectivity for linking remote LAN sites to each other or to a central site. Bridges and routers with ISDN interfaces can effectively use the fast dialing capabilities of ISDN to transfer data from LAN-connected client workstations to servers at remote sites, and vice versa. LAN interconnection and PC-to-PC communications over ISDN lines can cut transmission costs and greatly increase response time.
- *Telecommuting*—ISDN is an excellent medium for users working from home or small offices to access the corporate network over dial-up lines. The high speed nature of ISDN, compared to existing analog modem offerings, and its digital signal quality and reliability make it ideal for the growing number of telecommuting environments popular among corporations and users. The quick response time of ISDN links means calls are set up quickly, so that users can gain immediate access to corporate information, whenever and wherever they need it.

- Customized local applications—ISDN caller identification can screen an incoming call based on the callers phone number and accept or reject the call based on user-specified preferences. It can link to a directory and forward the call accordingly, or map to a database to pull the callers record. It can even bypass the local site and link the call to a remote site IP address for routing purposes, as in Figure 19.4.

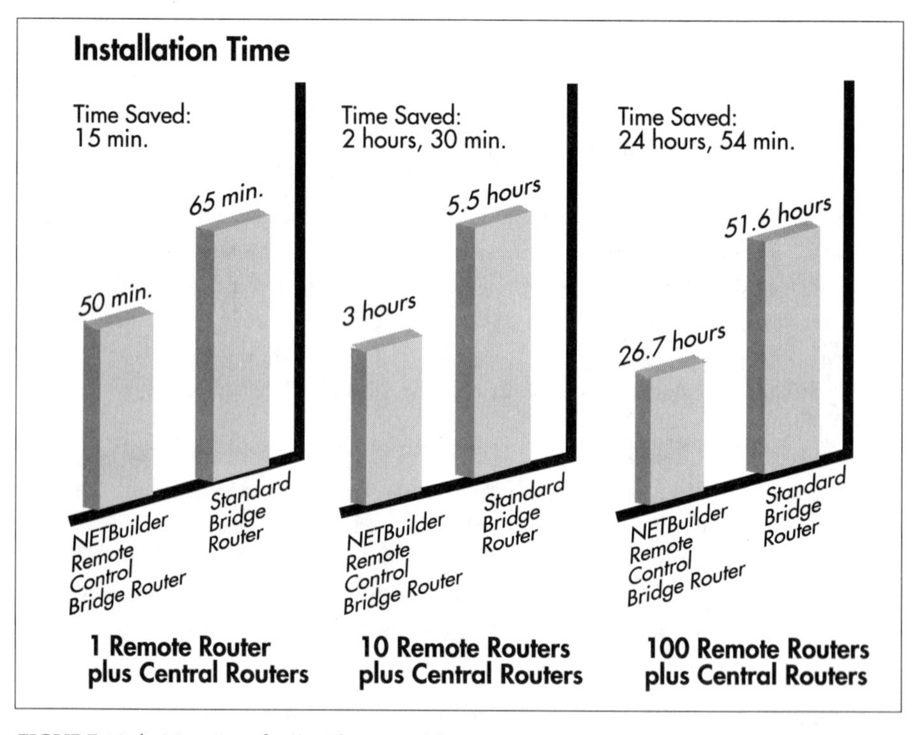

FIGURE 19.4. Mapping of caller Ids to IP addresses

- *Backup and disaster recovery*—ISDN is increasingly used to provide dial backup and disaster recovery facilities for networks. An ISDN connection acts as a backup for a leased line, offering the convenience and cost-effectiveness of a dial-up connection, together with line speeds comparable to current leased lines.

ISDN Deployment—A Status Update

Europe and Japan are leading the world in the deployment of ISDN. In the U.S., RBOCs are aggressively stepping up their efforts to deploy and test ISDN technology as a precursor to more advanced networks offering interactive multimedia services.

In Europe, the Post, Telephone, and Telegraph administrations (PTTs) have made ISDN technology widely available and are tariffing it attractively. The WAN cost savings are significant. Within Germany, for example, an ISDN BRI line that uses a single B-channel to connect across 100 km (62 miles) for six hours each day cost only U.S. $1,500 per month plus usage costs in 1993, whereas a 64-Kbps leased line cost U.S. $4,000 per month for the same distance. Companies that do not have constant traffic over the link can thus connect periodically using ISDN and maintain productivity at much lower cost. The pan-European standard ISDN service, known as Euro-ISDN, is being aggressively deployed across Europe with 85% coverage of the available ISDN telephone lines at the end of 1994, rising to more than 90% by the end of 1995.

In a 1993 study, Frost & Sullivan, a Mountain View, CA-based market research firm, estimated the total European ISDN market was worth $4.64 billion in 1992 and will grow to $5.91 billion by 1994, with a projection for $10.23 billion in 1997.

The following sections present the actual and projected deployment figures for the major worldwide implementors of ISDN.

Germany

The Deutsche Bundespost Telekom (DBT) has a virtual monopoly within the German telecommunications market. With more than 300,000 ISDN BRI lines installed at the end of 1993, Germany's ISDN market is expected to reach 800,000 BRI lines by 1995 (DBT sources). The number of ISDN PRI lines, around 11,000 in mid-1993, is expected to reach 33,000 in 1995. In the past, users in Germany ordering an ISDN line could choose between a proprietary protocol called 1TR6 or the Euro-ISDN version, but the 1TR6 version is available only by request for new installations. The Euro-ISDN implementation will be available in more than 90% of the German ISDN market by the end of 1994. This figure will reach 100% by 1995.

France

France Telecom, the French PTT, is the main supplier of voice and data services in France. With more than 150,000 ISDN BRI lines installed at the end of 1993, France's ISDN market is expected to reach 250,000 BRI lines by 1995. There were approximately 7,000 PRI lines installed in mid-1993. The French version of the ISDN protocol is VN3 and VN4 (a VN5 version in planned for the 1995-96 timeframe). The VN4 version will interoperate with Euro-ISDN for basic connectivity and with five other supplementary services. It will be available in 100% of the French market by the end of 1994. Today, France offers X.25 services to the Transpac network via ISDN B- and D-channels.

United Kingdom (U.K.)

Unlike the telecommunications markets in other European countries, the U.K. market is deregulated. This is encouraging a rapid adoption of a variety of services, including frame relay, SMDS, ATM, and broadband ISDN, in addition to ISDN BRI and PRI. BT, the largest ISDN provider in the U.K., had installed more than 17,000 ISDN BRI lines at the end of 1993, with a target of 120,000 for 1995. PRI lines numbered around 6,000 in mid-1993. The U.K. ISDN implementation is compatible with Euro-ISDN and is 100% available throughout the U.K.

Japan

Japan also has been very aggressive in the deployment of ISDN. Nippon Telegraph and Telephone (NTT) is the main supplier of telecommunications services in Japan. NTT has been encouraging the migration of telephone users to ISDN by means of price incentives. ISDN connections for large business users are available anywhere in the country and are priced at lower rates than comparable analog services. The ISDN market in Japan for INS Net 64 lines (CCITT-based ISDN BRI), represented in Figure 19.5, is expected to grow in the next few years at a rate of more than 20% per year and provide frame relay services on ISDN circuits.

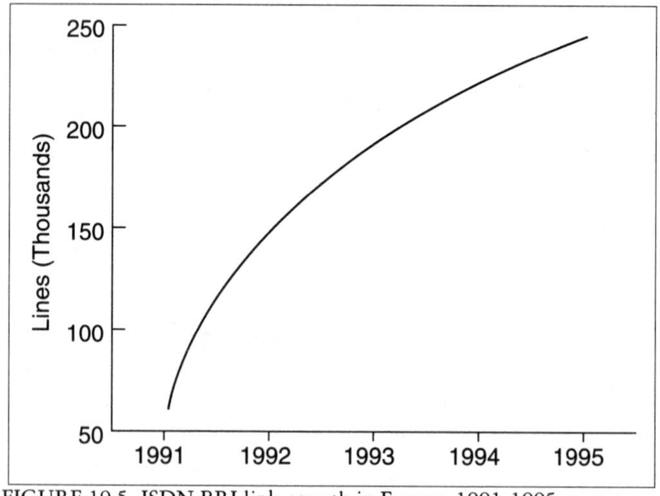

FIGURE 19.5 ISDN BRI link growth in Europe, 1991-1995

United States

Since the Regional Bell Operating Companies (RBOCs) standardized on a BRI implementation, called National ISDN 1 (NI1), ISDN penetration has made steady progress in the United States. Analyst group In-Stat, Inc. estimates that ISDN is currently available in 60% of the country, and most of the RBOCs are beefing up their ISDN efforts.

As shown below, the percentage of RBOC access lines that are ISDN ready is rapidly increasing in 1994 over 1993 figures. By 1995, according to Bellcore, which completed the study in November of 1993, RBOCs serving the largest population centers will have uniformly strong support for ISDN.

RBOC	1993	1994	1995
Ameritech	70%	80%	80+%
Bell Atlantic	59	87	90
BellSouth	46	53	64
NYNEX	31	55	76
Pacific Bell	60	78	87
Southwestern Bell	54	60	66
US West	43	57	59
Cincinnati Bell	57	73	83
South New England Telephone	27	34	44
Rochester	69	68	68
GTE	14	16	18

ISDN PRI deployment is more limited than ISDN BRI deployment in all countries because of high installation costs and tariffs. In Europe, only Germany has been promoting ISDN PRI, mainly for voice applications and PBX-to-PBX networking. In Japan, NTT is evaluating ISDN PRI with frame relay support. In the U.S., work is proceeding on the definition of a common standard, called NI2, which will be deployed nationwide in the next few years.

Attractive ISDN tariffs in Europe, Japan, and the U.S. allow customers to deploy the technology in a cost-effective way.

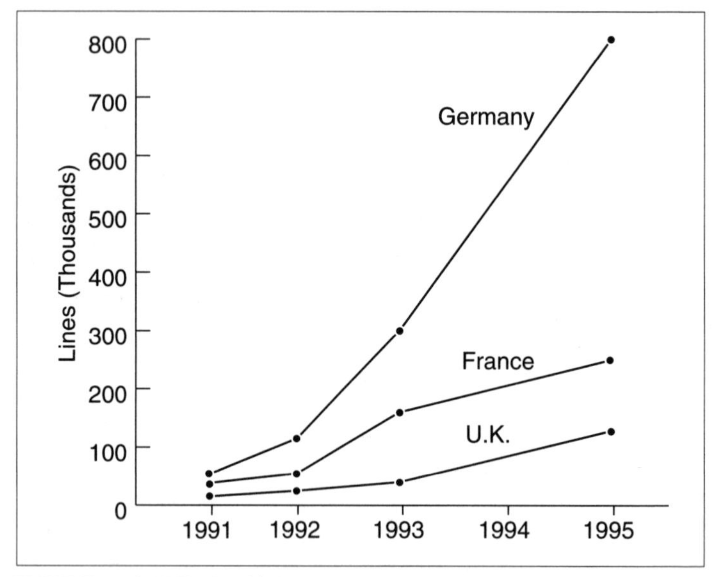

FIGURE 19.6. INS Net 64 Growth in Japan, 1991-1995

WAN SERVICES FOR DATA NETWORKING

Interconnecting end-users in a typical campus environment require hubs, bridges, and routers linked by privately owned wiring. Interconnecting networks in different locations into backbone configurations or connecting remote sites to a central site requires WAN links. Today, the majority of WAN connectivity for backbones and branch offices is provided over private, leased digital lines operating at 56 Kbps or 1.544 Mbps (T1 rates) in the U.S., and 64 Kbps or 2.048 Mbps (E1 rates) in Europe. For locations transferring controlled amounts of steady data traffic, leased 56 Kbps and 64 Kbps lines are popular because of their low charges for installation and monthly rental. Figures 19.7 and 19.8 show how the monthly cost of leased lines compares to the cost of switched (dial-up) services in the U.S. and Germany, respectively. In both figures, the horizontal lines show leased line costs, and the shaded circles show the usage points where leased lines become less expensive than comparable switched communications.

Interconnecting remote end-users to networks require the use of remote access servers and public telephone network services. Traditionally, remote users have used analog modems and public phone lines to transmit data to their corporate headquarters because of their low cost and high availability. With improvements in analog modem technology for line errors and data compression, high-speed analog modems are now used at speeds up to 28.8 Kbps, with an effective compressed throughput to 115.2 Kbps, depending upon the line quality.

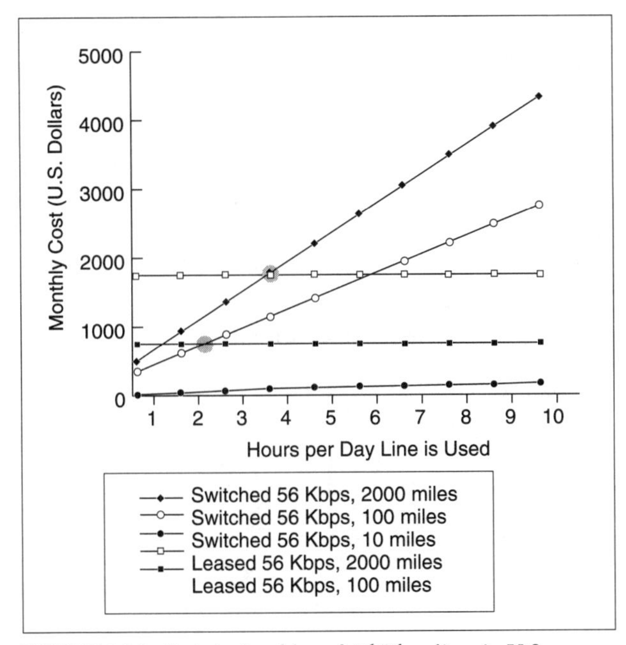

FIGURE 19.7. Switched and leased 56 Kbps lines in U.S.

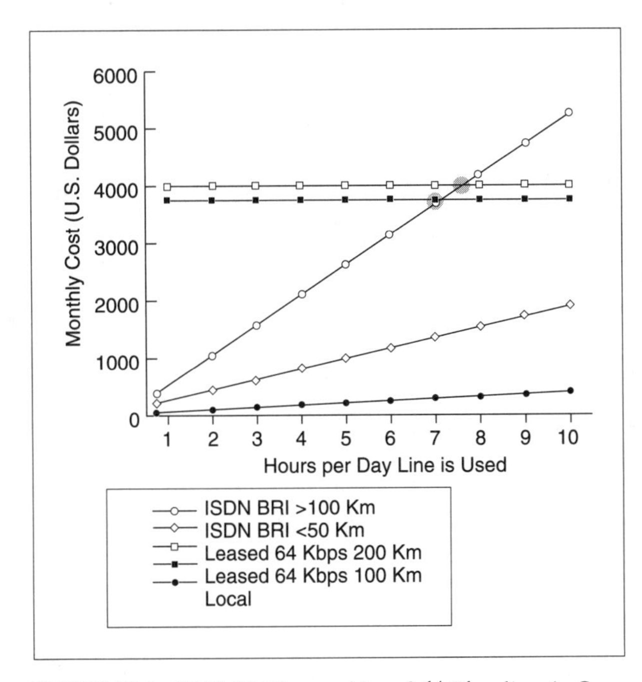

FIGURE 19.8. ISDN BRI lines and leased 64 Kbps lines in Germany

With the availability of switched digital services from the public telephone network providers, internetworking users have begun migrating from analog to digital technologies. The results are better quality, higher speeds, increased reliability, and more flexibility. In Europe and the Pacific Rim, digital technologies are appearing under the ISDN umbrella for both BRI and PRI. In the U.S., because of a slow adoption of a common ISDN implementation by RBOCs and long-distance carriers, digital technologies are offered on a variety of circuits:

- Switched 56 Kbps
- Aggregate Nx64 Kbps, based on ISDN BRI or PRI
- Switched 384 Kbps, based on ISDN PRI
- Switched T1 at 1.54 Kbps

ISDN has a clear advantage for switched WAN access because of its reliability as a digital technology, its fast call setup to support data applications, and its flexibility in supporting combinations of voice, data, and video services on a call-by-call basis.

DIAL-UP DATA NETWORKING APPLICATIONS

In addition to the cost savings they offer in comparison to leased lines, dial-up services like ISDN offer greater flexibility and allow more applications to be supported by internetworking products. Fast modems, inverse multiplexing, and a growing number of switched digital services offered by the public telephone providers including ISDN open new horizons for a host of new applications:

- Backup and disaster recovery for leased lines
- Bandwidth management
- Remote management
- Remote office internetworking
- Client-to-LAN personal office internetworking
- Dial-on-demand remote networking

Backup and Disaster Recovery

Applications that deliver mission-critical information are crucial to the functioning of an organization and have strong reliability and availability requirements that make disaster recovery an important criterion in network design.

Redundancy is usually the preferred way to achieve fault tolerance. In the case of WAN links that use a leased line as a primary line to ensure constant availability of a data path, another leased line can serve as a backup. This is a very expensive solution, since the backup line is needed only when the primary line malfunctions, yet companies must pay the monthly charge for the redundant line, whether it is used or not.

A dial-up connection, such as a switched 56 or an ISDN BRI line, is a much more affordable solution for backup of a primary leased line. The dial-up line is activated by the central internetworking device automatically when a failure is detected on the primary line, as shown in Figure 19.9, with no apparent degradation of service in the network. If the primary line is a high-speed pipe running at T1 or E1 rates, several lower-speed dial-up circuits can be combined through inverse multiplexing to afford comparable high bandwidth capacity.

Inverse multiplexing aggregates multiple dial-up data circuits so that they appear to the application as a single high-speed data connection in a process known as bandwidth-on-demand.

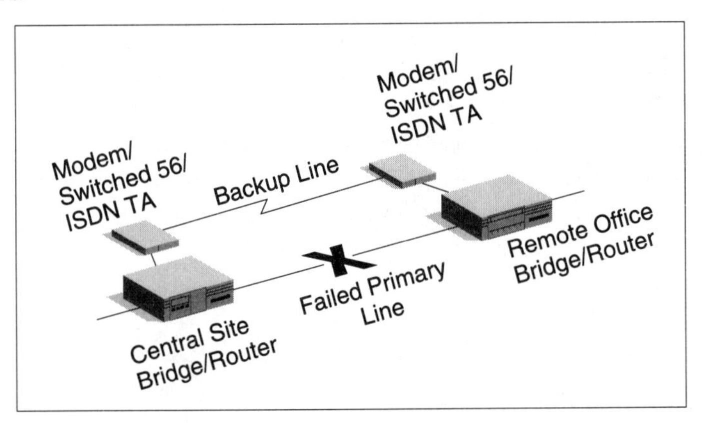

FIGURE 19.9. Backup and disaster recovery

Bandwidth Management

Dial-up lines can also be used as overflow data connections when the data load increases, represented by Figure 19.10. When the primary line reaches maximum capacity, the overflow traffic can be sent over one or more dial-up circuits on an as-needed basis.

Bandwidth-on-demand is an economical way to add WAN links between two locations through inverse multiplexing. The ISDN architecture allows the aggregation of all B-channels within BRI and PRI lines.

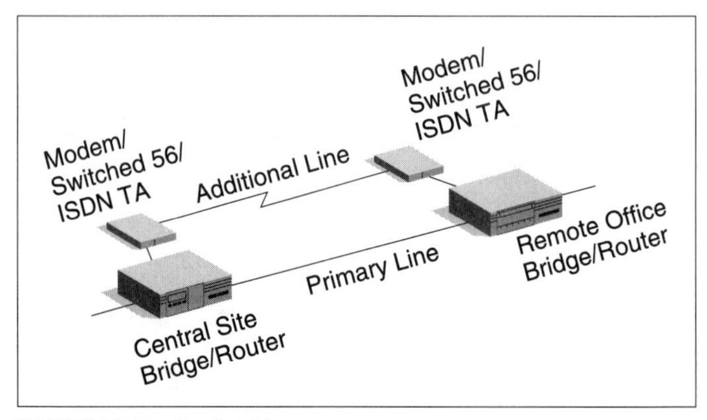

FIGURE 19.10. Bandwidth management

Remote Management

Dial-up internetworking greatly enhances remote network management by allowing a network manager to administer and control a geographically distributed network from a central location. There is no need to place a skilled, highly paid administrator in every remote site. Instead, network managers can configure, monitor, and control remote internetworking devices from corporate headquarters. Centralized management of remote connections, represented in Figure 19.11, is the most cost-effective way to connect a large number of sites.

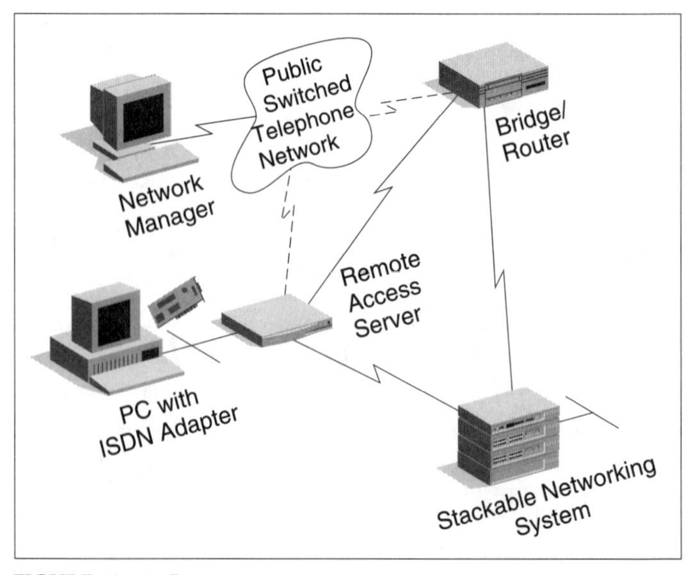

FIGURE 19.11. Remote management

Remote management can even extend its reach to single-user personal offices, with the aid of ISDN adapter cards. Software agents on plug-and-play adapter cards communicate with the centralized network management system for full visibility within the enterprise management framework.

Remote devices can dial in to the central site to report network status, alarms, and statistics. In addition, central site network management systems can poll remote sites for information on a periodic basis. Dial-up lines are well suited to these applications.

Remote Office Internetworking

A growing number of small to medium-sized remote offices are being connected to corporate backbones, as in Figure 19.12, to get access to centralized information. The nature of the traffic in low-population remote sites does not usually warrant having a dedicated leased line to connect it to the central site. Users at these remote offices typically are connected over a LAN and share some common resources, such as a printer and a server for storing computer data and programs. Occasionally, when electronic mail needs to be exchanged with the central site or when information needs to be uploaded or downloaded from central computers, a WAN connection is required for a short time.

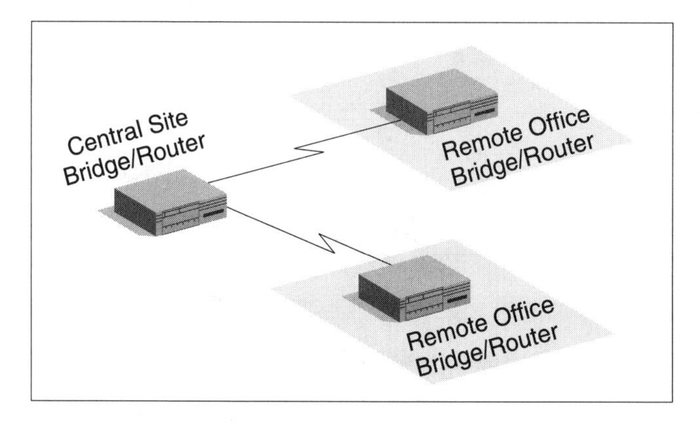

FIGURE 19.12. Remote office internetworking

Dial-up services are well suited for remote configurations since they provide a telephone circuit only when information needs to be exchanged.

Internetworking devices can be intelligent enough to schedule these connections when telephone rates are more economical, to achieve further cost savings. The devices also can dial a succession of numbers and connect to the first one available.

Client-to-LAN Personal Office Internetworking

The concept of remote office connectivity can be extended to a single user in a remote location, as shown in Figure 19.13. The user can call in to a remote access server from home or another location via analog telephone lines, switched 56 Kbps links, or ISDN BRI to connect to host computers on the enterprise network. The remote access server makes the remote user appear as a locally connected client on the LAN, with all the attendant services and privileges.

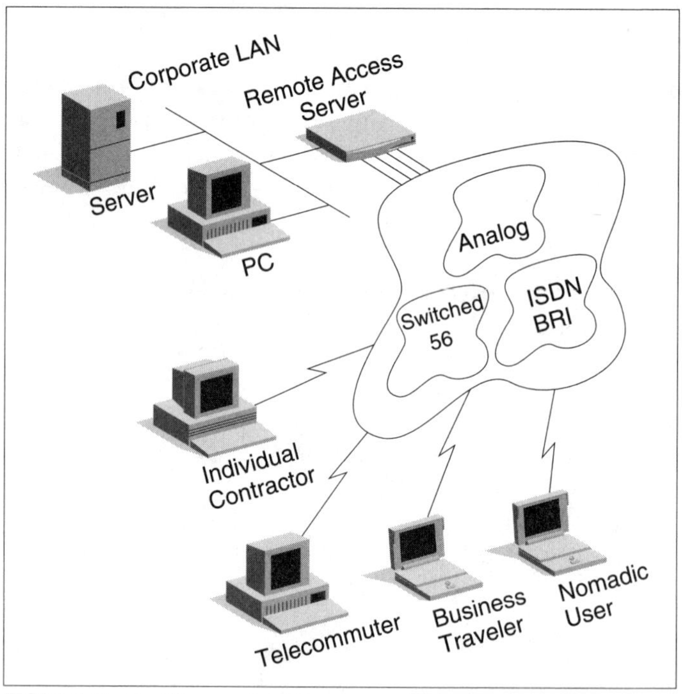

FIGURE 19.13. Client-to-LAN Internetworking

Personal office internetworking opens new avenues of productivity for corporate employees who travel extensively, work at home, or require after-hour access to the office. At the central site, internetworking devices receive incoming data calls from various locations through pools of dial-in lines, provide security authentication and validation of callers via login procedures, and route the calls over the corporate network. Alternatively, the internetworking device can hang up after identifying the caller and call back later, for purposes of security or economy.

Dial-on-demand Remote Networking

Remote users who need access to the corporate LAN might find it more economical to reach it through network links provided by a local LAN site. For example, to save long-distance access charges, business travelers might prefer to dial in to a local LAN and allow the network to route the traffic to the final destination, which might be in another location or even in another country. Depending upon the traffic pattern between the LANs, the connection between the local LAN and the destination LAN can be leased or dialed up. In the dial-up case, the local LAN can initiate the connection on demand and aggregate multiple circuits to offer the appropriate level of service, as shown in Figure 19.14.

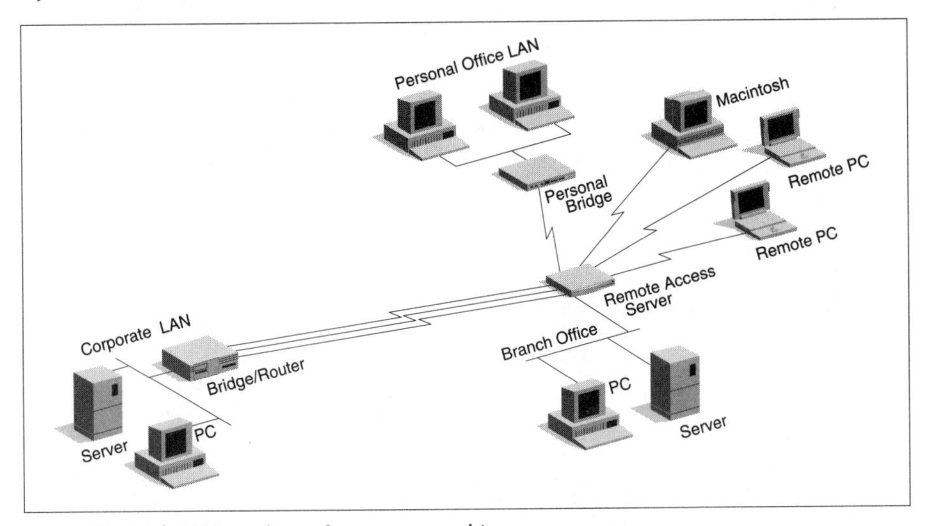

FIGURE 19.14. Dial-on-demand remote networking

With ISDN's fast call setup capability, client-to-LAN remote access can take advantage of dial-on-demand during an individual session. When there is no client data traffic destined for the LAN, the client can disconnect an ISDN WAN link, transparently to the running application, to save dial-up charges. The application continues to see a logical link to the LAN. When communications with the LAN are next required, the client can automatically reestablish a dial-up session and pass the data traffic over the WAN.

Dial-up Features for Successful LAN/WAN Integration

The flexibility that dial-up lines bring to the internetworking market needs to be kept in the perspective of basic LAN requirements for traffic flow and bandwidth access.

LAN protocols assume continuous path and capacity availability. Therefore fast call control and line capacity can be important considerations. In addition, dial-up access to LAN environments must be carefully controlled for security purposes. Moreover, WAN links are expensive resources that need to be monitored by accounting and management applications for cost-control purposes.

Three categories of features for internetworking devices are required to blend LAN and WAN capabilities:

1. Call Establishment and Termination
   Dial-on-demand
   Bandwidth-on-demand
   Load balancing
   Compression
   Automatic data call distribution and pooling for incoming calls
   Line hunting for outgoing calls

2. Call Security
   Caller authentication
   Call acceptance or blocking
   Callback
   Centralized security management

3. Call and Network Management
   Statistics
   Loopback
   Administration
   Centralized control of outlying sites
   Standards-based operation

Wide area links are valuable resources that have to be carefully managed and used. In contrast to LAN environments, where bandwidth is free, dial-up WAN access is charged on a usage basis. The diverse connectivity requirements of the multiple WAN solutions add extra complexity, from both a reliability and a manageability standpoint.

Remote Office WAN Connectivity

The 1980s saw the proliferation in campus environments of LANs, LAN backbones, and client/server computing architectures. The early 1990s focused on connecting outlying LANs to each other and linking them to central or regional sites. The next logical step is to extend LAN connectivity over the WAN and give access to central site data and resources to hundreds of remote users, thereby building virtual LANs over a geographically extended region.

Remote Office Growth

As shown in Figure 19.15, the percentage of remote LANs that are connected to central sites is expected to more than double by 1997. These remote LANs average around 20 users and one or two servers. More than 75% of these remote offices are managed from the central-site LAN.

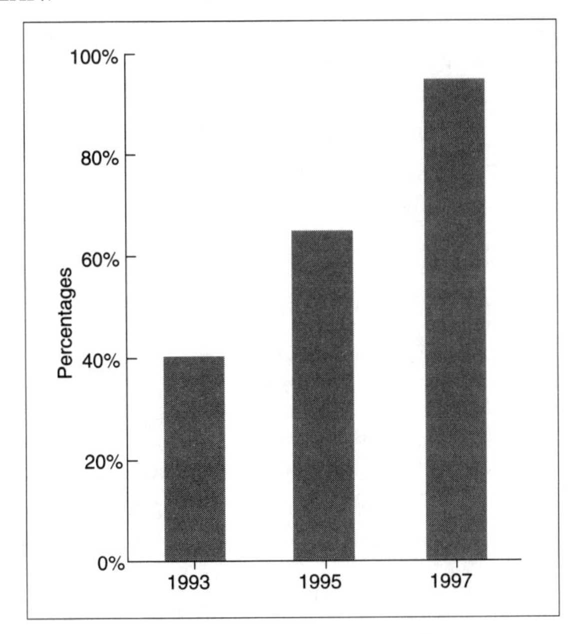

FIGURE 19.15. Percent of remote offices with network connection to the central site

The explosive growth of remote-office connectivity is focusing the attention of the internetworking vendor community on managing WAN links in order to lower transmission and other ownership costs and to provide efficient administration of large, dispersed networks.

WAN Challenges

WAN managers face problems that encompass all of the issues that LAN administrators must deal with, and many more. For example, LAN administrators usually assume nearly unlimited bandwidth for applications, as well as ownership of the physical media that connects computers and workstations. This is not so in the WAN environment where bandwidth management and transmission costs are important concerns.

A WAN manager must consider the impact each application will have on bandwidth utilization and look for ways to optimize the use of local and remote resources. WAN links typically are scarce and costly. Today, it is estimated that 35% of the expense of day-to-day remote network operations is devoted to line costs. Administration costs usually are higher. A recent survey conducted by Strategic Networks Consulting, Inc. (SNCI) confirms that WAN expenses are heavily concentrated in the people who keep the network running. In the case of remote networking, more than 30% of all expenses are allotted to day-to-day management and more than 10% to the administration of network changes, a total of 44% according to the study findings.

Today, leased lines are most widely used for WAN connections. X.25 is the most widely used WAN protocol, and frame relay is also popular. Analog dial-up connections and switched 56 and ISDN links are less well established but growing rapidly. The decision of leased over switched generally hinges on usage requirements. If data traffic is intermittent, such as brief exchanges of electronic mail or periodic file transfers, a dial-up connection is likely to be more cost-effective than a leased line.

PERSONAL OFFICE WAN CONNECTIVITY

More and more employees are accessing the workplace network from home or during business travels. While many remote access options are available to allow users to dial in to the corporate LAN, not all solutions offer the remote user the same productivity levels that local users enjoy. Personal office internetworking allows remote users full access to corporate resources by dialing in to an internetworking device, called a remote access server, through serial interfaces or channelized high-speed switched services (T1 or ISDN PRI). The internetworking device bridges the traffic directly to the enterprises backbone network. The remote user then appears as a local client to the LAN.

PERSONAL OFFICE FLEXIBILITY

Personal office internetworking is more flexible than dialing in to a modem pool using terminal emulation or remote control software applications. It provides tighter integration of the electronic workplace in comparison to a modem pool, and it offers much better support for graphical user interfaces and mission-critical applications in comparison to terminal emulation.

Personal office internetworking addresses the needs of two types of remote users: stationary and mobile users. Stationary users are telecommuters or very small personal offices that require connectivity to the corporate LAN to perform normal working functions, such as access to electronic mail and information databases. Mobile users are either business travelers or nomadic users, such as field personnel, who need some connectivity to the corporate LAN on an occasional basis to exchange electronic mail or to check technical or sales data.

The remote networking market is expected to explode in the next few years. The Yankee Group predicts that the market for LAN dial-up products and services will reach $4.4 billion by 1997. Telecommuters, nomadic users, and business travelers the remote users requiring personal office connectivity, comprise the largest share of this growing market, as shown in Figure 19.16.

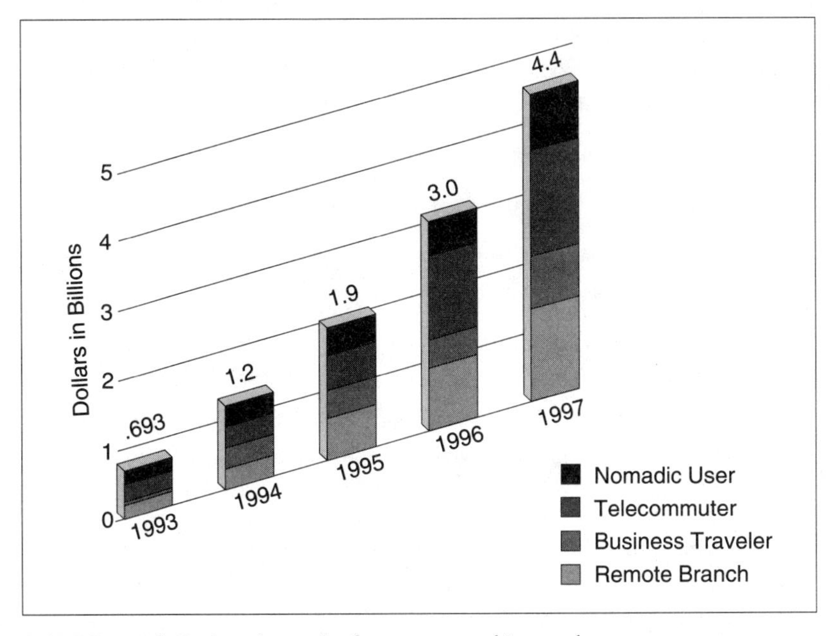

FIGURE 19.16. Projected growth of remote networking market

Personal office internetworking applications rely on dial-up lines to reach the enterprise network. Stationary users can choose from analog, ISDN BRI, and switched 56-Kbps lines to connect regularly to the network for extended periods of time. On the other hand, mobile users rely mainly on analog lines for their occasional sessions, because analog is ubiquitous. As ISDN availability spreads, it will become a more viable option for the mobile user. Today ISDN is beginning to be deployed in large hotel chains and in information kiosks worldwide.

WAN CHALLENGES

The remote user sees the following challenges as top priority:

- Ease of use

- Response time/throughput speed
- Speed of setup
- Ubiquitous and dependable access

Remote users want convenient, reliable, and high-bandwidth access to corporate resources. They want to reap the productivity benefits of working remotely without suffering the performance penalties often associated with dial-up technology.

The WAN manager faces a different set of challenges when implementing personal office internetworking solutions within the broader framework of the enterprise:

- The cost accounting in managing and maintaining the external phone lines needed to allow a large number of users to call in to the corporate network
- The security control in letting nonresident users have access to private data
- The centralized management of a widely dispersed user population whose locations can change radically from one session to the next

Often these functions reside in different departments within the enterprise, where the telecommunications group has ownership of the external WAN connections, the MIS group oversees corporate LAN access, and the network administrator monitors the traffic flow.

A large portion of the cost of running a remote network lies in the cost of administration and transmission. According to the survey conducted by SNCI, administration exceeds 44%, and transmission exceeds 34% of total costs. In contrast, for personal office connectivity, the transmission costs have a higher significance, given the number of remote users and their connection periods.

Security is a critical criterion in the acceptance of the personal office dial-in concept. When corporations open up their networks to external users, control over access to sensitive data has to be well defined and enforced. Certain security features that are available on a LAN network, such as filtering and logical groups, cannot be used effectively for many remote mobile callers. User identification, passwords, and resource managers must be used to limit and control access to the LAN.

ISDN Strategy and Solutions for Remote Office Internetworking

Several leading networking vendors are integrating ISDN into a remote office strategy through both a central and remote site router combination. For example, 3Com's NETBuilder II products will support both PRI and multiple BRI ports, to support larger populations at higher bandwidths. NETBuilder Remote Office products will focus on smaller remote office sites by providing native ISDN BRI capability.

European ISDN implementations vary from country to country, but are now converging to a common standard, Euro-ISDN. To support these different implementations, NETBuilder Remote Office products will connect to ISDN BRI networks through an integrated ISDN platform or through external ISDN terminal adapters, as it is done today. 3Com already produces an integrated ISDN BRI version of the NETBuilder Remote for the Japanese market and will expand this to include NETBuilder Remote Office.

In the U.S., a similar situation applies. American telephone companies have been slow in deploying ISDN because of the lack of standardization. However, the Regional Bell Operating Companies (RBOCs) are now embracing a standard BRI implementation through NI1. While the process is evolving, vendors will support ISDN using both integrated and external ISDN connections to various switch vendors and ISDN service suppliers. In this way, users can begin to take advantage of all ISDN features (such as caller ID and channel aggregation) in addition to the special features that optimize WAN connectivity, described later in this chapter. NETBuilder Remote Office with integrated ISDN BRI support is another option.

CONCLUSION

Organizations today are connecting more remote sites and remote users to their central networks. This expansion will continue, improving productivity and bringing users closer to the goals of the information age. WAN reliability, cost control, and ease of management will be key factors in remote office and personal office internetworking growth.

Narrowband ISDN offers promising capabilities in this respect, with digital access, fast call setup, bandwidth aggregation, and support for a rich set of services, including dial-up services to provide better use of WAN links.

A small group of vendors are leading the way in state-of-the-art solutions for the WAN marketplace by integrating end-to-end ISDN capability into remote and personal office products. By addressing remote connectivity from both ends, ISDN technology is finding a home in a transparent high-speed solution that satisfies network managers as well as end-users.

GLOSSARY

B-channel—The bearer channel in ISDN communication carries voice, data, or video signals.

Boundary Routing system architecture—Software algorithms and methodology let a router at a central site perform protocol-specific routing on behalf of a router at a remote site, greatly simplifying remote-site router administration.

BRI—The ISDN Basic Rate Interface offers 2 B-channels and a D-channel.

D-channel—The data channel in ISDN communication carries control and signaling information to set up and breakdown connections.

DSU/CSU—A Data Service Unit/Channel Service Unit connects an external digital circuit to digital circuitry on the customers premises. The DSU converts data into the correct format, and the CSU terminates the line, conditions the signal, and takes part in remote testing.

ISDN—Integrated Services Digital Network technology is able to deliver high quality digitally encoded voice, data, and other services such as video over telephone networks.

LATA—A Local Access and Transport Area is any one of more than 150 U.S. areas in which a local telephone company is authorized to offer communications services.

PRI—The ISDN Primary Rate Interface offers 23 B-channels and a D-channel (23B+D) or 30 B-channels and a D-channel (30B+D).

RBOCs—Regional Bell Operating Companies are the local exchange carriers that provide local telephone service and access to long-distance carriers and services.

X.25—A CCITT interface standard which lets computing devices communicate via wide area packet-switched data networks.

ABBREVIATIONS AND ACRONYMS

ATM—Asynchronous Transfer Mode

BRI—Basic Rate Interface

DS1—Digital Signal, level 1

DSU/CSU—Data Service Unit/Channel Service Unit

ETSI—European Telecommunications Standards Institute

IP—Internet Protocol

ISDN—Integrated Services Digital Network

ITU-TSS—International Telecommunications Union Telecommunications Standardization Sector

LATA—Local Access and Transport Area

MIB—Management Information Base

NFAS—NonFacility Associated Signaling

NI1—National ISDN-1

NTT—Nippon Telegraph and Telephone

PBX—Private Branch Exchange

PRI—Primary Rate Interface

PTTs—Post, Telephone, and Telegraph authorities

RBOCs—Regional Bell Operating Companies

TA—terminal adapter

TE—terminal equipment

Telcos—telephone companies

AUTHOR BIOGRAPHY

Najib Khouri-Haddad joined 3Com in December 1992 as the product manager for remote office products in the Network Systems Division. Today, he is responsible for the NETBuilder Remote Office for ISDN connectivity.

Previously, Khouri-Haddad worked at ROLM, a maker of business communications systems mainly for voice applications.

Khouri-Haddad holds a Master's degree in engineering management from Stanford University and a Master's in electrical engineering from the University of Michigan.

CHAPTER 20

Enterprise Multimedia Systems—A Case Study

by Dr. Schahram Dustdar

The convergence of four industries, namely the computer, TV, and broadcasting, telecommunications and print/publishing industries into a multimedia industry brings much turbulence into the organizational domain and leads to considerable organizational redesign. This study shows some evidence of how the deployment of multimedia information systems changes the nature of business processes within Barclays bank and leads to new products, and services and enables the organization to change delivery channels to reach their customers. Barclays sees video-on-demand as the model for a range of new multimedia services which will transform retail and wholesale banking over the next years. Among the possible implications of multimedia IS are a virtual end to branch banking and its replacement by multimedia kiosks and interactive television.

The 1990s have brought and are bringing a number of evolutionary changes to the organizational and technical environments of organizations. The changes in IT are to be found in advances in speed and quality of hardware and software, the rise of powerful and transparent networks, the ease of access to distributed hardware and software, the rise of powerful and transparent networks, the ease of access to distributed databases, and the advent of multimedia IS and multimedia services. One category of strategic information systems, multimedia information systems, can be an enabling technology for Business Process Redesign, Business Process Reengineering (BPR), Business Reconfiguration or however else it is termed.

THEORETICAL FRAMEWORK FOR ANALYSIS

In this chapter we use the MIT 1990s framework[7] as a conceptual frame of reference to understand the implications of multimedia IS usage regarding organizational redesign. The theoretical perspective and the normative orientation for the analysis of the case study is "segmented institutionalism," to use Kling and Scacci's term,[4] in the domain of the "substance of theory."[5] The "causal agency"[5] used for analysis of the case is the "emergent perspective" of action.[6]

WHAT ARE MULTIMEDIA INFORMATION SYSTEMS?

The IT industry seems to be in the midst of an evolutionary "leap forward." PCs are being transformed from computation-intensive to communications-intensive devices that use in addition to text and graphics the media types of audio, animation and video. But what is multimedia, and what are multimedia IS? In this chapter we use the term multimedia IS as a new category of IS which also includes components of a communication system, such as multimedia conferencing. There is no agreed definition in the literature of what multimedia or multimedia IS are because there is so much turbulence in this area.

To discuss multimedia IS, we have to develop a working definition. The two main characteristics of a multimedia IS are being a real-time system not a "batch" system and the degree of interactivity of the system. Through the digitalization of information in all media types and the emergence of ubiquitous information networks, sometimes referred to as "information highways," some organizations are moving from their traditional organizational structure and strategy into a new and redesigned organization.

MULTIMEDIA INFORMATION SYSTEMS AT BARCLAYS BANK

Barclays bank is a UK-based leading financial institution, the 3rd largest bank in the UK and the 5th largest bank in Europe with a market capitalization of $12,749.7 million in 1993, ranking 18th of the European Top 500 companies.[1] The commitment to IT can be measured through the total investment in IT throughout the Group of £800 million during 1993. There are plans to invest £2.5 billion into new technology for the branch infrastructure over the next 3 years in addition to the £800 million that the Group already invests each year.

These investments in IT have had a major effect in both domestic and global banking, with a consequential reduction in staff numbers whilst service levels have improved. As Andrew Buxton, chairman at Barclays, says: "Much has been achieved through rethinking business processes, economies of scale and the introduction of new technology."[1]

The recently (September 1993) established Multimedia Department, operating from the BNS (Barclays Network Services) headquarters, has been formed by merging the Interactive Video Unit and Divisional Research Unit of Central Retail Services Division (CRSD). Over the years, this reorganization shows the changing and evolutionary understanding of multimedia IS—from a separated Video Unit to integrated multimedia IS development—and the growing need to integrate video and audio technologies and information systems into networked multimedia IS. The head of the Multimedia Department reports directly to the director of alliances at Barclays, since multimedia is seen as a strategic opportunity to broaden the business network and to redesign the business. The head of Multimedia at Barclays has formulated the multimedia motto: "Our objective is to have the information available when and where the customer wants it." Barclays Multimedia has a staff of over 15 and a budget of over one million £ a year.

The main objectives of the Multimedia Department at Barclays are:

- Evaluating Multimedia Information Systems through pilot studies
- Developing electronic catalogues for retailers and Barclays Group Companies
- Developing a specification for a generic Multimedia application framework for retailers
- Developing a Software environment for Multimedia Information Systems for retailers
- Developing and testing online loan services
- Developing and testing online plastic card transactions

STRATEGIC MULTIMEDIA IS AT BARCLAYS

Barclays' exploitation of multimedia IS shows a growing need to change the delivery channel from the branch infrastructure to a digital delivery channel. The potential target market served through a digital delivery channel is a mass market, although multimedia IS will be used to customize the financial products. Hence, multimedia services are highly customized but delivered through a mass delivery channel, such as interactive TV or multimedia kiosks. The challenge of Barclays regarding design issues of multimedia services is to find a balance between customized IS and mass IS. The commitment to a customized product becomes clear in the following statement: "We recognize that customers are not alike, and we are becoming more focused in order to meet each customer's needs and desired level of service."[1]

In the past, selling of financial services was done solely through bank branches and traveling business bankers. Barclays' multimedia IS try to target the potential customers through redesigning its organizational boundaries as it is shown in Figure 20.1. The changing process of a "store-based shopping" to a "shopless shopping" paradigm is being implemented with multimedia IS. Financial services can be transmitted to the workplace, to the home, and to public access ATM multimedia kiosks basically via the same technology and software.

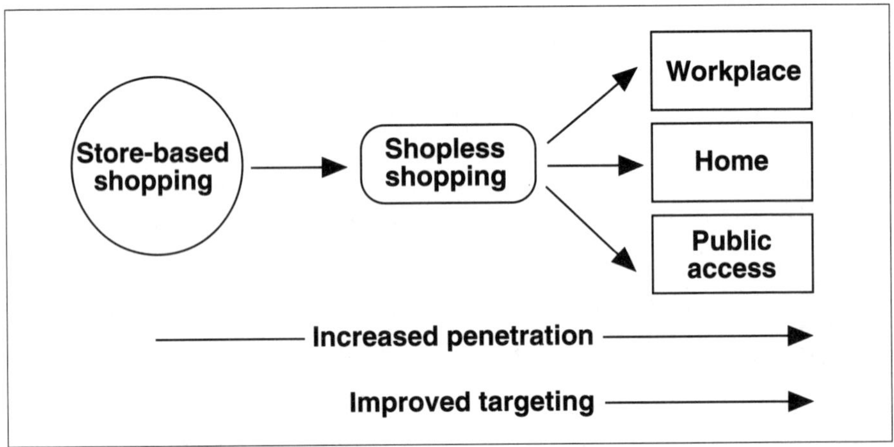

FIGURE 20.1. Barclay's store-based shopping

The process of implementing multimedia IS is given highest priority, as Joseph De Feo, head of information technology, says: "We thought we had time. We planned to have the program motoring by 1996-97 based on our estimates of the investment in infrastructure the telcos would have to make to deliver these services. They made faster progress. Here and in the US, the telcos can provide video-on-demand over the existing telephone network. We don't have five years now, and this is putting pressure on developments."[3] The main expected benefits of multimedia IS at Barclays are:

Customer

- Improved service
- Effective needs analysis
- Impartial advice

- Convenience
- Financial modeling allowed

Staff

- Sales aid
- Free sales staff to concentrate on selling
- Training
- Branch office procedures
- Fraud/security issues via video and audio in scenarios
- Professionalism

Operational

- Branch infrastructure
- 24-hour availability
- Interactive sales Terminals (IST)
- Recruiting new customers in public places
- Online networked services/information

Barclays develops multimedia IS for "internal" use and for "external" use. Internal multimedia IS are systems such as "Career Builder," which is a Point of Information (POI) system. It contains a database of relevant job information and career opportunities to which users can interactively match their own skills. Multimedia training IS are used in Barclays sixty-three "Learning Centres." External systems are targeted at a mass market, although the products and services are being customized via multimedia components. External systems are the Homestead projects, the Business Needs Analyser, and Touchbank. Figure 20.2 gives an overview of the internal and external multimedia IS developed in Barclays.

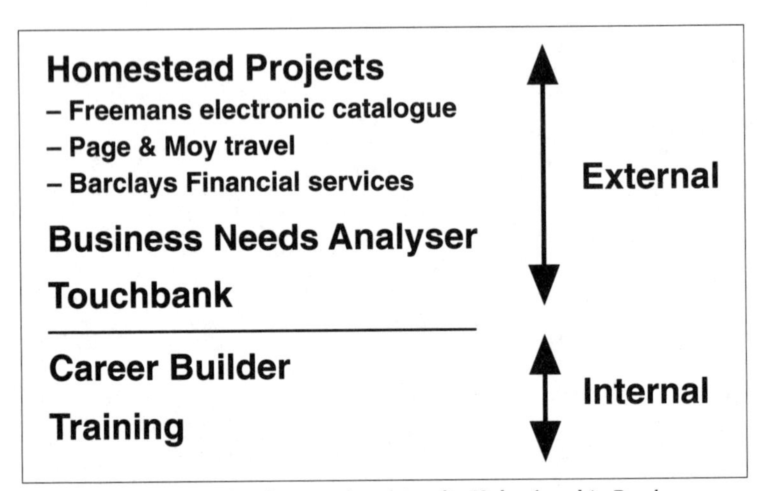

FIGURE 20.2. Internal and external multimedia IS developed in Barclays

CONCLUSION

As stated earlier, Barclays is now at the stage of exploiting multimedia information systems, hence the answer to the research questions are based on the early usage of multimedia IS. Regarding the first question, which multimedia IS play an important role for business process redesign, it is important to note that the emphasis is on the word "role." We agree with Ciborra and Jelassi[2] that the source of advantage cannot lie only in the possession of a unique sophisticated system, in this case a multimedia IS, be it internally or externally. Economic and technological forces push organizations to develop such systems jointly and even open them to competitors as the classical cases of strategic IS, such as McKesson, American Hospital Supply (now Baxter), and American Airlines show. The new challenge is to harness IT to tap the core competencies of the organization to create new information and knowledge or new content, as suggested in Figure 20.3.

In the case of multimedia IS, Barclays is pushed to ask itself the main strategic question: Which business are we in, and what are our core competencies? The answer to these questions is changing, as we can see through the deployment of CD-I and in the near future interactive TV as a delivery channel and the "information highway" as a networked infrastructure for the delivery of services and new products, through a multimedia mass IS.

Networked multimedia systems, such as a public access kiosk, multimedia PCs, or interactive TV devices at home, need to have a two-way interactive link to Barclays' staff via Desktop-Videoconferencing. Via Desktop-Videoconferencing, Barclays staff can offer advice and consultation services to clients. John Malty, deputy director of delivery channel automation at Barclays, has reflected the concern about multimedia systems video-quality in his comment: "For staff support we can work with current technology, but for direct customer facing it needs to be of much higher quality."

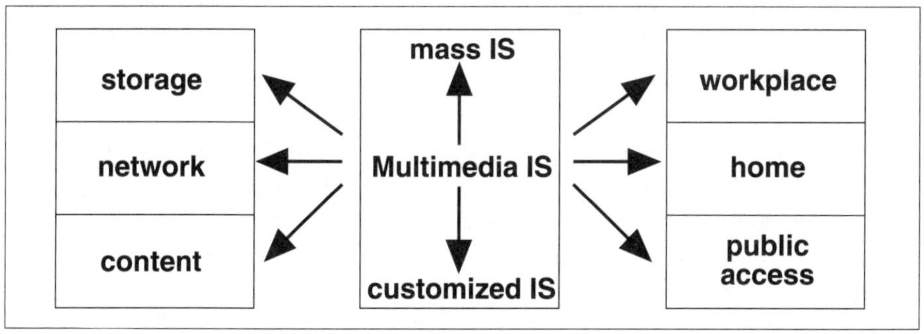

FIGURE 20.3. The new multimedia

Multimedia IS plays an enabling role at Barclays in the process of redesigning and rethinking core competencies and skills. The most important multimedia IS used in Barclays exploitation is interactive TV. Interactive TV will be the successor to CD-I systems and will redesign Barclays fundamentally. The delivery of customized multimedia services to a mass audience via CD-I, interactive TV or however else the technology is termed in future, is essential. The delivery channel will be changed from a branch infrastructure to a direct digital customer link into the home, public places and the workplace. The second key multimedia technology involved is the integration of Desktop-Videoconferencing capabilities into kiosks and other applications. This provides a direct link from the Point of Information or Point of Sales to Barclays staff, regardless where the staff is located—in offices, at home, or traveling.

However, as stated in the theoretical framework for analysis of this case above, MIT's 1990s framework shows the importance of organizational and social issues and their interconnection with technology.

Multimedia IS, and particularly networked multimedia IS, can be seen as an enabling technology to move Barclays into the state of a networked organization. The organizational challenge lies in the increasing need to manage distributed complexity. Through strategic alliances, Barclays is redesigning its business scope and developing new services and targeting customers differently than before. The problems of networked multimedia IS and the development of new services are closely linked with parties other than Barclays. Hence, Barclays is in close discussions with major multimedia companies such as, BT, AT&T, Mercury, and with computer suppliers. An internal program is looking at possible alliances and joint ventures. To offer truly global—or at least European—services, Barclays depends heavily upon regulatory situations since telecommunications and the associated network infrastructure (information highway) is still more or less regulated. Within the theoretical frame of "segmented rationalism",[4] the case shows some evidence that through the redesigning of organizational boundaries (see Figure 20.3) and the move to direct digital customer links to workplace, home, and public places, multimedia IS will increase the impact on social and organizational life tremendously. However, to measure and study the scope and degree of the impact, research of multimedia IS depends on its advanced status in organizational and social realities.

In answer to how multimedia IS redefine business processes, we can summarize that delivery channels shift from branch infrastructure to digital direct linkage to the customers. Hence, the way in which and the procedures by which customers learn about products and new services changes radically. Currently the shift from one-way communications to two-way, which is enabled by networked multimedia IS, has and will have strong effects on consumer marketing. Furthermore, remote Barclays staff gives advice via Desktop-Videoconferencing links to customers and potential customers.

The third research question asks what new products and services, markets, and delivery channels will evolve. The following statement by the chairman of Barclays makes quite clear what structural consequences multimedia IS will have at Barclays branch infrastructure: "I still believe that branches perform a valuable function at the interface between the bank and its customers, but technology and changing shopping habits mean that there will be fewer branches in the future."[1]

During 1993, the number of people employed in the Group was reduced by 7,200 to 97,800. In 1988, the Group employed 118,400 people, so there has been a steady reduction over the last few years. New products and services currently are being explored by the external multimedia Homestead projects. Through electronic alliances Barclays is bundling products with its financial service, as Freemans catalogue and Page & Moy travel multimedia IS evidence. If the travel catalogue of Page & Moy could be delivered via interactive TV to the customer. Customers can choose a vacation press the desired continent, a list of special requirements and dates—in fact everything they would normally do in a travel agency, and then call up video presentations of the resorts, including restaurants, hotels, and sights. The key services which Barclays would offer are payment services, personal loans, possibilities to change foreign currencies, travel insurance, and so forth. Similarly, the customer can select a car from the visuals, pick the upholstery, book a test drive, and arrange financial details.

Regarding the core competencies and skills, it is very important to mention that some 30,000 employees at Barclays have been through the multimedia training courses which were delivered at the sixty-three Learning Centres, but there are plans to deliver them through networked multimedia PCs to the employees desktop very soon. As the chairman puts it: "Changes in work practices, particularly in the UK Domestic Bank, have meant significant extra resources being devoted to training. We also respect the changing nature of our workforce. We allow increased flexibility in working arrangements—job sharing, a career break scheme for child care, responsibility breaks to care for elderly or sick people."[1]

Figure 20.3 shows the possible relationships between a service provider on the left and its client on the right. In the case of Barclays, we can see an exploratory move of the organization into the three domains. Joseph De Feo says: "Multimedia will make possible a range of services, but the customer will use an agent to gain access to them. The agent will be the brand. We would not want to get into agent work outside things we know about. But we believe we have to be a partner with the telcos, or technology companies, in developing agent activities. We could find that we will get as much in the way of revenues out of our participation in the agent business as we do out of banking."[3]

To cope with the changing nature of work and the workforce it requires the support of multimedia IS. This is an evolutionary development of IS as an enabling technology, first a competitive advantage to some organizations who deploy it, later a necessity for survival of the organization.

Organizations will have to think radically about how multimedia IS will affect them. As De Feo confirms, banks of the future will no longer control the delivery channel because customers will no longer physically go to branches. Multimedia IS fundamentally will change the way organizations do business. Moreover, multimedia IS have significant impact on social life and work itself.

ACKNOWLEDGMENTS

I wish to thank Anwer Shah, Paul Gold, Grant McKee, Prakash Jani and Peter Woods of Barclays Multimedia for their cooperation in this study.

REFERENCES

[1]*Barclays Bank: Reports and Accounts*, London 1994.

[2]Ciborra, C.; Jelassi, T. (Eds): *Strategic Information Systems: A European Perspective.* John Wiley & Sons, Chichester 1994.

[3]"Barclays In-house Revolution," *Financial Times.* 20. June 1994.

[4]Kling, R, W. Scacchi. "The Web of Computing: Computer Technology as Social Organization," *Advances in Computers*. 21. 1982. pp. 1-90.

[5]Markus, M. L., D. Robey. "Information Technology and Organizational Change: Causal Structure in Theory and Research," *Management Science* 34, 5. 1988. pp. 583-598.

[6]Pfeffer, J. *Organizations and Organization Theory.* Pitman, Marshfield, MA. 1982.

[7]Morton, Scott, M.S. (ed.). *The Corporation of the 1990s: Information Technology and Organizational Transformation.* Oxford University Press, New York. 1991.

AUTHOR BIOGRAPHY

Dr. Schahram Dustdar is head of the Department of Central Informatics (ZID) at the University of Art at Linz-Austria, where he heads the Austrian National Support Centre for MICE (Multimedia Integrated Conferencing for European Researchers), an EU-Project which supports the implementation and use of MBONE Multimedia conferencing tools on the Internet. His research interests are organizational impacts of new information technology, multimedia information systems, and the Internet. He was visiting research fellow at the London School of Economics, Department of Information Systems during 1993/94. He can be reached at dustdar@khsa.khs-linz.ac.at.

CHAPTER 21

Automated Tinker Toys for Developers: The Paradigm of Development Before the Fact

by Margaret H. Hamilton, Hamilton Technologies, Inc.

What would it take to automate software development? What would truly cost-efficient and cost-effective development look like? What if changes to the requirements of a system could be made at any point in the process—without having to change the code manually?

Recently the National Test Bed of the U.S. Department of Defense sponsored an experiment in which it provided a development problem to each of three contractor/vendor teams chosen from a large pool of vendors. The application was real-time, distributed, multiuser, client/server, and needed to be defined and developed under government 2167A guidelines.

All teams were able to complete the first part, the definition of preliminary requirements. Two teams completed detailed design. But only one team was able to generate complete and fully production-ready code automatically; a major portion of this code was running in both C and Ada at the end of the experiment.[1]

The team that was able to generate the production-ready code was using the 001 Tool Suite, which is based on a paradigm called *development before the fact*. A preventive approach, development before the fact means doing things right the first time. Systems

[1]*Software Engineering Tools Experiment-Final Report, Vol. 1*, "Experiment Summary." Table 1, p. 9, Department of Defense, Strategic Defense Initiative, Washington, D.C., 20301-7100.

based on development before the fact are carefully constructed to minimize development problems from the outset. The result is systems with properties that control their own design and development—reusable systems that promote automation. Each system definition models both its application and its life cycle.

Development before the fact systems are constructed like tinker toys, reusing effective components. Effective reuse is a preventive concept: reusing something with no errors to obtain a desired functionality avoids both the errors and the cost of developing a new system. It allows one to solve a given problem as early as possible, not at the last moment. To make a system truly reusable, however, one must start not from the customary end of a life cycle, but at the beginning.

THE TRADITIONAL APPROACH TO DEVELOPMENT

To understand development before the fact, it is necessary to know something about today's traditional system engineering and software development environments. Simply put, they are slow and inefficient. In government in the 1980s, an average developer using C was expected to produce only 10 lines of code per day; today the benchmark is more like 2 to 5 lines a day. Support to users ends up being the norm, which is fixing broken things, rather than building things that work. Innovations come too late, if at all. The result is mediocre systems that cost more than they ought—and more than they are worth to develop. One can never know whether a design is a good one until successful implementation.

Designers must think and design this way because of the limitations of available methodologies. In defining requirements, designers use mismatched methods to capture aspects of even a single definition. Data flow, state transitions, dynamics, data types, and structures all are defined using different methods. After definition, there is no way to integrate them.

The problems multiply. Integration of object to object, module to module, phase to phase, or type of application to type of application becomes even more of a challenge than solving the problem at hand. The mismatch of products used in design and development compounds the issue. Integration of all forms is left to the devices of a myriad of developers well into the development phase of the project. The resulting system is hard to understand, objects cannot be traced, and at best there is little correspondence to the real world. A consequence is to promote systems defined as ambiguous and incorrect. Interfaces are incompatible and errors propagate throughout development.

Definitions of requirements in traditional development scenarios concentrate on the application needs of the user, without consideration of the potential for the user's needs or environment to change. Porting becomes a new development for each new architecture, operating system, database, graphics environment, language, or language configuration. Critical functionality is avoided for fear of the unknown, and maintenance is both risky and the most expensive part of the life cycle. When a system is targeted for a distributed environment, it is often first defined and developed for a single processor environment.

System definitions contain insufficient information about a system's run-time performance, including that concerning the decisions between algorithms or architectures. System definitions do not consider how to separate the system from its target environment. Design decisions thus depend on analysis of outputs from *ad hoc* implementations and associated testing scenarios.

The customary focus for reuse is late into development, during the coding phase. Requirements definitions lack properties to help find, create, and make use of commonality. Modelers are forced to use informal and manual methods to find ways to divide a system into components natural for reuse. Modules do not lend themselves to integration, and they are error-prone. Because systems are not portable or adaptable, there is little incentive for reuse. Redundancy becomes a way of doing business.

In fact, though, automation itself is an inherently reusable process. If a system does not exist for reuse, it certainly does not exist for automation. But most of today's development process is needlessly manual. Today's systems are defined with insufficient intelligence for automated tools to use them as input. In fact, automated tools concentrate on supporting the manual process instead of doing the real work. Typically, developers receive definitions which they manually turn into code. A process that could have been mechanized once for reuse is performed manually again and again.

Under this scenario, even when automation attempts to do the real work, it is often incomplete across application domains or even within a domain, resulting in incomplete code such as shell code. The generated code is often inefficient or hard-wired to an architecture, a language, or even a version of a language. Often partial automations need to be integrated with incompatible partial automations or manual processes. Manual processes are needed to complete unfinished automations.

DEVELOPMENT BEFORE THE FACT

The environment in which the new paradigm developed taught some important lessons. In fact, the development before the fact approach was derived from the combination of steps taken to solve the problems of traditional systems engineering and software development. What makes development before the fact different is that it is preventive rather than curative. Consider such an approach in its application to a human system. To fill a tooth before it reaches the stage of a root canal is curative with respect to the cavity, but preventive with respect to the root canal. Preventing the cavity by proper diet prevents not only the root canal, but the cavity as well. To follow a cavity with a root canal is the most expensive alternative; to fill a cavity on time is the next most expensive; to prevent cavities is the least expensive option.

Preventiveness is a relative concept. For any given system, be it human or software, one goal is to prevent, to the greatest extent and as early as possible, anything that could go wrong in the life cycle.

The philosophy behind development before the fact is that it is inherently reusable: reliable systems are defined in terms of reliable systems. Only reliable systems are used as building blocks, and only reliable systems are used as mechanisms to integrate these building blocks to form a new system. The new system becomes reusable for building other systems.

From the very beginning, a development before the fact system inherently integrates all of its own objects (and all aspects, relationships and viewpoints of these objects) and the combinations of functionality using these objects; maximizes its own reliability and flexibility to change (including reconfiguration in real time and the change of target requirements, static and dynamic architectures, and processes); capitalizes on its own parallelism; supports its own run-time performance analysis; and maximizes the potential for its own reuse and automation. It is defined with built-in quality and built-in productivity.

The development before the fact technology includes a language, an approach, and a process (or methodology) based on a formal theory.

LANGUAGE

Once understood, the characteristics of good design can be reused by incorporating them into a language for defining any system. The language, meta-language really, is the key to development before the fact. It can define any aspect of any system and integrate it with any other aspect. These aspects are directly related to the real world. The same language can be used to define system requirements, specifications, design, and detailed design for functional, resource, and resource allocation architectures throughout all levels and layers of seamless definition, including hardware, software, and people-ware.

The language can be used to define missile or banking systems, as well as real-time or database environments. It even can be used to define and integrate function-oriented decompositions (control hierarchies) with object-oriented decompositions. The language also can be used to define and integrate these decompositions with networks of functions and objects, and to define systems with diverse degrees of fidelity and completeness. Such a language can always be considered a design language, since design is relative; one person's design phase is another person's implementation phase. Semantics-dependent but syntax-independent, the language has mechanisms that are themselves used to define mechanisms for defining systems. Although the core language is generic, the user "language," a by-product of a development, can be application-specific.

APPROACH

Development before the fact is a function and object-oriented approach based on a unique concept of control. The foundations are based on a set of axioms and on the assumption of a universal set of objects. Each axiom defines a relation of immediate domination. The union of the relations defined by the axioms is control. Among other things, the axioms establish the relationships of an object for invocation, input and output, input and output access rights, error detection and recovery, and ordering during its developmental and operational states. Table 21.1 summarizes some of the properties of objects within development before the fact systems.

Object-Oriented Properties of Development Before the Fact

CONTROL:	*The ability to have an object control and be controlled throughout all phases of its development and its birth, life, and death of operation.*
CREATION:	The ability to create an object.
REFERENCE:	The ability to use or mention an object.
UNIQUENESS:	The ability to select or find an object.
DESTRUCTION:	The ability for an object and all of its influences to be destroyed.
IDENTIFICATION:	The ability to identify an object with respect to its structure, its behavior, its relationships in development and in operational real time and real space with respect to an integrated set of all aspects of control.
CLASSIFICATION:	The ability to belong to the same class with other objects, each of which shares a common set of properties
TRACEABILITY:	The ability to trace the birth, life, and death of an object and its definitions, as well as its transitions between definition and instantiation; trace changes and *know* their effects; trace patterns (e.g., distributed patterns), control, and function flow (including data, priority access rights and timing).
ACCESSIBILITY:	The ability to safely access an object in all of its states of existence.
BOUNDARY CONDITION:	The ability to safely exclude invalid states of an object.
SECURITY:	The ability for an object to keep its behavior and structure secure (e.g., communication of one function with another function always takes place at the same level in a hierarchy and a given object has no knowledge of a higher level object.)
BELONGING:	The ability of an object to belong to, for example, a parent or a set of values.
HAVING:	The ability for all object to have for example, a child, sibling, a set of proper values, and an architecture.
TIMING:	The ability to instantiate an object at a given time or a given event.
MODULARITY:	The ability for an object to be portable, flexible and reusable.

HEALTHY EXISTENCE:	The ability for an object to live a full life throughout all of its states of being and doing (persistence).
COMPLETION:	The ability to determine when an object is completely defined, used, or instantiated.
INHERITANCE:	The ability of an object to derive behavior in terms of other objects.
REAL TIME and SPACE CONSTRAINTS:	The ability to realize an object in terms of its physical existence.
MINIMALITY:	The ability to define necessary and sufficient information about an object.
CONTAINMENT:	The ability of an object to have an inside and an outside (encapsulation). The outside view of an object may be completely replaced by the inside view of the object in question.
REPRESENTATION:	The ability for an object to have a natural correspondence to the desired aspects of the real-world object of which it is a model.
ORDERING:	The ability to establish a relation in a set of objects so that any two object elements are comparable in that one of said elements precedes the other said element.
PRIORITY:	The ability to determine when a particular object is more important than any other, given constraints such as time, priority, order, events, and architecture.
RELATIVITY:	The ability for an object to change roles depending on how it is being used (polymorphism) or viewed. This includes being vs. doing, controller vs. controllee, parent vs. children, requirements vs. implementation.
STRUCTURE:	The ability to distinguish between properties of dependence, independence, and decision making within, between and about objects.
PREDICTABILITY:	The ability for the behavior and structure of an object to be understood in terms of its relationships without ambiguity.

TABLE 21.1. Object-Oriented Properties of Development Before the Fact

This approach is used throughout a life cycle, starting with requirements and continuing with functional analysis, simulation, specification, analysis, design, system architecture design, algorithm development, implementation, configuration management, testing, maintenance, and reverse engineering. Its users include end users, managers, system engineers, software engineers, and test engineers.

The development before the fact approach had its beginnings in 1968 with the Apollo space missions. Research for developing software for man-rated missions led to the finding that interface errors accounted for approximately 75% of all errors found in the flight software during final testing (in traditional development, the figure is as high as 90%). Such errors include data flow, priority, and timing errors from the highest levels of a system to the finest detail. Each error was categorized according to how it could be prevented just by the way a system is defined. This work led to a theory and methodology for defining a system that would eliminate all interface errors.

The first technology derived from this theory concentrated on defining and building reliable systems in terms of functional hierarchies.[2] Having realized the benefits of addressing one major issue, such as reliability, just by the way a system is defined, research continued to evolve the philosophy by addressing other major issues the same way. Now systems are designed and built with development before the fact properties integrating both functional and type hierarchies.[3]

PROCESS

The first step in building a before the fact system is to define a model with the language. This process could be in any phase of development, including problem analysis, operational scenarios, and design. The model is automatically analyzed to ensure that it was defined properly. This includes static analysis for preventive properties and dynamic analysis for user intent properties.

In the next stage, the generic generator automatically generates a fully production-ready and fully integrated software implementation for any kind of application, consistent with the model, for a selected target environment in the language and architecture of choice. If the selected environment has already been configured, the generator selects that environment directly; otherwise, the generator is first configured for a new language and architecture.

It then becomes possible to execute the resulting system. If it is software, the system can undergo testing for further user intent errors. The system becomes operational after testing. Changes are made to the requirements definition, not to the code. Target architecture changes are made to the configuration of the generator environment, not to the code. If the real system is hardware or people-ware, the software system serves as a simulation upon which the real system can be based. Once a system has been developed, the system and the process used to develop it are analyzed to understand how to improve the next round of system development.

[2]Hamilton, M. *Zero-Defect Software: The Elusive Goal*, "IEEE Spectrum," Vol. 23, No. 3. March 1986. pp. 48-53

[3]Hamilton, M. and R. Hackler, *001: A Rapid Development Approach for Rapid Prototyping Based on a System that Supports its Own Life Cycle*, "*IEEE Proceedings, First International Workshop on Rapid System Prototyping*," Research Triangle Park, NC, June 4, 1990.

THE STRUCTURE OF DEVELOPMENT BEFORE THE FACT SYSTEMS: AN INTEGRATED MODELING ENVIRONMENT

The beauty of development before the fact systems is their logic. In a development before the fact system, students of logic or philosophy may appreciate watching, as one parent after another decomposes into its children, then inherits back from them. Most people probably will not react that way, though. Where possible, in what follows, then, there has been an effort to define terms and to proceed logically from one piece of the system to the next.

RMaps, TMaps and FMaps, OMaps and EMaps

Although development before the fact is an object-oriented approach, it can be said to integrate function-oriented and object-oriented development. The definition space is a set of real-world objects defined in terms of both functional hierarchies, FMaps, to capture time characteristics and type hierarchies, TMaps, to capture space characteristics (see Figure 21.1).

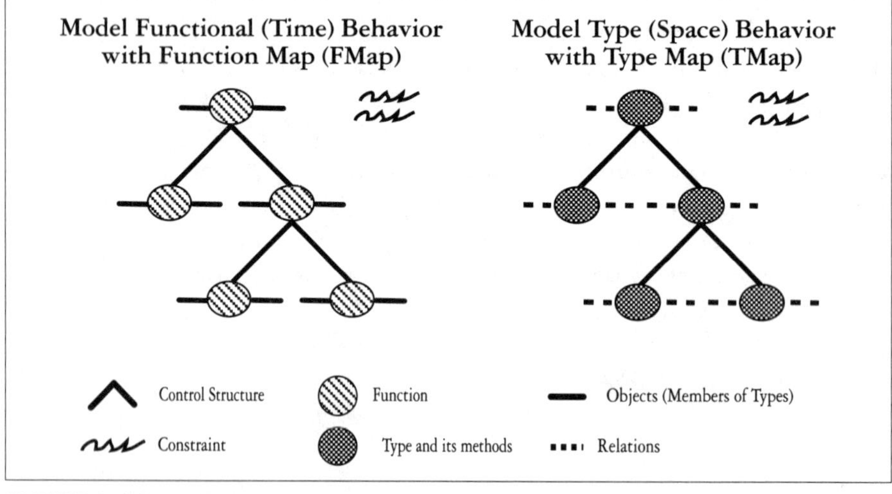

FIGURE 21.1. 001 Maps

A map is both a control hierarchy and a network of interacting objects. FMaps and TMaps guide the designer in thinking through concepts at all levels of system design. These hierarchies make available everything a developer needs to know, no more, no less. All model viewpoints can be obtained from FMaps and TMaps, data flow, control flow, state transitions, data structure, dynamics. Maps of functions are integrated with maps of types.

On each node of an FMap is a *parent function* that is defined in terms of and controls its *children functions*. For example, the function *build the table* could be decomposed into and control its children functions, *make parts* and *assemble*. The simplest of these decomposed functions is called a *primitive function*. Similarly, on each node of a TMap there is a parent type that is defined in terms of and controls its children types. For example, the type *table* could be decomposed into and control its children types *legs* and *top*. The simplest of these types is called a *primitive type*. Primitive types reside at the bottom nodes of a TMap. Every type on a TMap owns a set of inherited primitive operations. Similarly, primitive functions on types defined in the TMap reside at the bottom nodes of an FMap. Each function on an FMap has one or more objects as its input and one or more objects as its output.

Each object is a member of a type from a TMap, and immediately resides in an object hierarchy, an *OMap*. In fact, OMaps are instantiations of TMaps. FMaps have a counterpart in *EMaps*, or execution hierarchies, in which every action is a member of a function from an FMap, and which contain all of the object values plugged in for a particular performance pass.

FMaps are inherently integrated with TMaps through objects and their primitive operations. FMaps are used to define, integrate, and control the transformations of objects from one state to another (e.g., a table with a broken leg to a table with a fixed leg).

Typically a team of designers begins to design a system (hardware, software, peopleware, or some combination) at any level by sketching a TMap of the application. This is where they decide on the types of objects and the relationships between these objects in their system. Often a road map, or *RMap*, is sketched in parallel with the TMap. An RMap provides an index or table of the contents of the user's system of definitions and supports the managers in the management of these definitions, including those for FMaps, TMaps, defined structures (described below), primitive types, objects brought in from other environments, and other RMaps.

Once an agreement has been reached on the TMap, the FMaps begin almost to fall into place for the designers because of the natural partitioning of functionality (or groups of functionality) provided to the designers by the TMap system. The TMap provides the structural criteria from which to evaluate the functional partitioning of the system. For example, the shape of the structural partitioning of the FMaps is balanced against the structural organization of the shape of the objects as defined by the TMap. With FMaps and TMaps, a system and its viewpoints is divided into functionally natural components and groups of functional components which naturally work together.

Primitive Structures. All FMaps and TMaps are ultimately defined in terms of three *primitive structures*: *Join*, which corresponds to a dependent relationship; *Include*, which corresponds to an independent relationship, and *Or*, which corresponds to a decision making relationship. A formal set of rules is associated with each primitive structure. If these rules are followed, interface errors are "removed"—that is—prevented, and all interface errors are eliminated at the requirements phase.

A hypothetical system for making a table demonstrates the use of primitive structures (see Figure 21.2).

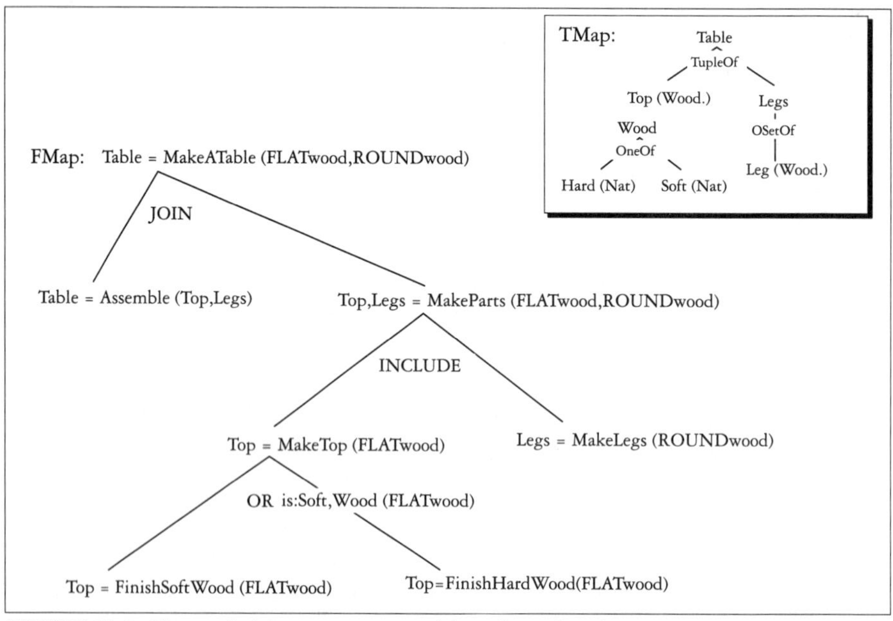

FIGURE 21.2. Three primitive structures are ultimately used to decompose a map

The *FMap* itself is called *MakeATable.* The top node of the function has *FLATwood* and *ROUNDwood* as its inputs and produces *Table* as its output. The parent *MakeATable* is decomposed with the primitive structure *Join* into its children functions, *MakeParts* and *Assemble.* MakeParts takes *FLATwood* and *ROUNDwood* as input from its parent and produces *Top* and *Legs* as output. *Top* and *Legs* are given to *Assemble* as input. *Assemble* is controlled by its parent to depend on *MakeParts* for its input. *Assemble* produces *Table* as output and sends it to its parent.

MakeParts as a parent is decomposed into children, *MakeLegs* and *MakeTop*, who are controlled to be independent of each other with *Include. MakeLegs* takes in part of its parent's input, and *MakeTop* takes in the other part. *MakeLegs* provides part of its output (*Legs*) to its parent, and *MakeTop* provides the rest. *MakeTop* controls its children, *FinishSoftWood* and *FinishHardWood*, with Or. Here, both children take in the same input objects and provide the same output objects, since only one of them will be performed for a given performance pass. *FinishSoftWood* will be performed if the decision function *is:Soft, Wood* returns true; otherwise, *FinishHardWood* will be performed. Notice that like all other objects in development before the fact systems, input (e.g., *FLATwood*) is traceable down the system from parent to children and output (e.g., *Table*) is traceable up the system from children to parent. *MakeATable*'s TMap, *Table*, uses non-primitive structures, a concept discussed below.

Each type on a TMap controls its children with structures called *parameterized types*. A parameterized type is a defined structure that provides the mechanism to define a TMap without its particular relations being explicitly defined. The TMap associated with the MakeATable example illustrates some parameterized types. *Table* as a parent type controls its children types, *Top* and *Legs*, in terms of a *TupleOf* parameterized type; *Legs* controls its child, *Leg*, in terms of *OSetOF*; and *Wood* controls *Hard* and Soft with a *OneOf*. A *TupleOf* is a collection of a fixed number of possibly different types of objects; *OSetOf* is a collection of a variable number of the same type of objects (in a linear order); and *OneOf* is a classification of possibly different types of objects from which one object is selected to represent the class. These parameterized types, along with *TreeOf*, can be used for designing any kind of TMap. *TreeOf* is a collection of the same type of objects ordered using a tree indexing system.

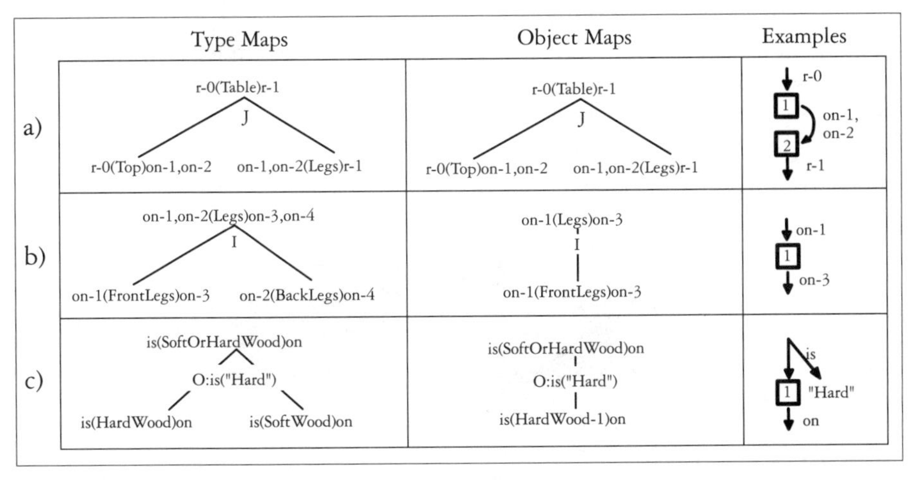

FIGURE 21.3. A TMap (and its corresponding OMaps) can be decomposed into its explicit relationships in terms of three primitive structures

A TMap and its corresponding OMaps can be further decomposed into explicit relationships. In Figure 21.3, *Table* as a parent has been decomposed into its children, *Top* and *Legs*, where the relations between *Top* and *Legs* are on1 and on2, the relation between *Table* and *Legs* is r1, and the relation between *Table* and *Top* is r0. Notice that *Top* depends on *Legs* to make a *Table* (see Figure 21.3a). In contrast, the front legs and the back legs of the *Table* are independent (Figure 21.3b). The *Table* may have *FrontLegs* or *BackLegs*, or both *FrontLegs* and *BackLegs* at once.

Figure 21.3c illustrates a decision structure for objects. The pattern of the OMap is different from the dependent and independent structures that characterize the TMap, since only one object is chosen to represent its parent for a given instance.

Defined Structures and Parameterized Types. Any system can be defined completely using only the primitive structures, but non-primitive structures can be derived from more primitive ones to accelerate the process of defining and understanding a system. For example, user-defined structures and parameterized types can be created for asynchronous, synchronous, and interrupt scenarios used in real-time, distributed systems. Similarly, retrieval and query structures can be defined for client/server database management systems. Non-primitive structures can be defined for both FMaps and TMaps.

CoInclude is an example of a system pattern that happens often when using primitive structures (Figure 21.4a). Its FMap was defined with primitive structures. Within the CoInclude pattern, A and B are the only leaf node functions that change. The CoInclude pattern can be defined as a non-primitive structure in terms of more primitive structures using *defined structures*. This concept was created for defining reusable patterns. Included with each structure definition is the definition of the syntax for its use (Figure 21.4b). Its use (Figure 21.4c) provides a "hidden repeat" of the entire system as defined, but explicitly shows only those elements subject to change (that is, functions A and B). The CoInclude structure is used in a similar way to an Include structure, except CoInclude provides more flexibility in repeated use, ordering, and selection of objects. Like primitive structures, defined structures have rules. Rules for non-primitive structures are derived ultimately from the rules of the primitive structures.

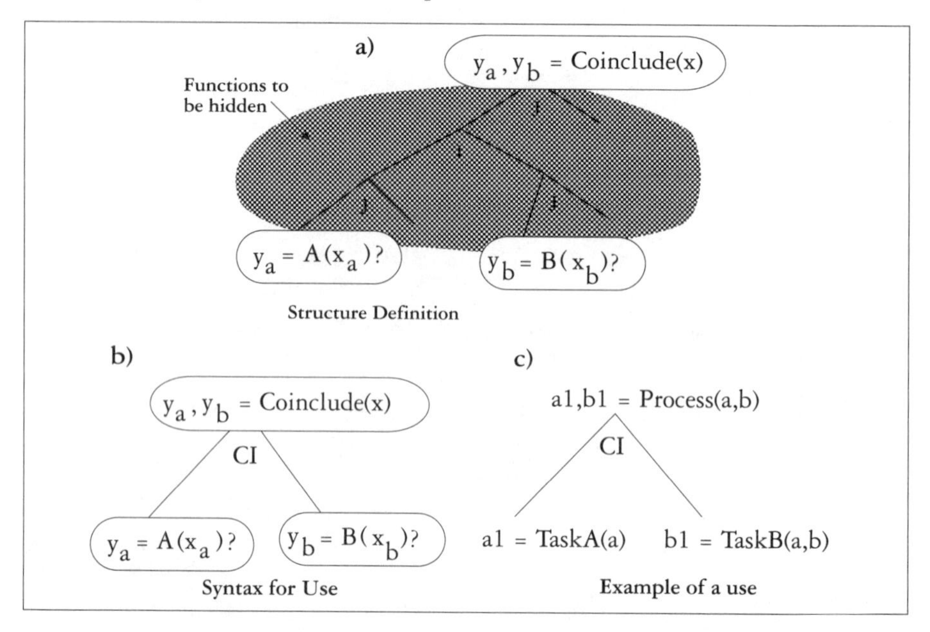

FIGURE 21.4. Defined structures are used to define nonprimitive structure reusables in terms of more primitive structures. CoInclude is an example of a system pattern that has been turned into a defined structure.

Async (Figure 21.5) is a real-time, distributed communicating structure with both asynchronous and synchronous behavior. The Async system was defined with the primitive *Or*, *Include*, and *Join* structures, and the *CoInclude* non-primitive structure. It cannot be further decomposed, since each of its lowest-level functions is a primitive function on a previously defined type (Identify2:Any and Clone1:Any under End), a recursive function (Async under DoMore), or a variable function for a defined structure (A and B under Process). If a leaf node function does not fall into any of these categories; it can be further decomposed or it can refer to an existing operation in a library or an external operation from an outside environment. Coordinate uses Async as a reusable where two robots, DecideNextStep and PerformTask, are working together to perform a task such as building a table. Here one phase of the planning robot, Master0, is coordinated with the next phase of the slave robot, Slave0.

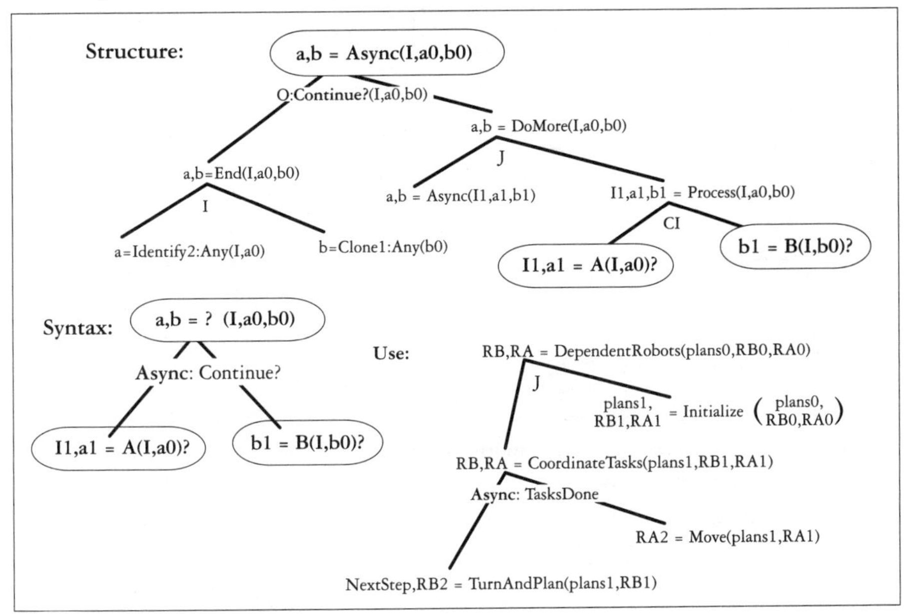

FIGURE 21.5. Async is a defined structure that can be used to define distributed systems with both synchronous and asynchronous behavior.

INTEGRATING FMAPS AND TMAPS: A SYSTEM

Figure 21.6 shows a complete system definition for a manufacturing company defined using an integrated set of TMaps and FMaps. This company could be set up to build tables with the help of robots to perform tasks using structures such as those defined above. Since this system is completely defined, it is ready to be automatically developed to end up with complete, integrated, and fully production ready to run code. This system's FMap, FullTime_Employee, has been decomposed until reaching primitive functions on types in TMap, MfgCompany. MfgCompany has been decomposed until its leaf nodes are primitive types or defined as types that are decomposed in another TMap.

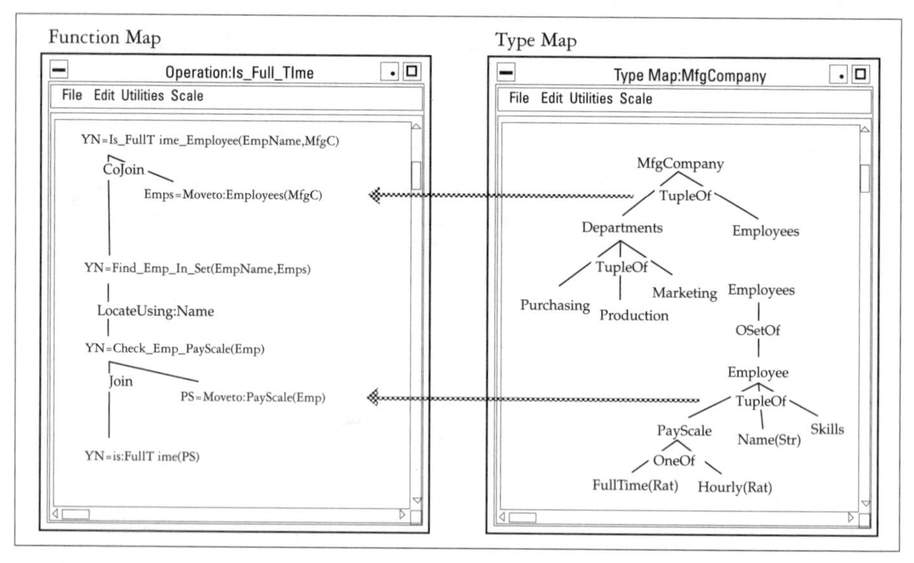

FIGURE 21.6. A complete system definition is an integration of FMaps that have been decomposed until reaching primitive functions on types in the TMaps and TMaps that have been decomposed until reaching primitive types

This system uses objects defined by its TMap to see whether an employee is fulltime or part-time. The first move is from the company type object to a type object for employees. A defined structure then locates an employee using a name. Once the employee has been found, the next move is to an object representing payscale. A primitive function then determines from the payscale object whether the employee is full or part-time.

The use of parameterized types gives this TMap model reusability. Each parameterized type assumes its own set of possible relations for its parent and children types. In this example, the TMap MfgCompany is decomposed into *Departments* and *Employees* in terms of *TupleOf*. *Departments* is also decomposed in terms of *TupleOf* into *Purchasing*, *Production*, and *Marketing*. *Employees* is decomposed in terms of *OSetOf*. One of the children of *Employee*, *PayScale*, is decomposed in terms of the parameterized type *OneOf*.

Abstract types decomposed with the same parameterized type on a TMap inherit the same primitive operations. So, for example, *MfgCompany, Departments*, and *Employees* inherit the same primitive operations from parameterized type *TupleOf*. An example is in the FMap, where both types *MfgCompany* and *Employee* use the primitive operation *MoveTo*, which was inherited from *TupleOf*.

A type may be non-primitive (Departments), primitive (FullTime as a rational number), or a definition which is defined in another type subtree (Employees). When a leaf node type has the name of another type subtree, either the child object or a reference to an external object will be contained in the placeholder controlled by the parent object. If it contains a reference to an external object, that means the parent is forming a relation with that object.

As experience is gained with different types of applications, new reusables emerge. For example, a set of mechanisms has been derived for defining hierarchies of interruptible, asynchronous, communicating, distributed controllers. This is essentially a second-order control system (with rules that parallel the primary control system of the primitive structures) defined with the formal logic of defined structures that can be represented using a graphical syntax. In such a system, each distributed region is cooperatively working with other distributed regions and each parent controller may interrupt the children under its control.

AUTOMATION OF THE DEVELOPMENT BEFORE THE FACT TECHNOLOGY: THE 001 TOOL SUITE

A look at an operational before the fact development environment gives a picture of some of the possibilities. The 00l Tool Suite, an automation of development before the fact, is a full life cycle systems engineering and software development environment encompassing all phases of development. It is the system that, in the opening example of the National Test Bed experiment, automatically generated production-ready code that was fully operational.

Developing a system using the 001 Tool Suite begins with the definition of the meta-process and the definition of requirements (see Figure 21.7). The tool suite builds on the characteristics of development before the fact by having a means to observe the behavior of a system as it is being evolved and executed in terms of OMaps and Emaps.

The tool suite, a development before the fact system, was used to generate itself. Every system developed with the tool suite is also a development before the fact system. The tool suite can coexist and interface with other tools as well, and it can be used to develop a prototype or fully production-ready system.

As with any development before the fact system, the definition space is defined in terms of FMaps and TMaps and the execution space in terms of OMaps and EMaps. Different parts of the 001 Tool Suite highlight the importance of different properties of objects in development before the fact systems.

The tool suite has evolved over years based upon user feedback and a continuing direction of capitalizing more on advanced capabilities of development before the fact.

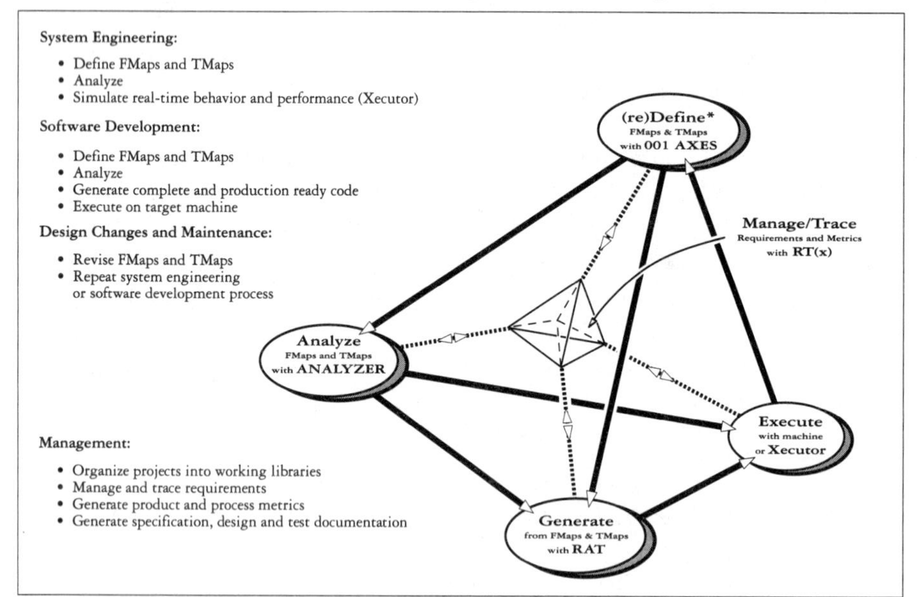

FIGURE 21.7. The 001 tool suite is an integrated systems engineering and software development environment

The many systems that have been designed and developed with this paradigm now reside within manufacturing, aerospace,[4] software tool development,[5] database management, transaction processing, process control, simulation, and domain analysis[6] environments. Some of these systems are systems engineering and some are software development efforts; others are a combination. The definition of these systems began either with the definition of the original requirements or with requirements provided from others in various forms.

The 001 Tool Suite automatically generated 800,000 lines of code for each of five platforms to create itself. The tool suite has generated over 8 million lines of code to generate its three major versions on these platforms. Contained within the integrated tool suite are many kinds of applications automatically generated as integrated systems.

[4]B. McCauley, *Software Development Tools in the 1990s*, "AIS Security Technology for Space Operations Conference." Houston, Texas. July 1993.

[5]*The 001 Tool Suite Reference Manual, Version 3*. Cambridge, MA: Hamilton Technologies, Inc. January 1993.

[6]Krut, B. Jr., *Integrating 001 Tool Support in the Feature Oriented Domain Analysis Methodology*, (CMU/SEI93TR11, ESCTR93188). Pittsburgh, PA: Software Engineering Institute, Carnegie-Mellon University. 1993.

They include database management, communications, client/server, graphics, systems/software development, and scientific/engineering programming systems.

Ongoing analysis of the results demonstrates more fully the impact of properties of a system's definition on the productivity in its development. Compared to a traditional C development environment, in which each developer produces 10 lines of code a day, the productivity of systems developed with the 001 Tool Suite varied from 10:1 to 100:1. Unlike traditional systems, the larger a system, the higher the productivity, partly because of the higher degree of reuse. Figure 21.8 shows the development before the fact systems engineering and software development environment focusing on the open architecture aspects of this approach.

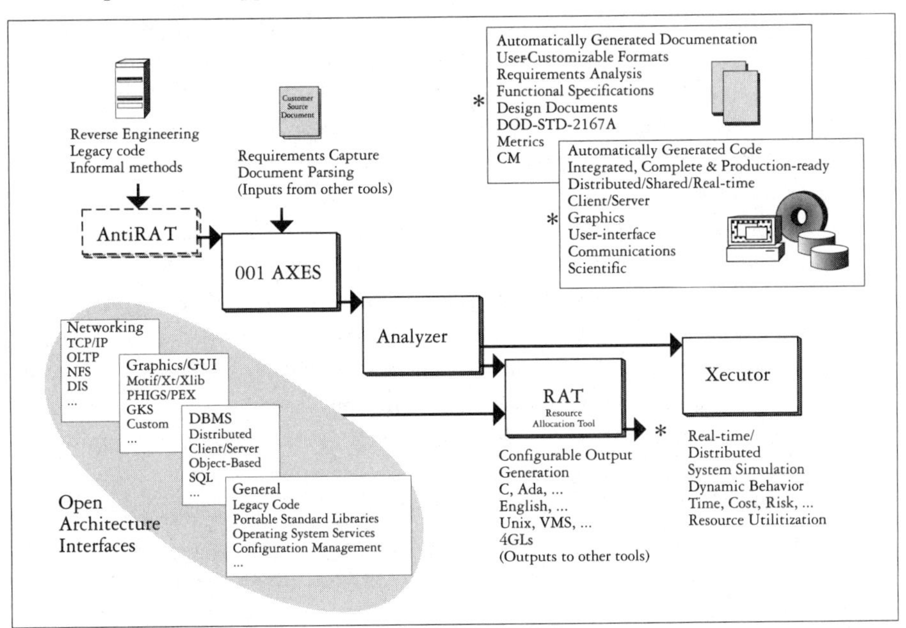

FIGURE 21.8. A seamless, open architecture environment, the 001 Tool Suite inherently supports an integration of function- and object-oriented development

CONCLUSION

Often the only way to solve major issues or to survive tough times is through non-traditional paths or innovation. One must create new methods or new environments for using new methods.

Innovation for success often starts with a look at mistakes from traditional systems. The first step is to recognize the true root problems, then categorize them according to how they might be prevented. Derivation of practical solutions is a logical next step.

Iterations of the process entail looking for new problem areas in terms of the new solution environment and repeating the scenario. That is how development before the fact came into being.

With development before the fact, all aspects of system design and development are integrated with one systems language and its associated automation. Reuse naturally takes place throughout the life cycle. Functions and types, no matter how complex, can be reused in terms of FMaps and TMaps and their integration. Objects can be reused as OMaps. Scenarios can be reused as EMaps. Environment configurations for different kinds of architectures can be reused. A newly-developed system can be safely reused to increase even further the productivity of the systems developed with it.

The paradigm shift occurs once a designer realizes that many of the old tools are no longer needed to design and develop a system. For example, with one formal semantic language to define and integrate all aspects of a system, diverse modeling languages (and methodologies for using them), each of which defines only part of a system, are no longer necessary. There is no longer a need to reconcile multiple techniques with semantics that interfere with each other.

Techniques for bridging the gap from one phase of the life cycle to another become obsolete. Techniques for maintaining source code as a separate process are no longer needed, since the source is automatically generated from the requirements specification. Verification too becomes obsolete. Techniques for managing paper documents give way to entering requirements and their changes directly into the requirements specification database. Testing procedures and tools for finding most errors are no longer needed because those errors no longer exist. Most tools developed to support programming as a manual process are no longer needed.

In the end, it is the combination of the methodology and the technology that executes it that forms the foundation of successful software. Software is so ingrained in our society that its success or failure will dramatically influence both the operation and the success of businesses and government. For that reason, today's decisions about systems engineering and software development will have far-reaching effects.

Collective experience strongly confirms that quality and productivity increase with the increased use of properties of development before the fact. In contrast to the just-in-time philosophy, the preventive philosophy behind development before the fact is to solve a given problem as early as possible. Finding an error statically is an earlier process than finding it dynamically. Preventing it by the way a system is defined is even earlier. Not having to define (and build) it at all is earlier yet.

The inherent reuse in development before the fact systems culminates in the ultimate reuse—automation itself. The answer continues to be in the results: automated development is here, and the systems of tomorrow will inherit the best of the systems of today.

AUTHOR BIOGRAPHY

Margaret H. Hamilton is the founder and CEO of Hamilton Technologies, Inc., which markets the 001 CASE Tool based on the tenets of the "development before the fact" productivity-enhancing architecture discussed in this paper.

From 1965 to 1976, Hamilton was director of the APOLLO on-board flight software program and director of the software engineering division of The Charles Stark Draper Laboratory at MIT. There she created a theory for defining and developing reliable systems. She then founded and was CEO of Higher Order Software from 1976 to 1984 where the first CASE product in the industry was developed. Its emphasis was to support the development of reliable-software-based systems. She founded HTI in 1986 to create methods and tools for developing cost-effective quality software.

Hamilton has authored numerous articles in the field. She is the 1986 recipient of the Augusta Ada Lovelace Award for Excellence from American Women in Computing.

CHAPTER 22

The Repository: Key to the Information Goldmine

by Barry Brown and Lewis Stone

The trend towards the distribution of data to multiple and heterogeneous platforms—PC, mini, mainframe—brings with it a built-in problem of enormous proportions. That is, if valuable information assets are widely distributed, the organization needs to find a way to organize and control those assets. Or the intellectual property those assets represent will markedly decrease in value.

This chapter explores the trend of organizations—both large and small—towards implementing a repository solution to this extraordinary dilemma.

THE ROAD TO PRODUCTIVITY

There are many roads to productivity. The one least traveled, but perhaps most profitable, is the one where software tools are integrated in a manner producing accessible and timely information.

The three keywords here are information, tools, and integration. Information, a buzzword of late, really is the most important asset a company owns. With the proper utilization of it, information becomes a potent competitive force. And in today's very global—and very competitive economy—information may, in fact, be the deciding factor in determining the color of the organization's bottom line.

Understanding that information is a resource to be valued, organizations made a heavy investment in information technology. This investment, in the billions of dollars, included development of new systems as well as purchase of a variety of software tools.

Software tools are decidedly two-flavored. On the one hand there are the end-user oriented tools which include report writers, 4GLs and the new breed of data mining tools. On the other hand there are the tools which specifically target the development function. These tools run the gamut from compilers to data administration to data warehouse tools. What was common among all of these tools, is the decided lack of interconnectiveness—or integration.

Lack of integration is a subtle defect with a powerfully negative impact on the productivity—and competitiveness—of an organization. It translates to the inability of information to be managed in a consistent and non-redundant fashion. Because software tools had seams information cannot flow easily from one tool to anther, forcing organizations to either manually move the information between tools—or worse, to create redundant and conflicting information stores.

The industry, recognizing the ramifications of these problems, began to move in the directional of what is often referred to as a development frameworks. IBM's AD/Framework, DEC's Cohesion, and Unisys' ASD framework are three prime examples. The goal of these frameworks is a pointed orientation towards consecutively. That is, IBM's or Unisys' goal is to provide a boundary-less environment to spur the free-flow of information through the use of standards and guidelines for the development of software tools. While development frameworks are most certainly a laudable goal, the major drawback is that, for the most part, these frameworks are philosophies only.

Software vendors began to stake claims to components of the development framework almost immediately. But overall, the integration problems persist. The missing component, that which enables a seamless environment, is a repository workbench.

A repository workbench has three functions. It is a repository. It provides tools. And it forms the "connecting glue" of the development framework—in other words, integration.

A short, standard definition of a repository is "an organized reference to the data content of something. That something could be a system, a database, or a collection of all the files, program databases, and manual records maintained by a large organization." While the definition of tools should be self-evident, in this context it is not.

Tools in a repository workbench environment encompass a broad spectrum of functionality that goes beyond what is commonly available. The last component of the repository workbench equation is integration. It is this component that meshes the repository and the repository-based tools into an organization's environment. The net sum of the repository equation is the ability to better leverage the skillset of a wide range of the organization's staff—from data administrators to programmers to analysts to the end-users themselves. This leveraging of skillsets leads to a dramatic increase in productivity.

THE REPOSITORY

The repository is the heart of the repository workbench. It is much more than a data dictionary. It stores information about objects—whether those objects are file definitions or process rules. The sections below itemize the major attributes of a repository. An effective and robust repository, should meet the objects presented in this section:

Initial Data Capture

For the most part, objects that will be required to be entered into the repository are objects that already reside in catalogs, files, databases, CASE encyclopedias or as part of a program (i.e., working storage, as well as the procedure division). These objects may reside on the mainframe, on a non-mainframe server or on a client PC. Scanning enables an organization to quickly populate the repository through the importation of objects from a pre-exiting source.

Scanning, however, should be robust. That is, a good scanning facility needs to have the ability to scan and understand multiple data types. For the most part, today's large enterprise shops continue to be COBOL-based even though a migration is underway towards more modern technologies. Whatever language or languages an organization uses, the repository must have the facility to scan and understand its structure of file definition. For COBOL-based systems this information is contained within the Working Storage section, the File Section as well as in the Procedure Division.

Since a great many COBOL programmers store definitions in what is referred to as copybooks, the scanner must be able to interpret these copybooks as well. Since copybooks are often stored serially in a large partitioned data set, the scanner could speed up the process by being smart enough to scan multiple copybooks all at one time.

Language is only one part of the definition problem, though. A scanner should have similar prowess in handling definitions stored in databases, their catalogs, as well as CASE encyclopedias.

Single-instance scanning is insufficient to meet the needs of most enterprises. Therefore, the scanner must permit the organization to re-populate the repository as many times as necessary through versioning. Since there is always the likelihood that the same element will be scanned from multiple locations, especially in client/server architectures, the scanner must provide collision resolution as well.

Tracking

A repository should have the ability to keep detailed information about objects. The repository defines an object as more than the traditional data definition. An object may be a field, a file, a procedure, a system. Because the repository maintains detailed information about objects, the organization has an excellent opportunity to "track" the status of many of the formal processes that form the underpinnings of IT. A robust repository should be able to keep track of jobs, and programs as well as systems wherever these jobs, programs, and systems are distributed.

The repository should also serve as a documentation medium as well keeping track of programs, jobs, systems, and their corresponding data and processes. It is not a total stretch of the imagination to expect the repository to document reports, screens, schedules and backup and handle maintenance responsibilities as well.

Source and Use

All organizations are different in the policies, methods, and procedures of their IT processes. The repository workbench must integrate itself, as well as act as an integrator of these policies, methods, and procedures. The repository workbench must be flexible enough to support both the data model(s) the organization chooses to use and its life cycle methodology.

Since few organizations today are at least examining client/server as a viable option, the repository chosen should have the ability to support distributed processing however the organization decides to configure it.

One of the side benefits of going the repository route is that the repository itself can serve as its own documentation medium. What better than a repository to keep track of a widely distributed communications network, the data security planning that needs to support it, and the hardware that is currently proliferating throughout the organization?

User Access

In studies on productivity it has been shown that the user interface has the greatest impact on the usability of the system. For the function of data administration, a flexible user interface is mandatory if the organization is to leverage the resources of skilled professionals. The repository workbench product must have the ability to support the user in the environment to which he or she is accustomed. There is no better way to decrease productivity than to force developers to adopt a new access mechanism which is totally at odds with the way they are accustomed to doing things.

Therefore, the repository must "run in" (at least the client version) the environment the user is accustomed to using. If the user is a mainframer, the mainframe repository environment must provide productivity-enhancing access techniques to make productive use of the mainframe. This includes providing pull-down menus, pop-up windows, customizable screens, fast-path commands, as well as a contextual help facility. In essence, the user must be provided a SAA/CUA-compliant user interface.

If the user chooses a PC as a client, then the repository workbench should take full advantage of that user interface by exploiting such things as graphical user interfaces, point-and-click, and drag-and-drop.

Perhaps the best test of a true client/server repository is its ability to scale itself to as many diverse clients as there are diverse organizations. Your repository vendor or how many platforms does his or her repository run.

Dialog

A robust repository dialog should provide a simple, intuitive means for maintaining and querying information assets, as well as accessing tools. Features should include contextual menus, context-sensitive feedback, full customization of windows, including ability to anchor particular attributes during scrolling and the ability to definite meaningful labels, spreadsheet-like displays, and full scrolling capability.

Because the power of client/server means that new tools are "plugged" into the client and/or server at any time, the repository must be "plug-compatible" as well. When the organization adds a new tool, the repository workbench menus should have the intelligence to rebuild the workbench automatically.

An even smarter repository workbench would go one step better. It would provide the user with the ability to reuse its component parts. That is, if the user decides that he or she wants to build something new, he or she should have the ability to reuse any part of the repository necessary.

Finally, a robust repository workbench will provide the user with the tools needed to do the job. This means the ability to perform ad hoc SQL queries, can their queries for re-use and even multi-level impact analysis.

Extensibility

A robust repository workbench is not rigid. It should support growth. This growth should not be limited to data definitions. In an object-based environment a repository workbench should have the flexibility to add new sources of information as well as new tools, reports, and procedures. Each of these is defined as an object. Extensibility features should include automatic creation of repository table spaces, automatic rebinding, recreation of repository indices, reorganizational capability, and full error handling.

The repository must provide full data/database administration tools which provide ease in defining security, validation rules, integrity rules, derivation rules, and domain constraints.

Extensibility means more than just data extensibility; it also means process extensibility. What good is a repository that cannot be integrated with the tools (i.e., editors, utilities, etc.) the organization uses today? And what good is a repository that will be unable to integrate the newer tools of tomorrow?

Project Control

A repository workbench must provide facilities to automate the enforcement of corporate and project standards and procedures and control distribution of repository resources. Capabilities should include: project-oriented security, clone function for rapid project definition, access/update/migrate privileges, life cycle phase authorization, and project parameterization.

Versioning

The repository workbench must provide a comprehensive set of facilities for supporting, monitoring and auditing the evolution of repository definitions. This feature makes it possible to plan and implement the maintenance procedures that become necessary as systems mature and require modifications. A robust repository workbench provides the capability to use aliases, an unlimited number of variation names, and the ability to perform set-level operations such as delete and copy. Just as importantly, a robust repository must support a revision number attribute.

Life Cycle Phase Management

Supporting an organization's methodology(s) is an essential role of a repository. A robust repository workbench provides an organization-extensible means for defining the various stages of object evolution. These stages are referred to as life cycle phases. Transition rules define the movement of an object from one phase to another. Relationships between entities based upon their respective life cycle phases should be verified to ensure proper migration results. Managing life cycle phases and object migration is a vital function within a repository if it is to control and participate in an organization's development and maintenance methodology. Features should include customizable controls, project controlled life cycle phases, versioning within those life cycles, as well as the ability to add or remove life cycle definitions. The repository also needs to be smart enough to support transition rules, migration paths, and relationship-state rules.

INTEGRATION

Developmental frameworks are philosophies. For the most part, software engineering tools such as CASE maintain key positions within this framework but do little to integrate themselves effectively to other tools in other quadrants of the framework—or even other tools within the same quadrant. The objectives in this section, if met by the tool being evaluated, will assure the organization that the repository will be seamlessly integrated with repository tools as well as in-house developed and third-party tools.

Architecture

A repository workbench is a unique hybrid of repository, tools and an integrative vehicle. In order to support this three-fold functionality, the underlying architecture of a repository workbench product must provide both openness and an extensible framework. The organization must be able to integrate easily into—and expand upon—the framework.

In a client/server environment, this can only be accomplished if the repository's architecture is based on an object-oriented approach. It is only in this way that a repository can provide both data and function extensibility.

An object-oriented architecture also serves as the foundation for a repository workbench that is not only easily configurable, hence modifiable, but easily integrated into the organization's current architectural environment.

Standards

The basis of any open framework is the standards on which it rests. For a framework to be fully integrative with an organization's environment, the framework must conform to and support the standards and guidelines that the industry has embraced. Additionally, the repository workbench must provide the organization with the ability to support the standards that it has developed as a part of its policy and procedures.

Unfortunately, the advent of client/server has added many new acronyms to our vocabulary list. A good rule of thumb in selecting a repository is that whatever standards we speak, our repository must speak as well. This includes: DB2, IRDS, SAA/CUA, LU6.2, DRDA, CPI-C, organizational naming conventions, organizational keywords and abbreviations, and organizational custom rules.

Gateways

The basis of a repository product is information. Information, however, is not confined to a single source. A repository product must provide the organization with a series of gateways which allow the organization to export and import information between these information sources (e.g., CASE tools, various databases, and files). Since it is expected that the organization will have multiple requirements for gateways, the most robust of repository workbenches will define generically a gateway bridge that provides a commonalty of approach across diverse products. Features should include: a bi-directional bridge between the CASE encyclopedia and the repository, upload/download facilities, check-in/check-out, collision resolution, impact analysis, import/export capabilities, bulk population ability, re-population through versioning, variable name mapping, catalog import, source import from multiple catalogs, as well as support for the file and database systems used by the organization.

CASE and Application Development Bridge

A very specific gateway is the one required by CASE and application development tools. The gateway allows CASE objects to be integrated into the repository with the goal of permitting CASE users to have a more efficient way of controlling, securing, reporting, and distributing specifications captured at their workstations. A robust repository can be thought of as a clearing house between workstations and CASE products. The repository workbench should provide management tools that enable the organization to share data resources. This includes the ability to share the same model between two different tools. This is an extremely important feature. Too often organizational productivity is greatly lowered by the utilization of more than one CASE tool. Having the ability to move freely between these two or more CASE tools is a function that the repository should serve.

Client/Server

Our discussion thus far has alluded to client/server in each of the categories mentioned. While organizations are most definitely moving in the client/server direction, the unfortunate fact is not all repositories are true client/servers. Those selecting a repository should ask themselves the following questions when selecting a repository:

1. Does the repository have a portable GUI? In other words, does the repository run in the environment the organization is running in (i.e., PM, Windows, etc.)?

2. Does the repository use standards? For example, if the underling repository database is proprietary, how can the repository be called standards-based?

3. Is the repository scalable? In today's enterprise environments, systems run in a three-tier architecture: on the mainframe, on the server, and on the PC. Can the repository be distributed to one, two, or all three of these platforms?

4. Is there a "cross object capability?" In other words, is an IMS object equal to a DB2 object equal to a CASE object? The answer should be yes.

5. Does the repository sport a tool server which permits distribution of functionality to distributed platforms?

6. Can these tools be encapsulated into brand new tools? In other words, can you reuse any and all objects, data, and functions to create a new tool?

7. Does the repository support the range of communications protocols that you need today—and those that you might need tomorrow?

8. Given the number of platforms, has the vendor used common code libraries or new code for each platform? Common code libraries are far superior.

9. Does the vendor provide APIs?

Workbench Integration

The repository workbench creates a productive environment where repository information is integrated with an extensible toolset. This approach offers an organization the flexibility to incorporate both its existing tools and those which it may consider in the future.

Integration includes user-defined tools as well as third-party tools. A good repository should go one step further, though; it should also "enable" these tools through the repository workbench interface. This requirement means that the repository dialog should be extremely, robust providing an extensible end-user interface, security, as well as customizable help dialogs and messaging.

TOOLS

A robust repository workbench needs to supply a series of tools that takes advantage both of the repository and its integrative prowess. The features described in this section are those of a robust environment.

Tool Development Environment

Being able to integrate tools to the workbench is only one side of the coin. The other side is in being provided with the facilities to develop in-house tools. A tool development environment should possess the following capabilities: vendor supplied shell programs, vendor supplied subroutine libraries, ability to encapsulate data and function into new objects, ability to invoke in-house tools through the repository dialog, and reusability of every repository object.

Groupware

Productivity is greatly enhanced when a facility is provided for project teams and users to communicate with each other. This is often referred to as groupware. Within a repository environment, this can be accomplished through the use of electronic mail which provides messaging to both project members as well as users. Messaging should include more than just text; it should also include batch output, as well.

Reporting

Various levels of the organization requires access to the repository for reporting. On one level, the end users require access to the repository to find out information about the types of information available within the organization. On the other hand, data administration staff has a real need to control the transition of information within the repository. Both levels of user-access needs to be supported. Reporting features should include: QMF, SQL, FOCUS, and other 4GL tool sets. A good repository should anticipate reports the organization will require and provide them as canned procedures.

Impact Analysis

In non-repository systems, a large percentage of non-productive time is spent in determining the impact of change. Analysts and programmers must manually review documentation and program source listings to evaluate the extent of change necessary as well as the length of time it will require to make those changes. This can be a lengthy process. A repository-based system automates this process through the function of impact analysis. Automatic impact analysis deconstructs the repository to determine the level of change required.

Scripting

Database administrative procedures are extraordinarily complex. The complexity of many of these tasks implies that the staff member involved must have the highest degree of skill and exercise the utmost level of care. In organizations that wish to leverage the skill set of the average user, increase the speed at which a task may be completed or wish to deploy vast functionality across differing layers of the organization, what is required is the means to decrease the complexity level of the activity and thereby reduce the risk of error. A repository-based scripting facility provides this functionality. Capabilities should include the ability to invoke any vendor-supplied tool, report, or script, as well as any in-house tool, report or script.

Forms

Forms provide the ability to establish external layout definitions which serve to present a modified view of the objects within the repository without altering the object itself.

Although the definition of objects in the repository are not altered, the user view can be modified to afford the greatest expediency in utilization of the repository without having to write code.

Generation

Since the repository acts as the central clearing house for corporate information resource management, the repository must have the ability to act in concert with definitions used by application development and end-user tools. To enhance productivity, consistency, and security, the repository workbench must have the ability to generate syntax including DDL, DML, DBD, PSB, DCL GENd, as well as copybooks.

Managing Relational Tables

A repository workbench needs to be more than just a repository. Facilities to manage the underlying database should be fully integrated into the toolset. These tools should provide the ability to do such things as unload/reload databases, create and drop objects, and support referential integrity.

Data Warehouse

A repository is also a perfect vehicle to provide the maximum support for organizations that are planning, building, accessing, and maintaining the data warehouse.

Whether the organization stores its information in databases, CASE and visual programming tools, source code—or all three—a robust repository provides a transparent methodology by which the organization can implement and maintain this unique organizational strategy.

There are four main areas of data warehouse support: construction, warehouse maintenance, support for warehouse builders, and support for warehouse users. Each is described below.

Construction

A repository provides the ability to capture and analyze data sources, document and build target warehouse structures, as well as interface to warehouse tools that are on the market today.

The repository provides warehouse users an easy-to-use framework from which users can run standard ad hoc reports or develop new ones.

Warehouse Maintenance

A repository that provides automatic program language scanning, SQL-based queries, DBMS, and CASE interfaces may be used to continually synchronize the warehouse with up-stream data sources as well as to enable the more traditional change impact analysis associated with warehouse usage and maintenance.

Support for Warehouse Builders

Repositories have traditionally stored a rich set of information regarding an organization's systems, processes, and databases. This information is invaluable in enabling warehouse builders to:

- Discover and document legacy sources of data
- Describe and design warehouse structures
- Capture rules for source-to-target mapping
- Specify, configure, roll out, and maintain warehouse users' reporting and query capability across the enterprise
- Analyze the impact of change on warehouse extract processes and warehouse queries

Support for Warehouse Users

The repository provides a vehicle to:

- Be usable at the skill level of an analyst, and should be available in the analyst's graphical environment
- Provide a "table of contents" to an organization's warehouse
- Provide helpful information on what data means, and what valid values it can assume
- Guide users in understanding how and when data is accessed (i.e., usage rules)
- Enable data stewards or other authorized users to alter and maintain descriptive information about their warehouse data
- Enable users to query their warehouse through existing report writers, or by a repository dialog

DATA MINING

The organization's knowledge workers have long searched for a tool with which they could "mine" the vast information storehouses the organization now merely minds.

Organizational data has suffered from the problem of multiple definitions of the same piece of information. What these workers need is a way to tie these disparate pieces of information together into a cohesive whole and then report on the results. For example, the term "sales" can have multiple meanings. Does it mean "after returns" or before?

End-user interfaces to the repository provide such a tool. Using an intuitive interface geared for the knowledge worker, the repository becomes a base to search and pinpoint distributed information as specified in a free-format query language that does not require the end-user to learn and master a programming language.

In this way, the knowledge worker is free to "mine" information that has historically been unavailable. Once the information is located, the end-user repository interface provides the ability to format a report (or link to third-party report generation tools) for that user.

The Repository and the Knowledge Worker

For data/database administrators and developers, the repository has always provided a flexible yet powerful way to view and manage corporate information assets.

With access to the corporate repository, the knowledge worker can take advantage of all this power and information. Business users such as financial analysts, accountants, and sales/marketing staff, have long lamented their inability to know exactly what corporate information was available to them.

Since repositories have historically been used as professional data/database administration tools, the inclusion of data mining into the fold requires the addition of some specific end-user features.

Graphically View Available Data

Knowledge workers are accustomed to a graphical paradigm. The repository providing data mining features must adhere to this paradigm as well. For example, utilizing icons such as cabinets and folders is a natural business metaphor that makes the repository more adaptive to end-users.

Create Graphical Queries Based on Business Rules

Business is about business rules as well as data. Since the repository can capture processes, the end-user benefits by seeing how the data is processed by the organization.

Build and Execute Queries Without Programming

SQL and other programming languages may be popular in the IS community, but these languages are not for the business user. While SQL is used under wraps, the user must be provided with a free-format query language to create inquiries against the repository.

Build and Share Query Libraries

Reusability is a productivity-enabling device for knowledge workers as well as IS staff. The repository is built on the principle of reusability, and therefore it can be used to enable the knowledge worker to store his or her queries for future and group usage.

SUMMARY

It is becoming increasingly obvious that organizations are suffering from a severe information overload. With the advent of client/server computing, more and more information is becoming distributed—and therefore harder to get a grip on.

A repository can serve as an important focal point in the time of all this information madness. It can keep track of where information is located as well as what it means. Most importantly, it can become a tool for the entire organization. Both as a data warehouse for IS staff and as a data mining tool for the knowledge worker.

AUTHOR BIOGRAPHIES

Barry Brown and Lewis Stone are co-founders of BrownStone Solutions, Inc. (recently acquired by Platinum Technologies) which has been a leader in enterprise repository products since early 1987. At the time, their emergence onto the scene was heralded as a pioneering effort in bringing IBM-repository-compatible, repository workbench capabilities to the DB2 administration development environment.

Since that time BrownStone Solutions has expanded its product offerings to become a true manager and expediter of corporate-wide information assets, wherever these information assets are located and whatever hardware platform, operating system, database, or programming language they use. Understanding the value of information as a competitive resource, BrownStone has been a leader in providing what is referred to as data mining tools—the ability for a knowledge worker to locate and understand diverse corporate data sources.

The derivation of the BrownStone suite of products stems from the diverse "Big Six" consulting careers of its two founders. During Stone and Brown's tenure, where both were heavily involved in major database projects; it was noted that time and time again they were required to spend hours in building a framework for the efficient administration of database systems across their diverse client base. Over the years, across industries and applications, Stone and Brown noted that this framework was indeed generic and, if developed into a software workbench, could provide enormous benefits to utilizing the tool. The goal was to develop a highly efficient, high-performance object-oriented, and therefore truly extensible, repository workbench.

That accomplished, today BrownStone Solutions continues to add products to its open systems' architected repository workbench. A leader in the marketplace, BrownStone Solutions provides a total approach to information management—from data warehouse, to CASE and visual programming to language scanning.

CHAPTER 23

The Three-Tier Client/Server Architecture

by Christopher Lozinski

The market has finally agreed that two-tier relational database applications are a good idea. Oracle, Sybase, Informix, and others offer accepted relational databases. Powerbuilder, Gupta, and others offer excellent client tools for these databases. Data can also be accessed from Spreadsheets, Visual Basic, or other common tools. This two-tier market is mainstream, and growing. This is where the mainstream market is today. The rest of this chapter explains where the leading-edge market is going. But first, let us see what the limits of the two-tiered systems are.

While the two-tier approach works well for simple applications, it shows its limits in the more complex applications being demanded in this competitive market. After a while, the client/server relational model begins to limit the complexity and performance of the application. Heavy database queries overload the disk. Shipping large quantities of data back and forth overloads the network. Data updates must be propagated to many local caches. Functionality, which should execute on the server, is built using stored procedures, which locks the user into the database vendor.

The two-tier applications also suffer from being inflexible compared to object-oriented applications. In a typical two-tier application, the client process consists of presentation logic intertwined with application logic. It is rigidly separated from the server process by a functionally limited API known as SQL. The two-tier developers emphasis is on reading writing and processing data. Changes in the server require changes in the client. Changes in the client often allow little reuse of code. There is no efficient way to integrate various network resources, such as IMS databases into the application. Changes to the system are done by modifying the data processing and may take significant resources to implement.

So what is the market doing to overcome these limitations? A bewildering array of approaches are being tried. Messaging Services, Distributed Transaction Processing Monitors, Object Request Brokers (ORB), Object Linking and Embedding (OLE), Remote Procedure Calls (RPC), and Database Gateways are all being used.

Each of these approaches will have some success, but, in my opinion, the most promising technology is the convergence of client/server and object-oriented approaches.

I see the client/server market moving towards a three-tier architecture. Figure 23.1 compares the existing two-tier technology with the coming three-tiered approach. In the two-tiered approach, the server is a relational database, and various development tools are used on the client side. In the three-tiered architecture, an independent application-specific process, sits between the client and the server. The client still runs the PC-based personal productivity applications, such as Word, Excel, and possibly Visual Basic, that make the system very easy for the end-user. The database server is still the existing corporate database that offers reliability and robustness. But in between the graphical user interface and the database server, many innovative companies are adding application servers to provide custom functionality. In this architecture, all of the system components can be communicating with each other using a new class of communications products called middleware. The client productivity software, the application server, and the existing databases are connected together by the middleware to create a complete solution.

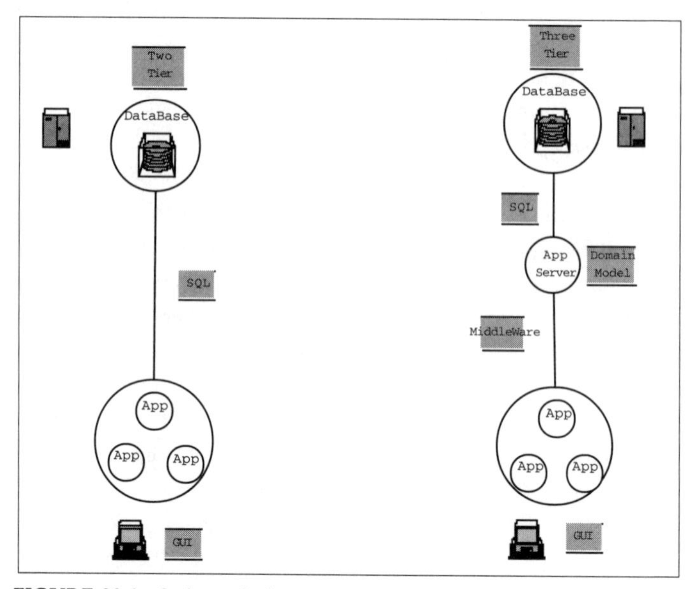

FIGURE 23.1. 2-tier vs 3-tier

Application servers provide additional custom functionality to the client/server system and are the place where object-oriented technology will have the most impact. Application servers are object-oriented representations of the systems being controlled.

They are becoming proper domain models, supported by rigorous object design methodologies. They are being built to be testable using the newest in object-oriented test methodologies. Their classes are real-world things like bank account, manufacturing workstation, or person. They model real world events such as transferring money, processing material, or assigning a person to a job. They are customized to accurately represent the business being served. As the business changes, they evolve to model the new business correctly . They are being treated as financial corporate assets.

The application server is connected to the other pieces of the system using middleware. Middleware is defined as "the network aware system software, layered between an application, the operating system, and the network transport layers, whose purpose is to facilitate some aspect of cooperative processing."[1] There are many different types of middleware. Message passing mechanisms provide reliable communications between computational processes. Directory services make it easy to locate network resources in a constantly changing environment. Transaction Processing Monitors allow nested transactions to span multiple databases. Database gateways make one database look like another. Object Request Brokers allow C++ programs to communicate with other applications. So the term middleware covers a dazzling array of products, many of which are difficult to understand. To clarify the use of middleware, let us look at some existing examples of three-tier architectures.

SOME GOOD EXAMPLES

In my work, I have run across many leading-edge companies building such three-tier applications. I will present several examples that illustrate the business issues driving companies towards three-tier architectures.

One of my clients sells custom applications to police. They need to do phonetic searches of the database, something which stored procedures do not support. To accomplish this objective, they acquired a phonetic search algorithm, used it as an application server to quickly search through hundreds of thousands of criminal records, and then only reported the matches to the client application. This is illustrated in Figure 23.2. Imagine if they had to transfer the entire database across the network to the client PC to

[1]The Gartner Group. "Middleware: Panacea or Boondoggle?," *Software Management Strategies, Strategic Analysis Report*. July 5, 1994.

perform the phonetic search. It would overload the entire network. Instead they did the phonetic searches on a high performance UNIX server, and only reported the results to the widely distributed PCs. I start with this example because it shows a simple case where an application server is essential.

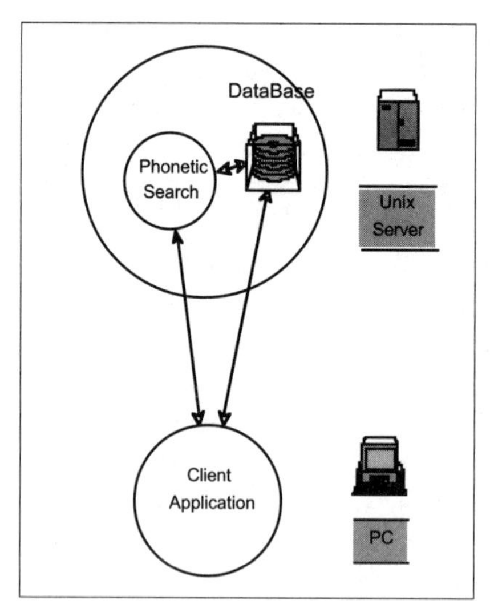

FIGURE 23.2. Quick searching on an application server

Another client sells production tracking software to manufacturing organizations. Production supervisors need to know where the material is in the factory in order to plan their work. This requires accessing all of the lot status information and displaying it visually so that they can make their decisions. Traditionally, such large queries would overload the system, particularly if multiple individuals were running them simultaneously. With an application server, the lot locations are stored in RAM, the aggregation is performed instantly, and the results are transferred to the PC for display. This is illustrated in Figure 23.3. Thus the supervisors have better knowledge of what is happening around them. They do not need to walk around to see where the inventories are; they can just check their computers. Time is saved, better decisions can be made, and factory throughput increases.

Another client competes in the commodities trading market. They literally need to be able to customize their programs overnight in order to compete against other traders. They are building an architecture where multiple servers provide different financial services, such as buying a commodity, or buying a future, or transferring money. See Figure 23.4. The client software runs EXCEL spreadsheets and accesses these various basic services to build new custom services for a particular client. The classic example is McDonald's, which, when introducing chicken McNuggets, needed to guarantee the price of chicken, and so they developed a future on synthetic chicken, designed to approximate the actual future price of chicken. It was made up of futures on corn, energy, and the interest rate. The individual futures were supported by several application servers. The complete synthetic chicken offering was developed on a spreadsheet that accessed these network application servers. Other customized offerings were developed as separate spreadsheet programs that accessed the same suite of application servers.

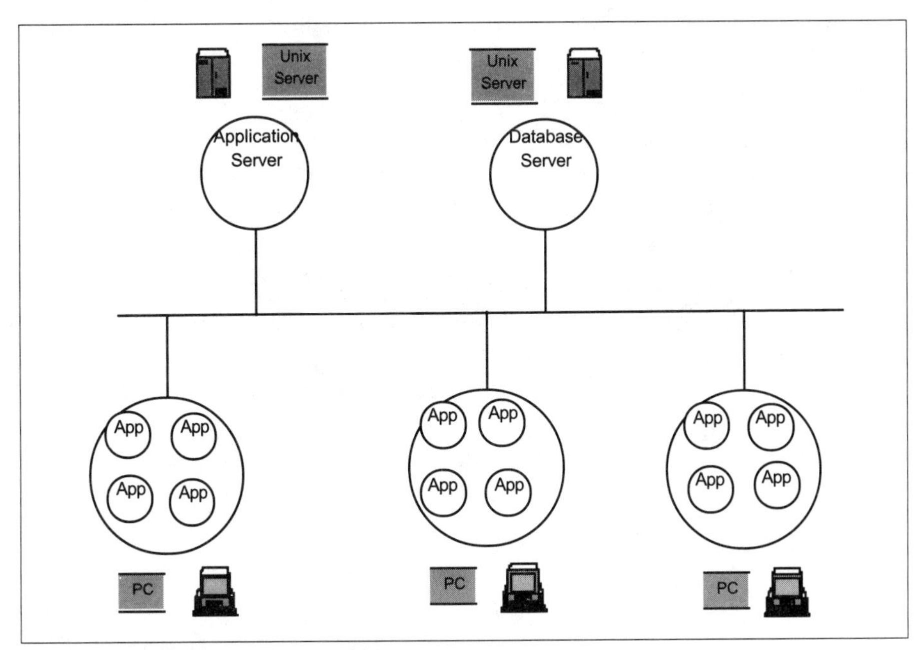

FIGURE 23.3. RAM to server searching

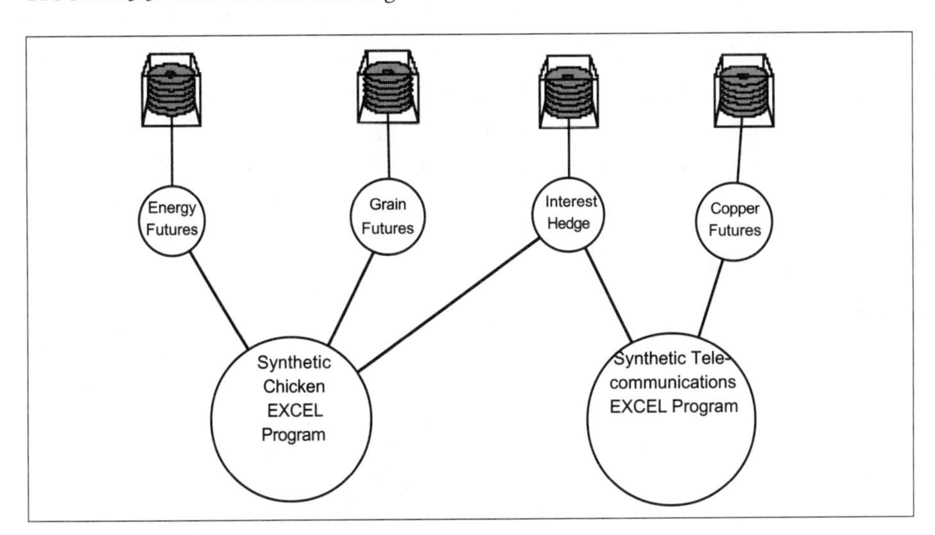

FIGURE 23.4. Customizing programs overnight

Several of my customers sell financial products either nationally, or internationally. They typically have a single central location for databases, and geographically distributed client applications shown in Figure 23.5. In the client office, several PC workstations access a local application server, which in turn accesses the central global database. In this example, the first two tiers are located in the client office, but the database server is located in the corporate office.

FIGURE 23.5. Geographically-distributed client applications

Perhaps the most sophisticated three-tier client server applications are being developed in the transportation industry. Typically, a transportation company has hundreds of hubs. Each hub has a three-tiered client/server system. Alternatively several hubs form a region, and share a single database. But all the hubs interconnect, so that these individual client/server sites all communicate with the other client/server sites. These projects are tremendously challenging and are still in their early phases.

Typically they also use another type of middleware, called a database gateway, or Transaction Monitor, that spawns multiple heterogeneous databases, creates a single relational view, and allows nested transactions. Symantec's Scale is an excellent example of such a product. For the transportation company, as material is shipped from one center to another, the record is automatically moved from one database to another based on programmable table fragmentation rules. In addition to managing multiple relational databases, the shipping companies have many server processes for each hub. It is understandable why these companies are still in the early phases of deploying such complex applications.

So we have just seen several examples of how the three-tier architecture is used in practice. We start with the simple phonetic search to illustrate the performance needs. We move to the manufacturing example, to show a more complex case, and finally we move to some financial applications, which is where some very sophisticated work is being done. The toughest applications are the ones where the complexities of the physical system (a shipping network) need to be supported by parallel client/server installations. Hopefully this series of examples helps you to understand your application options better.

So now let us abstract the advantages of the three-tier architecture.

ADVANTAGES OF THE THREE-TIER APPROACH

The advantages of the three-tiered approach are numerous. Database performance is improved. Frequently-accessed information can be cached in Ram on the application server, eliminating many disk accesses. For example, data validation information, such as valid choices, can be stored on the application server to allow diskless checking of transactions prior to processing. Network performance is also improved. Calculations can be performed wherever they best suit the application requirements. Only the results are shipped across the network, not the raw data. Thus the end-user sees a faster, more responsive application. Any arbitrary communication is allowed between client and application server, so the functional limitations of SQL are bypassed.

Not only does the system perform better at run-time, it also performs better during development. Individual components of the system can be updated with minimal changes to the rest of the system. Certainly, the database, and user interface are now completely decoupled, and so they can be independently modified. But even the application model can be evolved without changing the client application program. This is the power of the message-passing based approach to distributed systems. As long as the messages are the same, the components can be updated internally. The same cannot be said for the SQL approach to client/server computing.

Not only can the individual components of the system evolve separately, the application server can now be where all of the techniques of modern software engineering are applied. Formal Object Design methodologies can be used on the application server. Whereas previously relational databases were limited to Entity Relationship Modeling, there is now the opportunity to practice proper object modeling techniques. Rigorous object testing methodologies can be applied each time the model changes. The domain model, once built, evolves as slowly as the business it represents. Various different applications can now be built on top of the domain model, rather than building each one from scratch as a client application program. Software maintenance becomes much simpler.

THEMES AND VARIATIONS

Alas, one size does not fit all. In theory the three-tiered architecture is a very simple concept; in practice every system is somewhat different. There usually are multiple databases, not all of which are relational. They all have to be accessed, integrated and updated from the domain model. There are often multiple domain model servers. In addition to the application model, a separate financial model is often used. Another required server is the name server—given a name it finds a resource on the network. Additional computational resources are also frequently accessed, such as a network file server with graphic images. Finally the client application software packages come in all shapes and sizes. The key is that the whole system is assembled from different components held together by a flexible middleware package that allows easy interoperability.

The three-tiered architectures can run on one, two, or three platforms. For a portable demonstration all three tiers would run on a portable multi-tasking operating system such as OS/2. For production applications, the client would be on Windows, and the other two tiers would be on UNIX or a mainframe.

THE APPLICATION SERVER

What do I recommend for the application server? I believe that the majority of application servers are currently being built in C. Clearly the mass market is moving towards C++ application programs, with a CORBA Interface Definition Language (IDL). Personally, I believe that there are strong technical reasons for using a message-passing based language like Objective-C,[2,3] Smalltalk, or Eiffel. The most important of these reasons is that they are better for building business models than either C or C++. C++ confounds object modeling with performance enhancement issues like direct function calls. In contrast Smalltalk provides a very uniform domain modeling environment. This focuses the developer's attention on modeling the application and is thus very healthy for applications development. Smalltalk does come with a space and performance penalty, which is why I recommend Objective-C, a cross between Smalltalk and C. Objective-C is the basis of the NEXTSTEP development environment which NeXT recently licensed to Sun Microsystems as OPENSTEP. Sun plans to bundle OPENSTEP with the Solaris Operating System starting in 1995.

[2]Lozinski, Christopher. "Why I need Objective-C," *Journal of Object-Oriented Programming.* September 1991. pp 21-28.

[3]Lozinski, Christopher. *Why I still need Objective-C*, Current version published on Internet. For a copy, send email to info@bpg.com.

THE DATABASE

What do I recommend as the database? Certainly the mass market is moving towards relational databases. Elsewhere in this book, the advantages of object, or network, databases are extolled. I believe the ODI is the leader, but ODI is very tightly bound into the C++ world, which means that it is not the best choice for business modeling. Gemstone is an excellent database, but it only works with Smalltalk. Versant also is doing well, but its computational overhead is tremendous. Versant is really a combination of an object database engine, and a networking middleware. In contrast, I recommend a best of breed approach. Acquire the best stand-alone network/object database, and connect it to the other system components using the best middleware available. Currently my database recommendation is a database called Titanium.

THE MIDDLEWARE

What do I recommend as the middleware? The mass market is moving towards OLE or whatever Microsoft is doing this week. But there are far superior products on the market. For the purposes of building three-tier architectures with application servers, the message-passing middleware products are the ones to use. The developers I work with generally believe that large scalable systems can best be built with a message-passing based middleware product. Of these there are dozens on the market. The tools of interest to software developers are the message moving systems, and message queuing systems. The message movers require a direct connection, the message queuing systems can work in the presence of network interruptions. Although it is tough to generalize, message movers are in general more likely to lose a message. Message queuers often include transactional facilities to guarantee delivery. Message movers include products such as Peer Logic, Covia Integrator, Easy IPC, X-IPC, and Momentum. Message queuers include products such as System Strategies Inc.'s EZBridge Transact, DEC Message Queue, IBM Message Queue, and Dome Middleware. Whether you should you use movers or queuers depends on your trade off between speed and reliability.

Which one of these messaging products will dominate the market? No one knows. Choose whichever one runs on all of the platforms in your organization. They basically have very similar APIs and are reasonably interchangeable. The basic verbs are connect, disconnect, create message, send message, receive message. The particular function call arguments will change between products, but the resulting overall system design will not differ that greatly whichever of these products you use. Eventually, I expect that a standard message-passing based API will be developed by MOM, the Message-Oriented Middleware Association.

THE CLIENT SOFTWARE

What software do I recommend for the client? Which software your end-users want to run on their PCs is their choice. As long as it can call a DLL, or a DDE, these other tools can interoperate with it. One interesting choice is made when the developers are in charge of the desktop software selection. A lot of my developer customers want to run the same language on both Client and Server. This would either be Objective-C or Smalltalk. This is great for developers, because then they have a uniform object model. It is not so good for end-users, because they lack the ability to modify and extend object-oriented applications. Personally, I recommend Spreadsheets as the client software.

LIMITATIONS OF THE THREE-TIER ARCHITECTURE

I am sorry to say that the three-tiered architectures also have their limits. They are a more complex technology than the traditional two-tiered approaches. They require increased training, greater levels of education, and more skilled developers. On the other hand it requires a complex technology to manage a complex application. It is much like the difference between using a spreadsheet and using an object language like Smalltalk. The spreadsheet is great at building quick applications by unskilled workers, but it becomes unmanageable as the system complexity increases. Smalltalk takes a lot longer to learn, but it can handle tremendously complex systems and requires almost no maintenance effort, once the application has been properly modeled. Furthermore, existing Smalltalk applications can be very easily modified to track the changing business environment. Thus these three-tier architectures are only justified for complex and performance-limited applications.

MIGRATION STRATEGY

There is a simple, two-phased migration strategy from two-tier to three-tier model-based architecture. In the first phase, introduce objects into your organization on the client side of the two-tier architecture. In the second phase, migrate the objects into the middle tier of a three-tier architecture. This approach involves a combination of political, technical, and human resources strategies. Technically, the first phase starts by building new client applications and a domain model in an object-oriented environment. Politically, the objective is to have a successful project, in order to build political capital. It is preferable to address a problem that has defied solutions with traditional tools. This will build management's confidence in the importance of objects, and this confidence should carry over into political support for a three-tiered architecture.[4]

Once you have successfully deployed the two-tier client application, you are ready to move into the three-tier world. Choose an application that would benefit from having an application server. Choose the middleware products that meet your requirements. Spend the time to do a very careful design, because that is the basis for a failure modes and effects analysis (FMEA). The FMEA catches all of the things that can go wrong before they do go wrong. This step is critical because lots of things can go wrong in three-tiered architectures. This careful engineering may take time, but with the political capital built up during the first project, you should be given the time required to carefully engineer a successful three-tier project. If you engineer it carefully, the deployment will go smoothly.

This migration strategy, from two-tier to three-tier, also requires a human resource strategy. You need to have individuals skilled in the particular software technology. These skills are readily available on the market. You need expertise in object-modeling of your business environment. These people are all very busy, but you should be able to hire one for your organization. You need someone who understands distributed message-passing based architectures. These people are rare; you may be best off hiring a temporary consultant. Finally, you need a manager who is able to be a middleman between the traditional upper management and this new world of objects. This skill is the most difficult to find. The person needs to be able to represent the opportunities created with this technology accurately, and he or she must be able to sell the approach to a skeptical organization. This person also needs to be able to communicate the organization's legitimate business interests to the often frustrated developers. Too often this position is filled by a developer who has over-extended himself, or by a manager who does not understand the technology. Those organizations who hire the right person for this job develop wonderfully successful applications.

Once you have built the organization, you must continue to evolve it. In this industry training can never stop. To keep up to date you need to acquire a new skill every six months. I recommend that every individual should do this, but I insist that every software group do this. The penalty for failing to keep up, is that your organization will be making sub-optimal decisions, and wasting significant resources.

There are several ways to evolve your skills. As an organization you can buy training, hire skilled individuals, and have team members teach each other. Your human resource strategy needs to include all three elements. As an individual you should try to work for a company that invests in training. Otherwise you must invest your own time and energy to stay current.

[4] Bender, Robert, Guest Ed. *Communications of the ACM*. Special Issue devoted to Testing of Object-Oriented Systems. September 1994.

If you follow this technical strategy and human resource strategy, your chances of successfully implementing a three-tier client/server architecture are very good.

CONCLUSION

This is a book on technology trends, and I am blessed to be in a place where I can see what direction leading edge customers are going. These customers are clearly going to a three-tier architecture using one of the many different possible variations. They are clearly moving to a model-based approach to building applications.

Middleware is very hot. These changes are happening right now. Many new client/server installations are deciding between two and three-tier approaches. Three-tier applications are already running in highly educated organizations, typically the financial institutions where the salary and education levels are in the highest percentiles. One third of new client server installations in 1997 will be three-tier installations.1 Of those, probably one third will use an object model for the middle tier. Thus, I am confident that this is one of the more important directions that the industry is taking.

What should you do? If you have a simple application, keep it simple, and use the two-tier tools. If that does not meet your needs, you need to migrate to a three-tier architecture. Follow the migration path outlined in this chapter. Execute each step carefully, and your organization will be successful. If you are reading this, and your competitors are not, you are in great shape. Go implement a three-tier architecture before they figure out what hit them.

AUTHOR BIOGRAPHY

Christopher Lozinski (lozinski@bpg.com) is the President and Founder of Berkeley Productivity Group, a company that sells software components and subassemblies for building three-tier client/server systems. For product information send email to info@bpg.com. He has over 8 years of experience with Objective-C and was a columnist for two of the NEXTSTEP magazines before starting his own electronic newsletter. To subscribe, send email to newsletter@bpg.com. He has consulted extensively on strategies for building three-tier client server architectures. His applications background is in manufacturing and he is the co-editor of a new book, *Object-Oriented Models and Applications in Manufacturing*. He is the author of several of his company's software products, including the Smalltalk Interface to Objective-C, and the BPG BLOCKS workflow engine for manufacturing. In his free time he loves to travel, hike, and spend time with his family and dog.

CHAPTER 24

Virtual Reality: The Hot Ticket

by Richard V. Kelly Jr.

Most of virtual reality's essentials can be described in threes. There are three general forms of VR, for example. There are also three kinds of VR application, three levels of necessary VR software, and three general types of VR hardware peripherals.

The most common form of virtual reality, called "through-the-window VR," is already well known to the general public because of its use in arcade games and motion-based seat theaters. Through-the-window VR allows a participant to look into a virtual world from a seat in the real world. The "window" the user looks through may be as small as a home computer monitor or as large as a two-story movie screen. Motion-based seat theaters, the most common manifestation of through-the-window VR, allow for no true interactivity. The user is simply flown through a 2D film-based world, usually at high speed on a bumpy ride, without being given any chance to change the itinerary or to interact with objects in the world. Through-the-window arcade games, however, are usually based on computer generated images, not film, and so they are often more effective, usually allowing both 3D effects and some interactivity.

In a through-the-window theater experience the participant views scenes on the screen while the seat lurches and shudders in response to the images portrayed: roller coasters, swan dives off of buildings, and cliff-edge dune buggy rides. The images are almost always "real" images (that is, photographed with a motion picture camera) rather than "virtual" images (created in software). Any participant who looks away from the screen during the experience "falls out of" the world and back into the reality of the theater. But the sensations of speed and rapid movement while looking at the world can be convincing.

"Immersive VR," on the other hand, is done with a head-mounted display and emphasizes interactivity with a virtual (software-derived, not filmed) 3D environment. The head-mount allows the participant to enter and become immersed in the virtual world. The principal differences between immersive and other forms of VR include the fact that the user in an immersive system can turn around, look behind, and see something in the virtual world—swimming fish, exploding volcanoes, angry wasps, or the back door—not just the back of a theater seat. An immersive virtual world is genuinely three-dimensional and inclusive.

Immersive worlds are also usually interactive: the participant decides where to travel. This freedom may even extend to traveling outside the models built by the VR world developer, even allowing the user to fly up through the ceiling or into the walls. Immersive worlds are often manipulable; that is, the objects in the virtual world can be "collided" with and can respond with a behavior as a result of the collision. A virtual dog can, for example, back up or disappear or fetch a bone or attack the user after being contacted in the virtual world.

The head-mounted display actually has two components to it that allow the effect of immersion. One is the display itself, typically LCDs mounted in front of collimating lenses that straighten out the light rays from the LCDs, making them appear to emanate from optical infinity instead of four inches from the eyes. The other element is a tracking device that records the movements of the user and sends the coordinates it collects to the computer. Those coordinates tell the software rendering the images onto the LCDs where the user is looking. Without a tracker, the user's viewpoint would not change as the head's coordinates changed, and the effect of immersion would not be cogent.

The third type of VR, "second person VR," uses a camera to capture the image of the participant and insert it into the virtual world. Users then watch their own images on a television or movie screen interacting with objects in the virtual world. Virtual hockey rinks with virtual nets, skaters, and pucks that are deflected by a real keyed-in goalie are one popular game in second person VR.

In most second-person systems, the insertion of the participant into the virtual world is done via chroma-keying. This sometimes creates highlights around the participant or image resolution differences between the participant and the background (similar to what TV viewers see when they watch the meteorologist stand in front of weather maps on the local evening news). It often takes a minute of practice to correlate one's own body movements with what is happening on the screen. In time, the user can easily interact with objects and scenes in the virtual world.

A combination of second person and immersive VR yields a hybrid type known as a "cave." In a cave the participant walks into a room and is surrounded by a virtual environment. The environment may be displayed on multi-TV walls or on rear-screen projections. The user can interact with the projected objects because a camera, or an optical or magnetic tracker captures the user's coordinates and sends them to the system, notifying the computer where the user is in the world and thereby triggering responses from the virtual inhabitants who are displayed on the walls, floor, and ceiling.

Of the three main forms of VR, the one people most often think of as "real VR" is immersive. But virtual reality's definition has been stretched as widely as multimedia's definition has and is likely to be stretched farther as marketers begin to call everything from 3-D videogames to network visualization systems virtual reality.

VR Application Types

There are three general types of VR application, each with its own benefits, limitations, and uses. They are perambulation, synthetic experience, and realization/reification.

"Perambulation" involves walking or flying through some form of model. This may take the form of rolling through the CAD rendering of a hospital in a virtual wheelchair checking for architectural barriers to access. Or it may involve meandering down a human esophagus at the end of a virtual endoscope looking for lesions or ulcers in the stomach. In a perambulation, the user is mostly interested in observing aspects of the virtual world and discovering something about that world. Interactivity is generally limited. Interactivity may focus on moving objects around in virtual space (repositioning furniture in a virtual house), or removing objects from the scene (excising tumors from a duodenum). Mostly, however, perambulation applications center on observation rather than manipulation of the virtual world.

"Synthetic Experience," on the other hand, involves the training of muscle memory. Synthetic experience applications, such as virtual surgery or power plant control room operation, allow participants to easily practice skills that are dangerous or expensive to develop in the real world. In synthetic experience the most important aspect of the virtual world is not its appearance but its ability to allow direct interaction between a user and objects in the virtual world. A participant in such a world learns how to perform actions, usually with the hands, by practicing them (not just observing them) in the way a pianist learns to play a new composition.

The actions performed in a synthetic experience world are often those that may someday mean life or death to the participant, but which occur so rarely in real life that they cannot be practiced very often ahead of time. Firefighters, for example, can learn to direct water onto areas of a virtual chemical fire that will yield the greatest result the fastest. Police officers can learn to defuse virtual bombs. Plant workers can practice adjusting flow through pipes in a virtual refinery. The goal of such VR systems is to allow participants to practice critical actions over and over until they become second nature in hopes that the instinctive reactions the users develop will serve them well when those reactions are needed in the real world.

The third type of application, "Realization/Reification," allows users to see and manipulate profuse context-dependent data graphically. Reification means making a thing out of an idea. And Realization, an extension of Scientific Visualization and Visual Languages, refers to the representation of complex data in a graphical fashion. Such a representation, unlike most scientific visualizations or visual language icons, is usually both three dimensional and interactive.

Uses of realization are largely in industries that process prodigious quantities of data in real-time. These include network industries such as telecommunications, utilities, and financial services. The benefit that realization/reification offers is the presentation of directly manipulable data, one step beyond ordinary graphical displays.

In a traditional spreadsheet, for example, what-if analysis can be done by changing a single number in the spreadsheet and watching the other columns of numbers change in response. An ordinary graphical presentation of this data allows the user to see those changes occur on a bar chart or scatter graph. VR, however, allows not only the graphical presentation of the results, but graphical manipulation of the data; that is, not just graphical output, but graphical input.

A realization system allows the user to reach out into a seascape of livefeed data that the user sees as waves moving across the screen in realtime. The user can view, say, Spot FX data (foreign currencies) as they emerge from a Reuters feed and can perform what-if analysis interactively with the live data. By reaching out with a wand or glove, the user can press down on Yen prices to see the system reveal the concomitant increase in wave heights for the British Pound and Swiss Franc, or push profits from sales of a minor currency, such as the Swedish Krone, across the screen into purchases of a major one, such as the French Franc, seeing the resulting predicted effect on the whole portfolio.

The engine behind such as a system is the same as that behind any network predictive system: part expert system rulebase, part inductive algorithm, part neural net. What VR adds is the ability to reach out and touch data directly and see the results of those actions immediately displayed in the data-seascape virtual world.

VR SOFTWARE AND GRAPHICS

In order to run any of these VR applications, three pieces of software are necessary: device drivers, world model builders, and navigation tools.

"Device drivers" connect the unique devices needed to navigate in and interact with virtual worlds to the code that renders the images. They allow the user to communicate with the virtual world.

Typical devices include everything from optical coordinate sensors that track the movements and head orientations of a participant to spaceballs that allow unlimited flight through virtual worlds to the motion-based seats that respond to the user's movement in virtual space to soundboards that add a sonic component to virtual worlds. All of these devices capture signals from the participants and send them to the virtual world or take signals from objects in the virtual world and send them to the user. It is the device drivers that process and route those signals.

"World development kits," the next level up in the VR software hierarchy, allow the builder to construct stationary models to later be flown through, manipulated, or animated. Most standard CAD packages can act as development environments. CAD world development kits generally allow creation of objects in wireframe. These objects are then imported into the navigation engine where they can be filled, animated, imbued with lifelike behavior, positioned, and lighted. World development is the construction of the stationary model to be traveled through, as well as the dynamic objects that act in virtual space. But another piece of software is needed to allow the user to travel in that virtual space.

The "navigation/rendering engine" (not to be confused with the hardware "rendering engine," the graphics board that provides the processing power to draw the images on the screen) is a software tool used to navigate through, manipulate, and enliven the images created by the world development software. Navigation engines are the real workhorses of VR, the software components that make VR a possibility. Some are object based, some C++ based, some provide their own scripting language. Some even provide easy iconic graphical environments for object construction, making them combined world development kits and navigation engines. Others focus on navigation and rendering processing power alone.

When building a virtual world, a VR world designer has three levels of graphics with which to construct a new reality. The largest scale level is the geometry primitive. Common geometries are spheres, cones, and cubes. These can be variously shaded, lighted, colored, positioned, and animated to provide a world that conveys the impression of being filled with recognizable objects. The more detailed level below geometries is the polygon. Think of polygons as the faces of the cube or the sides of the pyramid. Onto these polygons can be mapped textures (scanned images such as photographs or zebra stripes or brick patterns) that add to the photorealism of the final world. At the lowest and most processing-intensive level are voxels, three dimensional pixels. This is the level at which medical imaging is done. Because the level of detail it can convey is huge—and, consequently, its CPU demands prodigious—voxel level work in VR is, at least so far, not readily attainable. But this is nothing more than a performance limitation, not requiring any fundamental breakthrough in the technology, only more horsepower, so it is likely to dissolve as a limitation over the next several years.

PERIPHERALS

Peripheral devices also fall into three categories: audio/visual, tracking, and navigation/manipulation.

The visual side of VR contains everything from the movie screens at theme parks to the standard VGA monitors that most developers use to build and test their worlds to the head-mounted displays that occlude the peripheral vision of users and make them feel as though they are inside the virtual world.

On the audio side, there are ordinary MIDI boards that release sampled sounds when buttons are pressed or virtual objects contacted. There are also audio boards that allow for "spacialization" of sound (the impression that a sound is coming from a given direction and altitude). They may even allow particular sound parameters to be tied directly to objects in the virtual world, such as airplanes whose engine roars doppler shift as they approach the listener. Also, because touch is still a difficult sensation to emulate in VR, many developers substitute sound cues for tactile ones. This sonic feedback collision announcement allows users to hear a sound cue when they have touched an object, rather than actually feeling the object.

Tracking devices, another major group of VR peripherals, are used to send the coordinates of the participant's head, hands, or body to the renderer, so the computer will know where the user is looking. Trackers generally work within "6 degrees of freedom"; that is, they handle movement in X,Y,Z space (position) and roll, pitch, and yaw (orientation). (Roll: turn your chin to the left and the top of your head to the right. Pitch: Lift your chin up and dip the back of your head back. Yaw: turn your head to the right or left. In all three instances, your body stays in the same X,Y,Z position, but your orientation changes.) Trackers come in optical, sonic, inertial, direct physical, and magnetic varieties.

Navigation/manipulation devices, the last major peripheral component of a VR system, include everything from gloves for picking up objects in the virtual world to globes for sailing through them to footpedals, joysticks, wands, and gyroscopic flying mice. The more advanced work in this area is now being done in the use of biological signals as inputs to the virtual world. Biofeedback signals such as EEGs, EKGs, galvanic skin response, and myography (muscle tension) are now being used as inputs that manipulate objects, viewpoints, or actions in the virtual world. This work may eventually lead to direct linkage between human bodily processes, even thoughts, and actions in the virtual world.

THE RANGE OF VR APPLICATIONS

Not unexpectedly, the most rapid development of VR applications is now being done in the entertainment realm. And within a couple of years VR entertainment in arcades and at home will be ubiquitous. Most arcade VR games, because they are relatively expensive and are time-fee-based, are testosterone-driven and involve little or no learning. The user simply dons a headmount or sits in a motion-based gondola and the experience begins, usually involving shooting opponents of some kind. Home-based VR tends to be far more involved and complex because home players spend far longer inside their games and prepay a lump sum for the experience rather than a by the minute fee. So, home games tend to be far richer, more learning-driven environments. However, they too tend to concentrate on merely thwarting villains.

In medicine, the Holy Grail is the virtual cadaver. Attempts to develop a virtual living body range through four different approaches. Film-based endoscopy allows a user to travel down a virtual esophagus cobbled together from several thousand photographs of real throats. MRI-based systems construct whole body sections organ by organ from scan data. Microtome reconstructions embed a real cadaver in wax, slice it, digitize the 2D slices, and arrange them into a 3D image in the computer. Software-based organ development builds models of living organs from scratch out of polygons and voxels.

My own work in this field has centered around animating virtual organs and organ systems using the principles of Artificial Life, a technology for modeling extremely complex living processes. This method involves combining 3D cellular automata, neural nets, genetic algorithms, and production rules, together with VR as the display mechanism, in order to portray processes such as tumor development or immune function, or to reveal approaches to gene therapy using engineered viruses as the mechanisms for site-directed mutagenesis (employing hand-changed viruses to insert their cargoes into malfunctioning cells).

In the network industries (telecom, utilities, financial services, and medical infomatics), VR work is concentrating on developing new metaphors for dealing with data: expressing relationships between vast amounts of related data, allowing users to interact directly with graphically displayed data, and representing data in three dimensions so as to elicit "intuitive understanding" of prodigious quantities of data.

The work here in these fields includes the exploration of novel metaphors for displaying profuse data: everything from Christmas trees the color and shape of whose bulbs represent types of problems in a network, to hydraulic models that show energy transmission lines as pipes and reveal capacity by the diameter of the pipe and usage by the height of water in the pipe, to rolling wavy seascapes of continuous livefeed data that can be processed and speculated against by pushing and pulling as if the data were interactive taffy.

In architecture, most VR work has focused on edifice prototyping and testing. Edifices range from rooms in a private residence to hospitals to vast chemical plant facilities. This approach allows the user to first construct the virtual building and then to test it for compliance with various regulations or for the comfort and well-being of future denizens.

Our work in this arena has concentrated on linking the construction of the virtual building to a spreadsheet so that design considerations can be visualized and their financial outcomes viewed simultaneously. Viewers can then make changes to the building in realtime to satisfy design versus finance constraints. They can then view the building to see if it looks the way they want it to look, and they can study the resulting spreadsheet created by the system to see if costs are in line with budget. We have also built in comments on the environmental impacts of design decisions—allowing a client to view the effects of building decisions on the environment as well as on the purse and on the eyes.

The area of VR development currently undergoing the greatest level of corporate and industrial exploration is the field of skill-based training. VR's unique benefit over film and ordinary 2D graphics is that it allows users to interact with its objects. This has allowed VR developers to construct worlds in which users can practice skills they use in the real world.

Synthetic experience investigations in my lab, for example, involve telephone line worker training, switching station training, and process control diagnosis training in large plants. And this has led to an interest in "living history" and "wraparound" educational applications that allow a user to become immersed in an historical environment (the wild west, a slave auction, traveling steerage across the Atlantic to America, under the bodhi tree with the Buddha, in Thomas Edison's lab, at a summer gathering of the nations of the Iroquois confederacy, on the battlefield at Gettysburg, in Timbuktu at the height of the Kushite empire). Beyond these real world applications lie the ability to practice viewing and working in environments that human beings have not yet experienced (a colony on the surface of the moon, en route to Mars, a journey to the center of the earth, and worlds that exist only in the imagination).

The Real World is Messy

In all of the commercial areas beyond entertainment, VR developers have discovered that pure VR applications seem to be relatively few. As a result, most commercial work in the field now involves either coordination with or direct integration to other technologies. Behavioral animation (the insertion of lifelike behavior into creatures in the virtual world), for example, may require half a dozen technologies and techniques outside of VR. In such an environment, VR becomes one technology among many. It is likely that in time, as VR entertainment develops, these same concerns will become important to game-builders as well as they broaden their scope from just 3-D graphics in a helmet to genuine artificially enlivened universes.

The future of virtual reality points in this direction: the commingling of other technologies with VR. For example, as optic fiber replaces copper, and bandwidth limitations recede, long-distance networked VR is becoming a reality. As CPU processing power increases, detailed voxel level images are being experimented with. As GUI designers begin to work in three dimensions, desktop metaphors will gradually be replaced with "landscape" and "interactive data-taffy" interfaces. As artificial life begins to play its role behind the scenes in behaviorally animating virtual objects, virtual worlds are beginning to take on the richness and complexity of the real world. As all of these advancements accelerate, it will become possible for participants in virtual experiences to feel for the first time as if they are natives in the virtual world, not just tourists.

AUTHOR BIOGRAPHY

Richard Kelly founded and runs the Artificial Life and Virtual Reality Applications Group at Digital Equipment Corporation, Nashua, NH. He has built commercial applications in advanced technologies with and for dozens of U.S., Asian, and European corporations; and he has taught extensively in the U.S. and Japan. His current research and application development work is focused on the conjoining of Artificial Life to Virtual Reality in order to enliven virtual universes. He is currently working on biomedical, telecomm, utilities, pharmaceutical, training, entertainment, and financial services VR applications. His latest book, in production, is entitled *The Future of Virtual Reality: Unexpected Consequences of a New Technology*.

CHAPTER 25

The Virtual Classroom

by Ken Gerlach

In a rapidly changing marketplace, corporate training programs can no longer deliver training in a responsive manner. Shorter product life-cycles and accelerated development places pressure on training employees whose skills need constant revision. U.S. companies spend less in time and cost to train workers than do their international competitors. A shrinking labor force must be multi-skilled and trained so they can quickly adapt to changing market conditions. Distance learning technology, delivered via satellite, can be utilized to quickly train geographically dispersed employees in a quick, cost-effective and efficient manner. In this Virtual Classroom, time and distance are collapsed as instructor and students interact in a simulated, traditional classroom setting. System configuration, instructor video skills and DL benefits are outlined. The future evolution of training promises desktop delivery of self-paced, individualized training when, where, and in as much depth as the student needs, to maintain a competitive position within and for the organization.

THE EDUCATION CHALLENGE

Corporate educational and training programs face serious obstacles and are not keeping pace with the reality and needs of a changing workforce. As John Naisbitt writes, "We have essentially the same education system we had for the industrial society and we are trying to use it to equip us for the information age."[1]

In the past decade we have seen constant, revolutionary change in products, technology, customer requirements, and in the skills and knowledge needed by company's employees. As a consequence, corporate America must confront the huge challenge of retraining its workforce to compete in a new and changing world economy. The American Society for Training and Development estimates that 55 million Americans, about 45% of the workforce currently employed, need to be retrained in the next decade in technical, executive management, supervisory, customer service and basic skills training. Moreover, "it is estimated that corporations will have spent in excess of $30 billion in 1992 to train employees" but that has not been sufficient to solve the problem.[2]

Training can be accomplished through traditional courses (over half of America's corporate training takes place in a traditional classroom), tutorials, or informal, "grapevine" learning. Structured, formalized training is more productive than informal training and much more effective when done in-house. The average U.S. company invests less than 2% of its payroll in formal training, reaching only one-tenth of the work force. The average amount of time corporations spend on training their employees each year varies by industry with up to five days per year for industries with established technologies and ten days a year for industries with rapidly changing technologies. At Hewlett-Packard Company the average for technical people is between 22-27 days per year. The surplus, trained, labor force of a few years ago has been considerably reduced, so each employee's current skills are especially crucial for a company to remain competitive.

The typical corporate training environment is comprised of a face-to-face class, either in a central facility or in field offices where a trainer is sent to each location. But traditional training cannot keep pace in an era where technical obsolescence of an employee's skills can occur within five years of graduation. Increased competition, driven by accelerated product obsolescence (decreasing product life due to shorter development and marketing cycles) also impacts the necessity for constant and on-going training. For instance, the average shelf life of a personal computer is less than nine months. At HP, 50% of revenues are generated with products introduced within the last 2 years—a very rapid turnover of products. With limited product life and rapid product development, the pressure to accomplish training goals within a limited time frame is enormous.

HP's field engineers became vulnerable during the 1980s when the accelerated development of new products was coming faster than technology updates could be transmitted. The interval between new product release and accomplishing marketing/training functions had to be cut dramatically. Training was struggling to keep up with the tech-

[1]Naisbitt, John and Patricia Aburdene. *Re-Inventing the Corporation*. Warner Books, New York, NY. 1985.

[2]Roelandts, Willem P. *Distance Learning Solution*, Keynote Address to Telecon XII. San Jose, CA. October 1992.

nology—field engineers were required to learn more in less time, but there was not enough time.

Traveling trainers, sent from corporate headquarters to the company divisions and field offices, can take months to deliver training to all sites. The optimal solution is to bring the training to all the employees at the same time—have the subject expert or panel of experts deliver the same information to all company sites simultaneously via some reliable and proven technological delivery system such as satellite. Tom Wilkins, an R&D Manager for Distance Learning products at HP says, "Between completion of course materials and delivery, a critical gap develops, sometimes as long as six months. The longer it lasts, the more likely it is that the training will be outdated before it can be delivered to the students."

Traditional classroom training will become increasingly less viable in terms of delivery time, cost, and user downtime since rapid changes in technology and product obsolescence are becoming the norm across industries. A company needs to train its people on new products or communicate new strategies in a way that collapses time and distance. The challenge will be to retrain employees while maintaining competitiveness, reducing soaring costs and managing limited time.

Training to Compete

For companies to remain competitive they must plan for and quickly identify opportunities in a world that is truly becoming a global village. Local, parochial markets are disappearing forcing organizations to either be responsive to changing market conditions or perish.

It is impossible for most corporations to be successful today without a worldwide strategy. In the 1960s, only 6% of U.S. businesses were exposed to international competition—today this figure is more than 70% and increasing. At HP more than half of its revenues come from overseas markets.

Frequent training and effective communication are fundamental for a company to be responsive to changing markets and products and to achieve a competitive advantage. Through evolving technology, collaborative management and market-responsive production, there is a premium placed on employees who are multi-skilled, multi-trained and who can be easily re-assigned as needed.

"America's competitors in other countries commit significant resources to building a well trained workforce," but the U.S. has the "...lowest percentage of spending on training compared to all the nations we...compete with" in Europe and in Japan. In the U.S. the amount spent on upgrading employees' skills is approximately only 3% of what is spent on capital improvements to plant and equipment. Yet "...the gains in productivity from workplace learning exceed the gains from capital investment by more than two to one."[3]

In today's global economy, the U.S. can no longer compete on the basis of mass production alone. We must respond faster to market needs and dramatically shorten product life-cycles bringing products to market in record time. Time-to-market will be crucial with companies dependent on a smaller work force that must continually revise skills and use more technology in order to compete.

THE TRAINING SOLUTION

In spite of the traditional classroom's inability to keep pace with growing training requirements, there has been a serious under-utilization of available technology which could be used to solve the training/education dilemma.

For a company that needs to update and quickly train geographically dispersed employees, delivering a consistent message worldwide, there is one technology that will enable them to quickly and efficiently respond to the training challenge...Distance Learning! Distance Learning (DL) is an electronically-assisted teaching/learning environment where teacher and student are geographically separated, rely on electronic devices and interact within a format that closely resembles the traditional classroom.

In the past, technology has focused on improving discrete instructional tools and access to outside information. The new generation of technology is aimed at creating a "Virtual Classroom" where artificial or physical boundaries are removed. Students in different locations across the country not only observe and listen but participate and become involved in subject matter that is delivered by the most knowledgeable and effective learning facilitators available and flavored by applicable, real-world case studies.

An effective DL environment creates a virtual classroom by using a telecommunications channel (i.e., satellite telecast), to link people at a number of locations. Interactive, two-way communication and real-time feedback is provided which encourages the users' active participation. Given this technology, class size is unlimited, time and distance boundaries are removed, and the student has direct access to experts. Corporate training can be conducted quickly and cost effectively.

Training via HP's Interactive Network grew out of a strategic need to provide a cost effective and rapid delivery system for continuous information transfer from the factory to worldwide field sales and support groups. Sixty percent of HP's technical field people now participate in DL.

[3]Portway, Patrick S. and Carla Lane. "Corporate Training," *Technical Guide to Teleconferencing and Distance Learning, Applied Business teleCommunications*. 1992. pp. 279-315.

In adopting a new training technology, a valid criteria to consider is whether it lowers cost and decreases time while maintaining or exceeding quality. HP has found that DL can deliver more training simultaneously to all locations, for lower cost with consistent quality. Instead of sending instructors to the field, one expert presents the same message to everyone, at the same time, across multiple sites. It certainly is more efficient than physically moving people across the country to training classes. The ability to share resources through technology is also a viable alternative to building more buildings and individual classrooms. DL can maximize resources and combine assets with other groups to produce programming.

DL is a strategic solution for corporations—an effective means to train employees quickly, cost effectively and with enhanced quality, for the training is delivered undiluted by an expert. DL can be a crucial part of an enterprise's strategy to better manage and continually improve its competitive position.

A Distance Learning System

In the Virtual Classroom that a DL system creates, specially designed hardware and software enable real-time interaction between an instructor at a host site, and students who may be located at multiple, global remote sites. It employs two-way interactive telecommunications and electronic devices to teach from a distance.

There are various delivery technologies for DL to choose from: satellite, fiber optic, cable, codecs over telephone lines and microwave. Computer response units, scanners, facsimile machines, CD-Rom, videodisc, and radio are also utilized. For purposes of focusing on larger numbers of students and widely dispersed receiving sites, satellite delivery best serves the objective.

Transmission, via traditional, analog satellite signals, broadcasts audio, and video to multiple receiving sites with return audio and two-way data communication supported over conventional land phone lines. High quality audio is paramount in learning environments—speech delivered without excessive reverberation, noise, feedback, distortion, or returning echo characteristic of a satellite transmission delay are important considerations.

Video compression eliminates the redundant part of the analog signal, taking the analog video transmission and converting it to digital. With digital, there is no degradation of signal as it is retransmitted; and an additional benefit is that it requires significantly less bandwidth. Advances in digital compression will lower transmission costs but more importantly will double or triple channel capacity (number of channels that can be sent over any transmission medium).

A total distance learning solution (production, video, audio and control system) is pulled together from various suppliers. A full integration of components requires expertise in audio, video, data, and control. This integration has been a barrier to market acceptance because users do not have all the necessary expertise. Outside services are usually required for hardware/software installation and test, design, production training, and prototyping of the specific training/educational programming.

Let us consider the various configurations of communication technology available and how they are applied.

Synchronous Mode:

1 way audio	Radio program
2 way audio	Telephone conferences, radio talk show
1 way audio/1 way video	Television program and video tape distribution

Asynchronous or Virtual Classroom:

2 way audio/2 way video	Video conferencing with compressed digital over telephone lines

The most economic and effective for training environments is:

2 way audio/1 way video	DL transmitted over a high quality analog satellite to multiple sites over a vast area.

The virtual classroom simulates the feeling of a large number of students being in one classroom together even though they are physically dispersed. Technology is applied seamlessly and is the most transparent to learners and instructors where the TV-quality image is maintained, and the two-way questioning/answering process is unimpeded. Interaction is supported through computer-assisted response devices and communication technologies that enhance further feedback.

SYSTEM CONFIGURATION

For simplicity, DL configurations include a host site, mode of signal transmission and receiving sites.

Host site: The program originates from the host site where the session is broadcast from a studio classroom. This teleclassroom would include a lectern/console where an instructor has the ability to either take complete control of the console features or have the production crew handle them. The lectern is equipped with a computer-based console that facilitates interaction between the instructor and students at remote sites through voice or electronic question and answer.

The broadcasting facility, or uplink, is equipped with a control room where a technician has duplicate controls to assist the instructor if needed. It may also include a director's console, various special effects generators including VideoShow, stillstore, animation, chromakey (for displaying and annotating one image in combination with another), and camera and sound system controls. Video editing and duplicating equipment could also be included in this facility.

Transmission chain: The first mile of communication link between the studio and satellite uplink usually consists of cable, fiber optics, microwave, laser or telephone land line. The program signal is delivered to an uplink antenna or dish, then transmitted to a geosynchronous satellite in a fixed position 22,300 miles above the equator. Satellites have transponders (video channels) that receive and retransmit the video signal. If a broadcast is encrypted, the encoder or scrambler would be co-located with the uplink and each receiver location would need a decoder to unscramble a broadcast.

Receive sites: A satellite downlink receives and decodes the satellite signal into channels. Equipment configurations may include satellite dish, low noise block converter, satellite video receiver, decoder for encrypted, scrambled code to maintain confidentiality and security, connectors, power supply, and video display.

Interactivity: Interactivity distinguishes DL from one-way, passive instructional television. Interactivity allows each student to participate with the instructor and with each other, giving students a sense of being connected with the remote class members and facilitating their active involvement. Interaction is accomplished via telephone, two-way audio, video, two-way electronic text, and/or graphics interactivity. Interactive Response Units (IRUs) or keypad devices, allow remote and local students to interact electronically with the instructor and among themselves during live classes. With IRUs, students can receive quick and meaningful feedback and actively participate when directly linked with the instructor. Instructors also receive immediate feedback from compiled student responses to numerical or multiple choice questions via the interactive keypad system, and are able to immediately assess the students' comprehension of the material, modifying the content, rate or presentation style in real-time. A database can be used to gather all the transactions for later evaluation and analysis.

THE INSTRUCTOR

Typically when under pressure to improve efficiency in training, the choices are either change the technology you use, change the way instructors teach, or some combination of the two. The task at hand involves putting the technological pieces together via system integration—bringing proven hardware and software into the mainstream of training.

The first obstacle for instructors to overcome is the paradigm shift from traditional, central training to DL, and one determining factor for acceptance is whether the technology is transparent to the learning. If the technology overwhelms the instructional process instead of serving as an effective teaching tool, then this approach will be resisted. Technology must be practical and easy for students and instructors. The goal is to take the face-to-face training environment and make the technology "invisible" so both the instructor and students can be actively involved in the learning process.

PRODUCTION STAFF

There are production staff who can aid the instructor and insure that the technology does not inhibit the process. The instructor is then free to concentrate on content and leave the production to the experts who might include:

Administrator: Scheduler of the facilities and satellite time, organizes timely distribution of materials to all remote sites.

Remote Site (field) Coordinator: Responsible for communicating directly with each receiving site coordinator. Reserves room and equipment, sets up a room for satellite signal receiving, trouble shoots technical problems, arranges participants' question procedure and follows up with student evaluations.

Producer: Overseer of the entire production. Conducts initial meeting to define what is needed and to advise the instructor on techniques, resources, media and production methods. The producer facilitates the process and makes sure the instructor's teaching aides, (i.e., live product demonstrations, videotapes) are properly integrated into the class.

Director: Works with instructor and directs the crew during rehearsal and broadcast to plan and select best way to "shoot" the program, timing and pacing, camera angles, placement of graphics.

Additional crew (when necessary): videotape recorder operator, additional camera operator, character generator/graphics operator, production assistant.

SKILLS NEEDED

Televised teaching requires skills different from traditional teaching. Instructors need to familiarize themselves with the studio environment, the control room and computer console, and learn new teaching skills for telecourses. An array of electronic educational tools including stored graphics, animation, videotape (roll-ins), drawing tablets and remote cameras will need to be mastered so they can be incorporated in the program effortlessly.

Since timing is critical, detailed organization of the class and explicit pre-planning of the presentation is essential. A script should be developed to insure broadcast time constraints are adhered to, and procedures established for course administration (i.e., student questions and distribution/retrieval of materials). New presentation skills, suited to the medium, need to be fostered such as responding to camera movement, cueing and speaking directly to the camera in an unhurried fashion occasionally varying the tone of voice. Select clothing with bright and bold colors, avoid plaids, stripes, and patterns that cause video interference. Orchestrate frequent interaction with students through predefined questioning strategies that stimulate discussion. Practice and rehearse to perfect timing and transitions, and to become comfortable with the technological tools. This is an extremely important objective—the instructor should be comfortable enough to focus on the student easily and to use the technology to modify the course pacing or content to meet students' needs.

BENEFITS OF DISTANCE LEARNING

Using technology for technology's sake can prove less than optimum, however a new training methodology or technology should be adopted if it meets two objectives:

- Reduces costs associated with conventional training by reaching more students in less time.
- Provides effective learning as well as, or better than, conventional face-to-face training.

Distance learning fulfills these needs in offering the following benefits:

A. *Saves time*—content presented by a single source is received in many locations simultaneously and instantly. Information is disseminated to all employees at once, not just a few at a time. Travel is reduced, there is less 'down' time, and productivity is maintained without travel disruption or the need for 'catch up' time. DL can greatly reduce the time to retrain a large, geographically diverse population and deliver time-critical messages immediately.

B. *Large audiences*—more people can attend. The larger the audience, the lower the cost per person. Delivery to the workplace at low cost provides the access to training that has been denied lower level employees. A company can increase the number of students trained without increasing training resources and can eliminate or minimize trickle down training, road shows and information "gate keepers."

C. *Lowers Costs*—travel, meals, and lodging costs are reduced by keeping employees in the office. In addition, lost opportunities for revenue-generation are diminished since field sales personnel participate in necessary training with minimal disruption to the time spent with customers. HP has continually experienced significant cost reductions in delivering DL (compared to centralized training) with an average return on investment of $20 million per year over the last 5 years. The initial cost to install the entire studio and control room was paid for with the first two-week teleclass delivered to 750 students worldwide.

D. *Improves learning*—HP surveys students to evaluate DL teleclasses and results show that consistently more than 70% to 80% select interactive teleclasses as the optimum method for instructional delivery compared with traditional types of training—students felt they learned more than in a conventional classroom. This is further supported by institutional and business studies which demonstrate that training at a distance is as effective as training face-to-face.

E. *Leverage limited expertise across locations*—the expertise of the best instructors can be delivered to thousands of people without sending trainers to conduct numerous training sessions at various locations. Topical experts spend more time in the office doing research and productive work instead of traveling. Each student, no matter how large the class, has direct access to the experts rather than another representative instructor "trained-by-the-trainer."

F. *Accessible and Flexible*—Accessible—through any origination site in the world. Flexible—with a remote transmit or receive truck, a transmit or receive site can be located anywhere; a benefit of the pervasive nature of satellite delivery.

G. *Adaptable*—useful for business, associations, hospitals or universities to present, inform, train, or educate. Company updates, training, meetings, and other events can be broadcast live to multiple locations.

H. *Secure*—encryption prevents unauthorized outside viewers and protects confidential data.

I. *Interactivity*—defeats the boredom and passivity of one-way television classes. DL requires greater emphasis on involvement strategies and methods than traditional, teacher-centered strategies. A condition of effective learning is an environment where ideas can be freely communicated. Students receive not only visual and audio stimuli, but through interactivity (i.e., quizzes, telephone dialogue, and surveys) they participate in active, experiential ways. With distance learning the students can interact with the instructor when they need more discussion, enabling the instructor to respond to questions in real-time which is critical for students' learning and retention. Interactive keypads provide for immediate response and feedback, a crucial element since a student who does not receive feedback in a timely fashion quickly becomes demotivated. The dynamic exchange of ideas and questions keeps students alert, makes them feel they are participating and involved, and increases information retention and comprehension.[4]

FUTURE DIRECTIONS

The next decade promises greater access and expansion of national and international education and training networks utilizing interactive telecommunications.

The traditional classroom at a central location, or alternatively "train-the-trainer" strategies, are being displaced by real-time interactive virtual classrooms at regional locations where instructor and students no longer need to be in the same place at the same time.

The next major step will be towards workstation-to-workstation training and education that will replace DL studio-to-receiving sites' classroom training. Multimedia, interactively manipulating text, images, animation, graphics, sounds, digitized voice, and video will enhance the desktop as the delivery vehicle.

Eventually, encapsulated training packages will be delivered on demand—individualized training that is self-paced and delivered when and where a student needs it. Students will have the ability to acquire new skills and knowledge when they need them, where they need them, and in as much depth as they need them. Responsive training, delivered to the workstation, will enable employees to meet the demands of an increasingly competitive global marketplace.

With rapid product development and changing worker skills requirements, the critical characteristics of a training system will be one that is distributed and just-in-time. Training will be responsive to the individual's needs, taking place when and where needed. Multimedia will provide the necessary visual and auditory stimuli and facilitate interactivity. The components will be modular so that the student can select only those training elements that are demanded for the immediate task.

The virtual classroom returns full circle, focusing once again on the individual student's needs. The future promises individual freedom and control where important skills required to efficiently compete in an ever-changing, new world economy can be easily revised and enhanced as needed.

[4]Lane, Carla. "What is Teleconferencing?" *Business Television and Distance Education: Partnerships in the 90s, An ITVA Industry and Education National Videoconference Resource Manual*. February 12, 1992. pp. 10-12.

AUTHOR BIOGRAPHY

Ken Gerlach is the Marketing Programs Manager of Hewlett-Packard's Distance Learning Systems Operation. He manages activities for the planning and introduction of integrated products and services for delivering interactive distance learning training and education. Products include multi-media computing integrated with HP-IN (Hewlett-Packard Interactive Network), one of the most advanced interactive distance learning television broadcast production and telecommunications services in the world. HP-IN incorporates HP's global business television network for interactive technical programming, utilizing three broadcast host sites and 117 remote downlink receiving sites in North America and Europe.

Gerlach's twelve years experience at Hewlett-Packard also span engineering project and marketing program management of computer hardware and software applications that include emerging datacom, remote conferencing and graphic presentation technologies.

Formerly, Gerlach provided project leadership as a Management Consultant at SRI International on techno/economic and market feasibility studies of computer automation and telecommunications for business and institutional clients in the U.S., Europe, and Asia. His early career began in project engineering design with the world's two largest engineering consulting firms, Skidmore, Owings & Merrill in the U.S. and Ove Arup & Partners in London.

Gerlach holds an M.B.A. in marketing from Santa Clara University, M.S. and B.S. in mechanical engineering from Pennsylvania and Portland State Universities. He currently is a part-time instructor of graduate and undergraduate business courses at the U.C. Berkeley Extension and other extended education programs. He is a member of numerous engineering and trade associations and a Chartered Engineer (Europe).

CHAPTER 26

Telecommunications in an Age of Convergence

by Kevin Brauer

The last decade of the 20th Century has spawned numerous prophecies about the Information Superhighway that is going to provide integrated broadband communications to every home and business in America. We have been hearing about all the new capabilities that will result from the convergence of computers, telephones, and television. The prophets also have created visions of billions in profits for the company that can be first to acquire all the technologies that are converging to create the superhighway.

To succeed in the era of the Information Superhighway, some believe we will need to create mega-companies that embrace—and control—as many of the converging technologies as possible. Others say that the small high-tech venture companies, particularly software firms, will lead us into the convergence era.

While most experts agree that consolidation is inevitable in the communications industry, no one large company can create the intense focus and agility needed to compete successfully across the breadth of the Information Highway. No small firm will be able to generate the intellectual and financial capital necessary to control a segment of the highway in the same way that Microsoft and Intel did in their respective industries. An alternative way to achieve the benefits of both large company scale and small company flexibility is by creating joint ventures with partners who have complementary technologies.

The promise of modern communications is vast, and so are the financial stakes. Sales of communications and information services in the U.S. totaled $310 billion in 1993 alone. And the bidding for new wireless Personal Communications Services (PCS) frequencies that began in December, 1994 reached into the billions of dollars. Why are the frequencies so valuable? They are licenses for frequencies in the 50 biggest United States cities, and they represent access to customers.

Building the Information Superhighway requires collaboration, and it requires totally free competition in the telecommunications market, including the so-called last mile of the Information Superhighway. In what is likely to become the industry model, Sprint recently joined an alliance of telecommunications and cable television that will revolutionize what can be delivered over telephone, cable television and computer.

The joint venture with three of the nation's leading cable television providers will create a new communications alternative for America's consumers and businesses. Local, long distance, wireless, and multimedia communications will be delivered in one package over an advanced network. Consumers can select from a menu of communication services as much or as little of what they want and all from a single source

The alliance will bring true competition into the local telephone marketplace. In that one area where competition does not yet exist, there has been insufficient incentive to invest and innovate and to build an integrated broadband communications capability. That technical deficiency threatens to delay all the advances of technology, and the flow of the many products waiting for connection to the Information Superhighway all standing ready to make imagination become reality.

Competition in the long distance marketplace has gone a long way toward making the Information Superhighway possible. Sprint and MCI entered the market and challenged AT&T with innovative marketing and competitive pricing, which in turn attracted more users of telecommunications services. Companies like Sprint had the incentive to invest in fiber optics and network technologies like Asynchronous Transfer Mode (ATM) and Synchronous Optical Network (SONET), which created the opportunity for an Information Superhighway. AT&T also benefited, pushed by competition to become more customer-focused and innovative in its approach to the long distance market.

By joining forces in the new competitive arena, the partners can deliver to consumers and businesses the promise of Information Age services sooner, rather than later. Switched broadband services, such as high-speed data transmission and video-on-demand, could be available in the near future. Competitive wireline services, such as full-featured voice telephone service, and next-generation wireless services, including digital cellular and PCS applications, soon will open opportunities for business.

For the most part, the barriers to integrated communications services are not technological. Today's local area network users, for example, already have much of the flexibility, capacity and transmission speed they need to make their communications more productive. What is required for simple, reliable and affordable access to integrated services are:

- Direct access to local customers on a national scale
- Enough bandwidth to transmit huge volumes of data

- A more flexible regulatory environment
- A nationally recognized brand to make marketing activities cost-effective
- A complete selection of end-user services, including local service, long distance, wireless, interactive/multimedia and entertainment

Sprint's cable partners have almost 300,000 miles of cable passing one-third of the homes and businesses in the United States. They are currently upgrading that infrastructure to fiber optic. The local cable infrastructure—and pending wireless service to additional homes and businesses—will be connected into a competitive access hub operated by Teleport Communications. Owned by several cable TV operators, Teleport's broadband metro network serves 19 major markets and will expand to 25 by mid-1995. From the local hub, communications will flow over the Sprint nationwide fiber optics network.

Customers will have cost-effective, one-stop shopping on the Information Superhighway and a complete package of local, long distance and wireless services in major market locations. Small business customers will have a single point of contact, which will help simplify the management of all their communications needs and allow them to invest more time in their business.

By integrating communications services, the venture will be able to provide a total-solutions approach for nationwide wireless applications, emerging "virtual office" applications and controlled business costs. This new partnership, coupled with Sprint's global alliance with France Telecom and Deutsche Telekom, provides global customers with a one-stop partner that can help them extend their businesses around the world.

The alliance answers the question of how the benefits of information superhighway will be available in the nation's homes and businesses. They will be coming through upgraded fiber optics and a digital, broadband communication system, a substantial portion of which will be in place in the next few years, and which will move communication into the 21st century.

What will new alliances provide? Cable television video offerings will expand to include a growing number of interactive, multimedia services. Local telephone service will be offered to subscribers served by alliance members nationwide. Wireless Personal Communications Services will give consumers full mobility, including voice and data capabilities. Consumers will enjoy a broader array of long distance services on a national and global basis.

The alliance makes sense for Sprint and its partners for a number of reasons. First, is the already wired base of the cable infrastructure. That wired base will be upgraded from narrowband to broadband fiber optics and coaxial cable, providing the same "pin drop" quality on cable television and local telephone service as is currently available on Sprint's long distance fiber optic network. Second, Sprint's growing expertise in wireless service is a strong foundation for partners in the venture, including the bidding for

Personal Communications Services licenses. Third, the breadth and depth of Teleport Communications Group, as the nation's largest competitive access provider, links the local cable provider to the long distance network. Finally, the communications services of cable television, Sprint local communications, Sprint long distance, and Sprint cellular and wireless all will be available to the consumer in a simple package.

Sprint's rapid commitment to building a nationwide high-speed, all-digital fiber optic network in the mid-1980s laid the groundwork for the new connectivity. Its strategy then involved creation of an innovative network design: Five north-to-south routes and three east-to-west routes across the United States formed a "ring" pattern that is inherently more survivable than point-to-point or "star" designs.

Recent news has focused on Asynchronous Transfer Mode (ATM) and Synchronous Optical Network (SONET). These are two technologies that, combined, will offer unprecedented levels of reliability in transmitting voice, video, images and data over a single fiber pipeline. Sprint now is rapidly moving toward a vast increase in transport capacity with the world's first all-SONET, broadband switching network—a transformation taking place from now into 1996. While SONET is the evolving transmission standard, ATM is the companion switching protocol. ATM accommodates high-speed, high density transmission, as well as burst data transport. How powerful is that? If the customer has a project with so much data that it will take the system 10 days working 24 hours a day to send the information, together the new technology can move that entire ten days of data in less time than it would take to tell you about it. The SONET ring configuration offers recoverability from a physical break in as little as 60 milliseconds, a lapse easily addressed by extant, error-correction schemes.

Customers will be at the center of the new age of communications, and world-class service is at the center of every technology Sprint deploys. Sprint is continually designing and testing the network so customers can select the communications products and services that best fit their needs, which are varied and changing all the time.

An "open systems" design makes it simpler and more efficient for customers to link directly to an intelligent network platform to better manage their business applications or to develop customized solutions from multiple product building blocks. This open system design means not being tied to proprietary hardware or software solutions. Instead, Sprint develops partnerships with multiple communications vendors, offering multiple solutions—giving customers the ability to select the best available solutions to enhance their current business performance and confidently plan to incorporate emerging technologies needed for future growth.

Today the global telecommunications industry has over 100 equity alliances, up from only 21 just three years ago. The primary factors in successful partnerships appear to be technological or strategic fit and most importantly a common goal and response to market needs. Sprint's pending agreement with France Telecom and Deutsche Telekom, in which they will have a 20% investment in Sprint, responds to the need for European and American multinational customers to have a full spectrum of communications linking their international operations. Those companies have discovered the productivity and competitive benefits of responsive communications, and the alliance will help domestic customers achieve those benefits in Europe and support Sprint's partners' customers in the United States. That, in turn, will open up opportunities for the North American telecommunications industry to serve European customers.

To reach less developed areas of the world, Sprint teamed with several firms in the Iridium satellite project led by Motorola. That will provide telecommunications any place in the world without the necessity of a wire or cellular infrastructure. When Iridium is completed later this decade, Sprint will have the first truly worldwide, universal telephone capability.

Sprint was the first company to build a coast-to-coast fiber optics network, which is an essential component—the interstate portion—of the Information Superhighway. And it was the first to deploy Asynchronous Transfer Mode (ATM) and Synchronous Optical Network (SONET)—key technologies that provide the switching speed and bandwidth to enable the Information Superhighway to work.

But while Sprint has the confidence of a leader it also has the prudence to know where it should let others lead. For example, Sprint believes it is good at building the roadway...but unlike some of the other telephone carriers, Sprint does not see the benefit in trying to produce or control the entertainment and information products that will travel on the Information Superhighway. In fact, Sprint sees an advantage in creating a network that will be totally open to carry all information and entertainment products.

The alliance between Sprint and leading companies of the cable industry is built on the principles of customer need and a common goal. It takes advantage of a force that has propelled the successes of the telecommunication industry—competition. Competition has brought Sprint closer to the dream of the Information Superhighway. Today, thousands of hardware and software companies are creating products for home and business, for entertainment and education, and for efficiency in business. Fiber optic lines are installed from coast-to-coast, even internationally, to connect those products.

AUTHOR BIOGRAPHY

Kevin E. Brauer was named president of Sprint Business in June 1994 and is responsible for the sales, marketing, and operational support of all voice and data communications products to Sprint's business and government long distance customers. He was promoted to his current post after serving since August 1993 as president, Sales, for Sprint Business. In that position, he was responsible for all of Sprint's long-distance sales efforts directed at business customers regardless of size.

Sprint is a company that has long been known for its technological prowess. Today, Sprint is known as well for its people and their consultative approach to business, helping customers become more efficient and productive by properly applying the vast array of communications technology available in the age of convergence. Sprint Business is providing customers more value for their communications dollar, more reliability from the network, and more services to help their businesses succeed.

CHAPTER 27

The Rise of Natural Intelligence

by Isidore Sobkowski

Professionals in the field have long known that AI was going to make its way out of the rarefied air of labs and smack into the business environment. Nowhere is this trend more useful than in Help Desk support; as PCs have proliferated for the last decade or so, organizations have seen the advent of the confused user. Thus sprang the Help Desk.

Help desks are interesting animals. People call them to get answers to such questions as, "My screen is blank, what should I do?" Given the complexity of today's PC hardware and software there is truly a combinatorial explosion of things that can be wrong with someone's PC at any point in time.

The Antioch California school district needed this genre of solution when they found that they were supporting more than 500 computers at 17 sites throughout the Antioch area. They must not only maintain and repair their existing wide area network, but they must also train and support a user base composed of 40% administrators and 60% students on a variety of computer applications.

The October 1994 issue of the *Client/Server Economics Letter* even goes so far as to say that Help Desks are a "killer app." And an article in that most respected of all computer rags, *ComputerWorld,* says that customer service software will be the next major wave of client/server applications. Writers Rosemary Cafasso and Julia King have found in their research that customer service needs are taking an ever higher priority in organizations. Cafasso and King go on to say that Help Desk personnel are involved in many more functions, and in more areas of their business than ever, requiring software that allows them greater flexibility and wider access to other company systems.

These Help Desk success stories do mount up. Satellite manufacturer Hughes Space and Communications Co. has signed a five year outsourcing agreement for Electronic Data Systems Corp. to manage their LAN environment and provide Help Desk support.

Color Tile, headquartered in Fort Worth, Texas, is a leading home improvement retailer with over 770 stores and 4,600 employees. After going live with an automated Help Desk in the summer of 1991, they found that time spent per call to the Help Desk was down from 10 minutes to 2.8 minutes. The Help Desk is staffed by six operators who provide service seven days a week and handle virtually every call that comes into Color Tile's corporate headquarters.

Help Desk productivity has jumped more than 40% at JCPenney since Help Desk software was installed by the Plano, Texas-based department store chain approximately three years ago. Besides saving time, the use of the system has enabled the chain to reduce its number of Help Desk support teams from eight to three. This, in turn, has made possible the installation of an automated call-routing system that allows the user to select the team that's needed to help solve the problem.

Traditional Help Desk software acts as a type of call manager. It logs the caller, records his or her problem and its resolution. Where most Help Desk software fails is in its inability to provide a timely solution to all but the most simple problems. If a Help Desk is only as good as its "knowledge base," then most of today's Help Desks are truly in trouble since they rely solely on the expertise of the Help Desk analyst. While even a great Help Desk analyst can solve some of the problems some of the time, it is statistically impossible for a human to solve all of the problems all of the time.

That is where Natural Intelligence comes in. If call management functionality can be coupled with AI technology, then you have the possibility of a system that can only get smarter.

WHAT IS NATURAL INTELLIGENCE?

Simply put, Natural Intelligence is nothing less than providing software with the ability to think. It is the fusion of the branches of Artificial Intelligence with "record and calculate" capabilities to achieve new heights of capability. Today's customer service software merely records customer problems and attempts resolutions of problems by utilizing minor database searches—hardly an intelligent approach.

A naturally intelligent customer support module would serve to amplify the abilities of the support technician using the software. Software, up until now, has only been as intelligent as its most intelligent user. But naturally intelligent software could serve to "lead" the support technician through the steps necessary to solve his customer's problem. Naturally intelligent software, then, should be profoundly more intelligent than the most intelligent user. This is true because its embedded intelligence is a composite of the most expert of information with the most expert of analytical techniques.

Natural Intelligence, then, is the confluence of the branches of Artificial Intelligence with the capabilities of commercial software—whether that software be customer support, financial or word processing.

While readers will readily understand the functional components of a commercial system (e.g., customer support consists, at a minimum, of call management, problem tracking and problem resolution capabilities), the AI aspect of Natural Intelligence requires some explanation.

Expert Systems

What are expert systems really? Let us answer this by first explaining what they are not. They're not data base management systems. The database contains facts and figures. For example, it shows that John earns $26,000 a year and is 36 years old. It also shows that Pete earns $67,000 a year and is 45 years old. There's no knowledge here, just facts, all the facts and nothing but the facts.

Knowledge bases are different. For example, a human resources knowledge base might show that if the salary of the employee is greater than $25,000 than the job title is manager. It may also indicate that if someone is 35 years old or older then they are vested in the company's pension plan. This is knowledge in the form of rules. There is no facts or figures, just knowledge about how this department works.

The secret behind expert systems is the "expertise." Instead of capturing the data surrounding a securities trade, you capture the knowledge of why someone makes that trade. Instead of just capturing data that describes a PC problem, you capture the knowledge that the Help Desk analyst uses in determining how to solve that problem. Just like databases were a better way to store data than just file systems, expert systems do a better job at storing and processing knowledge than do conventional systems.

Stored knowledge is easy to play back. A Help Desk analyst merely requests assistance and is immediately provided with a set of questions, procedures, and images that assists him or her in solving a given problem. Think of it as an expert-in-a-box. An expert that stores knowledge in a very particular way.

Rules are probably the easiest knowledge-recording structure to understand. Our lives are filled with decisions, so it is natural to turn these into decision rules.

A rule has two parts. A premise: if it is cold and it is creamy; and a conclusion: then it is ice cream. Therefore, a rule is a conditional sentence starting with an IF and ending with a THEN. A rule, or production rule as it is sometimes called, stems from a decision. Look at the decision tree below. Here we're trying to decide whether a patient's claim for medical services should be reimbursed.

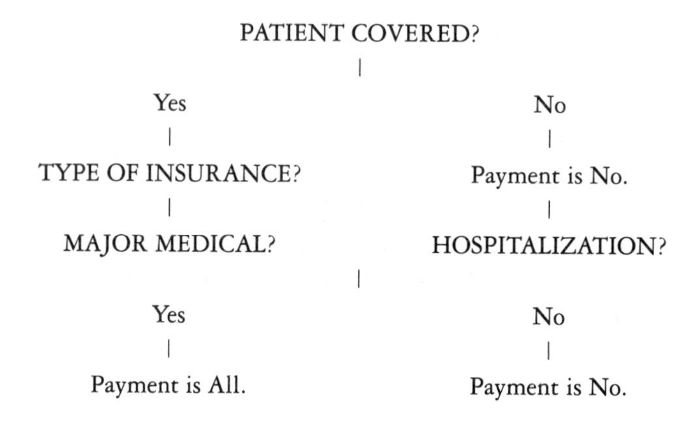

Once rules or decision tables are coded, there needs to be a mechanism to move around these rules picking and choosing the ones that are to be "fired." Expert systems do not operate in the same way traditional systems operate. Traditional systems execute one statement at a time, unless of course there is a branch statement.

Expert system rules execute according to a grand design called a control strategy which controls the method of searching through the knowledge base. There are two major categories of searching.

In Forward Chaining we move from a set of assertions or facts to one or more possible outcomes. The way rule application works is that the system searches for a value where the conditions in the "if" part of the rule are matched in memory (deemed to be true). If so, then the "then" part is put into working memory. Applying a forward chaining control strategy to a list of rules forces the execution of the rules from beginning to end. This is a sort of TOP DOWN REASONING approach. Here it reviews all known facts that were either entered at the very beginning of the system or became known as other rules triggered or fired. After a rule is reviewed, if the premise is true then it is fired. In forward chaining every bit of available evidence is examined since the final goal is not predetermined.

Suppose a CRT goes on the fritz. It just doesn't work. The screen is black. Well there can be many solutions to this problem. Perhaps the plug is not plugged in. Perhaps the power is out. Perhaps the VGA board is bad. We go from an event—a bad screen—to one of many possible outcomes. Our forward chaining system would wind its way through rules that dealt with this topic trying to assist in figuring out what is wrong.

The opposite of forward chaining is backward chaining. This strategy operates from the perspective that you already possess an outcome and are searching for the conditions or circumstances that lead to that result. Here the system tries to determine a value for the goal by identifying rules that conclude a value for the goal. Then the system backs up and tries to determine if the IF clauses are true. Again an example can serve to enlighten you. This time we know we have a bad VGA board and we leaf through our rules to find out if a black screen is indeed symptomatic of this problem.

Help Desks employing an expert system stratagem are providing a powerful method for improving productivity and a superior method for providing customer support. How many times have customers called with a problem only to hang up with no solution? The reason for this dilemma is that it is simply not possible to maintain a consistently high level of expertise on a corporate Help Desk. It is hard, often unrewarding work. As soon as an analyst reaches any level of experience, he or she opts for greener pastures. But an expert system-enabled Help Desk has the ability to capture the experience of senior level Help Desk analysts and maintain that level of experience long after that analyst jumps ship.

Neural Nets, Fuzzy Logic and Case-Based Reasoning

Neural nets simulate a network of hundreds of parallel processing interconnected units, shooting messages to each other at a rapid-fire pace. The job of a neural net is to receive the input and respond. This may first look like a task that can be handled as adequately by conventional means, but neural nets are computer programs with a difference. First, neural nets, and its sister paradigm fuzzy logic, have the capability of recognizing down-graded inputs. This capability is quite useful, it turns out, in dozens of commercial venues. From handwriting recognition, to finding a solution to a problem when the variables offered as evidence are not quite in the database—or expert system knowledge base (i.e., a user calls with a PC problem that has never been encountered before).

Neural Nets are not easy to understand. The simplest explanation demonstrates how a neural net might understand a handwritten letter. Suppose that we represent a letter as a 5 by 7 matrix of binary values as shown below.

```
0 0 1 0 0
0 1 0 1 0
1 0 0 0 1
1 1 1 1 1
1 0 0 0 1
1 0 0 0 1
1 0 0 0 1
```

The above matrix is a series of 35 ones and zeroes, therefore, we have 35 input units. Each one represents one of the 35 positions in the matrix. The output of the analysis of each matrix would be an ASCII letter such as "A" or "Z." (Draw a line through the ones in the matrix above and see what you wind up with.) Each ASCII letter is actually represented to the computer by an eight digit code. For example "A" would be 0100 0001. So what we have is input units and output units. We also need middle units since it is these units that actually experience the activation that triggers a response in the output units. All units, like the neurons in the brain, are joined by connections. Each unit has an activation value of either 0 or 1. Input activity triggers a pattern of activation across the middle units, which causes some sort of pattern of activation in the output units - sort of a ripple effect.

Say a perfect letter "A" is handwritten into the system. Activation of the units produces a numeric score which is calculated as a result of a weight being multiplied by the activation that passes through the units as shown below:

0.001 0.977 0.002 0.002   0.015 0.009 0.011 0.959

Using the rule that any output above 0.9 is considered a 1 and any output less than 0.1 is considered a 0 then we get the correct ASCII code for the letter "A":

0 1 0 0   0 0 0 1

Even if our handwritten letter "A" was less than perfect, that is even if one or more of the 35 input units was faulty, it would still be possible to produce the correct ASCII code.

This is an example of what is known as a back-propagation net which is probably the most popular and certainly the simplest. It is also one of the neural net strategies that can learn by itself. Simply stated the input is inputted and the output is calculated. This output is then compared to the desired output. Next the real output of each unit is subtracted from the desired output and thus the net slightly adjusts itself towards correcting the error. This is a repetitive process that is performed until the net has learned a new set of connections that produce the correct output.

If the letter is so poorly written as to be unrecognizable a more powerful net strategy could be employed to decipher even that.

Neural nets, case-based reasoning and fuzzy logic use intriguing statistical search, matching and retrieval techniques (similar to the example above) to sift through large amounts of data. The key, though, is the data. The more you have, the better the solution. It is not uncommon for users of this technique to refer to its "learning capabilities." That is because the quality of the solution that these technologies provide is directly proportional to the amount of information stored in the system—not unlike a human being. Compare this to a doctor, although possessed of a medical degree, one with 25 years of experience has more to base his/her decisions on than one directly out of medical schools. Experience really does count.

Although this genre of AI is probably the most difficult to understand, it is the easiest to use. There is really no work on the part of the organization other than installing the software and then using it. The more it is used the more it knows.

Where the expert system helps the more junior analyst give expert advice to users fuzzy logic "guesstimates" a solution to a difficult problem when there is no exact match. A real-world example illustrates how well this works.

Selling the likes of Jell-O Gelatin and Maxwell House coffee to literally thousands of grocery stores requires some crafty maneuvering to remain competitive. Kraft Foods does this in a high-tech way. Armed with laptops, salespeople are not only mobile but are able to keep their information right at their fingertips.

To keep them that way required Caroline Summit to find an equally high-tech way to manage her Help Desk. As Manager of Kraft Foods' White Plains, NY-based Sales Computer Help Line, Summit began a search through the dozens of Help Desk products that are on the market.

Summit needed something that would manage calls and problem resolutions. But many of the products on the market today do just that. So how do you differentiate between so many products that do much the same thing?

The great differentiator, according to Summit, was intelligence. Kraft Foods did not want anything that needed a heavily loaded database to begin with. What Summit and Kraft Foods wanted was a Help Desk system that sort of learned on its own. Most of Kraft Foods' problems are not exact replicas of problems that they had yesterday or the day before—they are variations on the theme. Kraft Foods wanted their Help Desk database to get smarter and smarter but did not want to have to build it before they used it. They wanted it to grow and learn with them.

The product Summit choose for Kraft Foods was PHD (Professional Help Desk), a smart Help Desk product from Greenwich-based PHD. According to Summit, when a problem comes in that nobody knows the answer to they click on PHD's fuzzy logic component which then makes suggestions as to what the answer might be (with percentage proximity to match). What's more, Summit recognizes that the use of fuzzy logic in a Help Desk means that the more problems she has the more adept at solving those problems her Help Desk becomes.

Natural Language Understanding Systems

That Kraft Foods so easily adapted to an AI-infused Help Desk was due to more than just the presence of fuzzy logic in the software. There also has to be a way for the software to interpret what the problem is.

Most software provides a rather rigid user interface. We enter data in hard-to-remember formats. How much better it would be to talk to the computer in our native tongue—our natural language. So instead of pull-down menus, lists or radio buttons to indicate a flickering screen how about just typing "The user has a flickering screen?"

Natural language processing, yet another branch of artificial intelligence, enables people and computers to communicate on an equal footing. But it is hard work. Perhaps the biggest problem computer scientists working in this area have had to tackle is syntactic ambiguity. Especially in the English language. Many words can have more than one syntactic category. That is, some words can be used both as nouns and verbs. For example let's look at the sentence WHY DON'T YOU GO OUTSIDE AND PLAY. Here the word play is used as a verb. In the following sentence the word play is used as a noun: I ENJOYED THE BROADWAY PLAY.

Along with syntactic ambiguity there is a host of other grammatical rules that must be factored into a natural language system. One of the harder to handle is pronominalization.

A example of how difficult it is for computers to understand the use of pronouns is the following request: GIVE ME A REPORT ON DIRECTORS WHO HAVE SUBORDINATES AND THEIR SALARIES. Whose salaries do we want? The directors or their subordinates? And added to all this confusion is the very human penchant for speaking ungrammatically. You know what you mean, your friends know what you mean, but does this computer understand PROFITS FOR LAST MONTH SCREWDRIVERS?

People refer to the same thing in a multitude of ways. Keyword searching systems, a poor substitute for natural language, force users to memorize exact formats or the system refuses to give up the information. A robust natural language system should allow for this type of ambiguity when processing the following requests:

HOW MANY FIRMS SUBMITTED THEIR TAXES FOR 03/13/95?

HOW MANY FIRMS SUBMITTED THEIR TAXES FOR 950313?

SHOW ME THE TAX RETURN OF FIRM 950313.

In the first two examples, the system needs to interpret the various formats of date. Easy as it looks, date and time are troublesome to all natural language systems, no matter how robust. In the last example we see that number again. Is 950313 a date, a tax identification number or what? It's obvious to us that we're referring to an identification number, but the natural language system must do some pretty fancy semantic interpretation to figure this one out.

Help Desks with natural language front ends enable the organization to solve problems more quickly, without the interference of rigid requirements which often serve to distort the meaning of the problem message itself.

How Natural Intelligence Confluence Works

Using the example of Help Desk software, Natural Intelligence becomes quite easy to understand. Aside from the traditional functionalities of call management, a naturally intelligent "Help Desk" would enable the rapid resolution of customer problems—whether or not the information provided by the customer matched what was in the database.

When a call is forwarded to the Help Desk, the customer support technician or Help Desk analyst uses Natural Language processing to type one or more English sentences into the computer describing the problem. At that point, a naturally intelligent Help Desk syntactically and semantically parses the English input and then passes control to what can be referred to as an experience-based processing module. Here it uses a combination of techniques such as fuzzy logic to "guesstimate" a solution to a difficult problem if there is no exact match and neural nets and/or case-based reasoning to sift through an accumulation of prior knowledge-base experience to determine the most optimal solution.

For those occasions when the Help Desk analyst is required to provide a step-by-step detailed set of procedures to solve a problem (i.e., changing a board or installing a new memory chip), the decision tree/expert system component, coupled with its multimedia and visual capabilities, provides the optimum fit. Here the customer, as well as the customer support technician benefits from domain expertise unmatched in any of the other AI disciplines.

It should, by now, be obvious that it is only due to the convergence of "smart technologies" that Natural Intelligence can be a viable alternative. But that Natural Intelligence is a method of making smart commercial software is not in dispute at all.

AUTHOR BIOGRAPHY

Isidore Sobkowski is Division Vice President of Greenwich-based PHD (Professional Help Desk), a part of DSSI—a 35 year old multi-million dollar, public company (800-4PHD-SALE). A widely-known expert in the field of Artificial Intelligence, Sobkowski was the chief architect of the WinExpert, an expert system development tool and WinBrain, a neural net development tool. Author of numerous articles and columns on the use of artificial intelligence, Sobkowski is currently finishing his book, "Intelligent Systems in the Service Economy." Prior to joining PHD, Sobkowski held the position of president of Stamford-based Applied Cognetics, Inc. from 1991 through 1994. Between 1985 and 1991, Sobkowski founded and managed a multi-million dollar high technology consultancy with clients including the New York Stock Exchange, Guardian Life Insurance Company, Dow Jones, Unisys, and Con Edison. Sobkowski received his Bachelors degree in Artificial Intelligence from City University of New York. He also received a CUNY Masters degree in Data Communications. Sobkowski is a frequent keynote speaker on the topic of Artificial Intelligence and the Service Economy.

INDEX

REGISTRATION CARD

Van Nostrand Reinhold
115 Fifth Ave., New York, NY 10003
Tel: (800) 842-3636
Fax: 212-254-9499

Thank you for your interest in Van Nostrand Reinhold publications. To enable us to keep you abreast of the latest developments in your field, please complete the following information.

WE WILL SEND YOU THE FOLLOWING UPON RECEIPT OF YOUR RESPONSE:

☐ Free exhibit pass(es) to select trade shows and updates about our forthcoming books.
☐ A catalog of titles.
☐ Information on becoming a VNR technical reviewer.

A. Title of this book:

____

B. How I first heard about this book:

☐ 1.Bookstore (specify):____________
☐ 2.Advertisement (specify):____________
☐ 3.Book Review (specify):____________
☐ 4.Colleagues
☐ 5.Catalog
☐ 6.Other (specify)____________

C. Number of communications/networking books I currently own:

☐ 1 ☐ 2-5 ☐ 6-10 ☐ >10

D. I purchased this book for my:

☐ 1.Professional use ☐ 2.Personal use ☐ 3.Both

E. I would be interested in new books on the following subject(s):

☐ 1.Internet for
 ☐ a. users ☐ b. developers
☐ 2.Media & Protocols
☐ 3.Networking Tools
☐ 4.Telecommunications
☐ 5.Wireless & Mobile
☐ 6.Information/Network Management
☐ 7.Other (specify)____________

F. My business or profession:

☐ 1.Communications
☐ 2.Consulting
☐ 3.Education/Research
☐ 4.Finance/Banking
☐ 5.Government
☐ 6.Healthcare Services
☐ 7.Information Industry
☐ 8.Insurance
☐ 9.Legal
☐ 10.Librarian
☐ 11.Manufacturing
☐ 12.Non-profit
☐ 13.Publishing
☐ 14.Other (specify)____________

G. My title:

☐ 1.Administrator
☐ 2.CEO/President
☐ 3.CIO/VP/Director
☐ 4.Engineer Specialist
☐ 5.General Manager
☐ 6.Independent Consultant
☐ 7.Librarian
☐ 8.Sales/Marketing
☐ 9.Other (specify)____________

H. My job function:

☐ 1.Computer Specialist
☐ 2.Datacom
☐ 3.Financial Analysis
☐ 4.Information Specialist
☐ 5.Management
☐ 6.Marketing/Sales
☐ 7.MIS/IS/IT
☐ 8.Network/Systems Management
☐ 9.Publisher
☐ 10.Researcher/Scientist
☐ 11.Systems & Software Design
☐ 12.Teacher/Professor
☐ 13.Other (specify)____________

I. I use a PC at:

☐ 1.Home ☐ 2.Work ☐ 3.School ☐ 4.Other

J. I have a CD-ROM player: ☐ Yes ☐ No

K. I have access to the following online services:

	Home	Work
1. America Online	☐	☐
2. CompuServe	☐	☐
3. Internet	☐	☐
4. Prodigy	☐	☐
5. World Wide Web	☐	☐
6. Other (please specify)	☐	☐

____

L. I participate in the following online activities ***(specify top one):***

☐ 1.Forums____________
☐ 2.Conferences____________
☐ 3.Newsgroups____________
☐ 4.Other____________

M. Name the top periodical you read for information in your field:

____

Comments:

Name:____________
Company name:____________
Street address:____________
____
City/State/Zip:____________
Country:____________
E-mail:____________
Phone:____________

For information on quantity discounts on ten or more books, please call our special sales department at 212-254-3232 or fax 212-254-9499.

Mail this postage-paid form or FAX to (212) 254-9499